Mechatronic Systems: Developments and Applications

Mechatronic Systems: Developments and Applications

Guest Editors

Nicola Ivan Giannoccaro
Ramiro Velázquez
Gianfranco Parlangeli

MDPI

Basel • Beijing • Wuhan • Barcelona • Belgrade • Novi Sad • Cluj • Manchester

Guest Editors
Nicola Ivan Giannoccaro
Department of Innovation Engineering
University of Salento
Lecce
Italy

Ramiro Velázquez
School of Engineering
Universidad Panamericana
Aguascalientes
Mexico

Gianfranco Parlangeli
Department of Engineering for Innovation
University of Salento
Lecce
Italy

Editorial Office
MDPI AG
Grosspeteranlage 5
4052 Basel, Switzerland

This is a reprint of the Special Issue, published open access by the journal *Machines* (ISSN 2075-1702), freely accessible at: https://www.mdpi.com/journal/machines/special_issues/U3Z6M4D357.

For citation purposes, cite each article independently as indicated on the article page online and as indicated below:

Lastname, A.A.; Lastname, B.B. Article Title. *Journal Name* **Year**, *Volume Number*, Page Range.

ISBN 978-3-7258-7512-2 (Hbk)
ISBN 978-3-7258-7513-9 (PDF)
https://doi.org/10.3390/books978-3-7258-7513-9

Contents

About the Editors

Nicola Ivan Giannoccaro

Nicola Ivan Giannoccaro is Full Professor in Applied Mechanics at the Department of Innovation Engineering, University of Salento (Lecce, Italy). His main research interests include mechatronic systems, control and modeling of mechanical systems, modal analysis, and damage detection by using vibrational data. He is a member of the Editorial Board of *Machines* for the section 'Robotics, Mechatronics and Intelligent Machines'.

Ramiro Velázquez

Ramiro Velázquez is Full Professor at the School of Engineering of Universidad Panamericana (Aguascalientes, Mexico) and currently serves as Vice Chancellor of this institution. His main research interests include mechatronic systems, assistive technology, haptic and tactile devices, and human perception. He is dedicated to comprehending how technology can help people with disabilities, especially visually disabled individuals. His research projects in assistive devices for blind and visually impaired people have been featured in IEEE Spectrum, CBS news, and BBC Horizons, among others.

Gianfranco Parlangeli

Gianfranco Parlangeli received an M.Sc. degree (with honors) in Electrical Engineering from the University of Pisa, Pisa, Italy, in 1999 and a Ph.D. degree in Information Engineering from the University of Lecce, Lecce, Italy, in 2005. He is currently Associate Professor at the Department of Innovation Engineering, University of Salento, Lecce, Italy. He is a member of the scientific board of the Interuniversity Center of Integrated Systems for the Marine Environment (ISME). His research interests include the analysis and design of multi-agent systems and robotic networks, fault-tolerant control, and variable structure control systems and robotics. He has published over 70 papers in the field and has contributed to several national and international projects in the area of robotics and industrial automation.

machines

Article

Self-Regulating Fuzzy-LQR Control of an Inverted Pendulum System via Adaptive Hyperbolic Error Modulation

Omer Saleem [1], Jamshed Iqbal [2,*] and Soltan Alharbi [3]

[1] Department of Electrical Engineering, National University of Computer and Emerging Sciences, Lahore 54770, Pakistan
[2] School of Digital and Physical Sciences, Faculty of Science and Engineering, University of Hull, Hull HU6 7RX, UK
[3] Department of Electrical and Electronic Engineering, College of Engineering, University of Jeddah, Jeddah 23890, Saudi Arabia
* Correspondence: j.iqbal@hull.ac.uk

Abstract: This study introduces an innovative self-regulating intelligent optimal balancing control framework for inverted pendulum-type mechatronic platforms, designed to enhance reference tracking accuracy and improve disturbance rejection capability. The control procedure is synthesized by synergistically integrating a baseline Linear Quadratic Regulator (LQR) with a fuzzy controller via a customized linear decomposition function (LDF). The LDF dissociates and transforms the LQR control law into compounded state tracking error and tracking error derivative variables that are eventually used to drive the fuzzy controller. The principal contribution of this study lies in the adaptive modulation of these compounded variables using reconfigurable tangent hyperbolic functions driven by the cubic power of the error signals. This nonlinear preprocessing of the input variables selectively amplifies large errors while attenuating small ones, thereby improving robustness and reducing oscillations. Moreover, a model-free online self-tuning law dynamically adjusts the variation rates of the hyperbolic functions through dissipative and anti-dissipative terms of the state errors, enabling autonomous reconfiguration of the nonlinear preprocessing layer. This dual-level adaptation enhances the flexibility and resilience of the controller under perturbations. The robustness of the designed controller is substantiated via tailored experimental trials conducted on the Quanser rotary pendulum platform. Comparative results show that the prescribed scheme reduces pendulum angle variance by 41.8%, arm position variance by 34.6%, and average control energy by 28.3% relative to the baseline LQR, while outperforming conventional fuzzy-LQR by similar margins. These results show that the prescribed controller significantly enhances disturbance rejection and tracking accuracy, thereby offering a numerically superior control of inverted pendulum systems.

Keywords: inverted pendulum; intelligent control; optimal balancing; fuzzy-LQR fusion; self-regulating control; hyperbolic tangent functions; disturbance rejection

1. Introduction

The inverted pendulum (IP) system is a well-known benchmark problem in control theory that has been extensively researched because of its nonlinear dynamics, underactuated configuration, and inherent kinematic instability [1]. Their balancing control is not just a theoretical challenge but a practical necessity in many mechatronic applications, especially where systems are required to maintain postural stability even in the presence of environmental disturbances [2,3]. The IP system needs an agile control mechanism

 https://doi.org/10.3390/machines13100939

to actively maintain vertical balance because the bounded disturbances can cause it to collapse when upright [4]. The solution to this control problem is extensively used in a vast range of applications, ranging from attitude stabilization of drones to gait control in self-balancing scooters and walk-assistive robotic mechanisms [5,6]. Similarly to IPs, these systems are also required to maintain stability while moving or interacting with their surroundings, which makes efficient balancing control crucial [7]. The traditional control methods lack effectiveness against large disturbances owing to their nonlinear characteristics [8]. Therefore, devising control procedures that concurrently deal with nonlinearity and disturbances can substantially improve the performance and dependability of these systems in practical situations.

1.1. Literature Review

Different control techniques have been explored to achieve stabilization and trajectory tracking in IP systems [9]. Among them, the ubiquitous PID control is simple, easy to implement, and computationally efficient, making it suitable for small disturbances and linearized models [10]. However, it struggles with nonlinearities, requires manual tuning, and lacks robustness [11]. Fractional Order PID (FOPID) improves upon PID by offering finer tuning flexibility, better disturbance rejection, and improved robustness, but it introduces additional design complexity and higher computational demands [12]. It also introduces additional parameters, making the tuning process more complex [13]. Moreover, implementing non-integer-order integrals and derivatives is computationally intensive and often requires specialized numerical methods or approximation techniques for practical situations [14]. The sliding-mode control is highly robust to disturbances and uncertainties, ensuring finite-time convergence and strong stability, but it suffers from chattering, requires a precise model, and may need additional smoothing techniques [15,16]. The model predictive control effectively addresses constraints and nonlinearities by optimizing control actions over a prediction horizon [17]. However, it is computationally intensive, demands an accurate system model, and may introduce latency issues due to solving optimization problems at each step [18].

The Linear Quadratic Regulator (LQR) has been a dominant approach because of its optimal state feedback formulation and ability to minimize a quadratic cost function balancing control effort and state deviation [19]. LQR offers a computationally efficient solution for controlling an IP by deriving an optimal gain vector by solving the Riccati equation. The method ensures stability under nominal conditions and is extensively used in balancing underactuated robotic and mechatronic platforms [20]. However, standard LQR assumes a linearized state space model, which limits its performance in highly nonlinear regimes, model uncertainties, and external disturbances [21]. Its ineffectiveness against parametric variations has led researchers to explore hybrid approaches that integrate nonlinear and intelligent control techniques to enhance its performance [22].

Among the intelligent control systems, the neural controllers provide an agile control effort [23]. However, they require large training datasets and put a recursive computational burden on the digital computer [24]. The Fuzzy logic control (FLC) scheme, on the other hand, has also gained traction in controlling nonlinear and underactuated systems due to its ability to handle imprecise and uncertain information [25]. Unlike model-based controllers, the FLC relies on expert knowledge and heuristic rules, making it adaptable to complex dynamic environments [26]. They need an extensive set of qualitative rules in order to create a precise control model. The calibration of the associated membership functions is a laborious process, necessitating their hybridization with traditional controllers to achieve improved performance [27].

Comparative studies indicate that hybrid FLC approaches, such as Fuzzy-PID and Fuzzy-LQR, outperform standalone fuzzy controllers in terms of design flexibility and adaptability [28]. These hybrid schemes leverage the robustness of fuzzy inference while maintaining the inherent optimal characteristics of conventional controllers, leading to improved balance control and disturbance rejection. Recent studies on the hybridization of FLC and LQR controllers have shown that fuzzy inference can compensate for LQR's sensitivity to parameter variations by adjusting the LQR's control yield dynamically based on system states [29]. The use of nonlinear scaling functions for control signal modulation and adaptive gain scaling has been extensively studied in robotic applications [30]. Hyperbolic functions, particularly, have been utilized for their smooth saturation properties, allowing for adaptive gain adjustments and dynamic self-regulation of state error magnitudes without abrupt changes [31].

Recent developments in data-driven and machine learning-based control strategies have shown significant promise for rotary inverted pendulum systems [32,33]. Reinforcement learning (RL) techniques have been successfully applied to trajectory tracking and stabilization tasks in underactuated pendulum systems, demonstrating improved adaptability to nonlinear dynamics [34]. Recent studies show that RL can achieve model-free swing-up and balancing of a quadruple inverted pendulum [35]. However, these techniques often demand substantial training data and significant computational resources, which can limit their practicality for real-world applications [36]. Moreover, their lack of formal stability reduces their reliability in safety-critical environments [37]. Finite-time continuous extended state observers have also been designed and experimentally validated on electro-mechanical and electro-hydraulic systems, demonstrating improved disturbance rejection and accurate state estimation [38].

1.2. Proposed Methodology

This paper presents a novel self-regulating compounded-state Fuzzy-LQR (FLQR) control strategy for stabilizing a Single-Link Rotary-Inverted-Pendulum (SLRIP). To overcome the standard LQR's performance constraints against modeling uncertainties and bounded disturbances, the proposed scheme is realized by decomposing the control law into two key components: a Compounded State Error (CSE) variable and a Compounded State Error Derivative (CSED) variable. These aggregated variables encapsulate the system's dynamic behavior and are fed into a Fuzzy Inference System (FIS) to generate an adaptive and robust control input. The integration of fuzzy logic enhances the LQR's disturbance rejection capability, providing improved adaptability to system uncertainties while maintaining optimal state feedback control.

The proposed scheme is further enhanced by introducing a cubic hyperbolic tangent function (CHTF) to preprocess the CSE and CSED variables before feeding them into the FIS. The hyperbolic function not only normalizes the input variables within ± 1 for improved fuzzy reasoning but also introduces selective attenuation and amplification regions to enhance control input efficiency. Additionally, the variation rate of CHTF's waveform is adaptively modified online as a function of state deviations, via a pre-calibrated model-free learning algorithm. The principal contributions of this study are listed below:

1. Formulation of the proposed CS-FLQR scheme for the SLRIP. The baseline LQR law is decomposed into CSE and CSED variables, which are processed through an FIS to enhance the control system's resilience against exogenous perturbations.
2. Integration of pre-calibrated CHTFs for adaptive preprocessing of CSE and CSED variables to normalize them within ± 1 and to create selective attenuation and amplification regions for improved control efficiency.

3. Augmentation of the CHTFs with model-free adaptive tuning principles to dynamically adjust the said function's variation rate and hence the magnitudes of compounded error variables to further enhance the controller's adaptability.
4. Performance validation of the proposed self-regulating CS-FLQR against baseline CS-FLQR and classical LQR via customized experimental trials conducted on the Quanser rotary pendulum platform [39].

This hybrid structure significantly improves control input economy, robustness, and adaptability in the presence of disturbances and uncertainties. In comparison to standard LQR, the fuzzy-augmented LQR allows for dynamic adjustment of control actions based on error conditions, resulting in faster convergence and lower steady-state error. The integration of CHTFs for the preprocessing of compounded error variables produces a smooth and bounded control input, which reduces excessive oscillations and economizes the control effort. Additionally, the online adaptive reconfiguration of the CHTF's waveform aids the control law in efficiently adapting to varying system dynamics, making it more suitable for highly dynamic systems like SLRIP platforms.

Despite advances in LQR-based, fuzzy-based, and hybrid control systems, existing approaches continue to encounter issues in terms of real-time adaptability and robustness against disturbances. The deployment of reconfigurable CHTFs as an adaptive error scaling mechanism is still underexplored. This study fills these gaps by providing a self-regulating CS-FLQR controller that dynamically modulates control signals using cubic error-driven CHTFs and model-free adaptive tuning principles. The proposed control procedure is intended to greatly improve tracking accuracy and disturbance compensation in IP-type robotic systems.

It is worth noting that earlier works addressed inverted pendulum stabilization control from different methodological perspectives. Specifically, ref. [12] introduced dual fractional-order PD controllers optimized via nonlinear intelligent adaptive mechanisms, ref. [16] developed an EKF-based fuzzy adaptive sliding mode controller with hardware validation, ref. [21] proposed a phase-based adaptive fractional LQR tailored for pendulum-type robots, and ref. [22] presented a complex-order LQIR framework with experimental verification. In contrast, the present study contributes a self-regulating fuzzy-LQR formulation that integrates LQR decomposition, nonlinear error modulation, and model-free variance adaptation and validates its efficacy on an SLRIP platform.

The remainder of the article is organized as follows: The SLRIP system's model description and the baseline LQR formulation are presented in Section 2. The development and stability analysis of the proposed self-regulating CS-FLQR scheme is detailed in Section 3. The controller parameter tuning process is outlined in Section 4. The simulation results and their analysis are presented in Section 5. The study is concluded in Section 6.

2. System Description

The SLRIP system consists of a vertically oriented rod attached to a horizontally rotating arm, which is driven by a geared DC servomotor. The schematic is illustrated in Figure 1 [22]. The system necessitates a closed-loop control algorithm for maintaining the pendulum's link in an upright position while ensuring effective tracking of the rotating arm's reference position. An input voltage signal produced by the control algorithm controls the angular displacement of the DC motor, causing the horizontal arm to rotate. The arm is pivoted at the motor shaft. This rotational motion provides the energy required to swing the pendulum upward and keep it in an upright position. The angular motion of the rotating arm is referred to as α. The angular motion of the pendulum about its pivot is represented by θ. These angular positions are measured using rotary optical encoders attached to the pendulum's pivot and the motor's shaft.

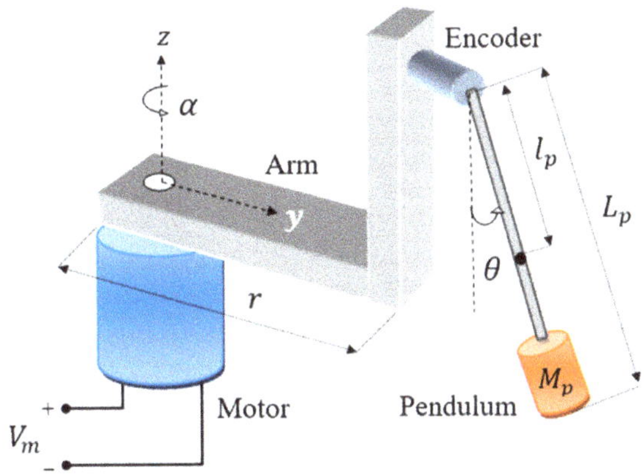

Figure 1. Schematic of the SLRIP system [22].

2.1. *System's Mathematical Model*

The Euler–Lagrange formula is used to derive the system's mathematical model, which takes into consideration the system's mechanical and electrical dynamics. The dynamics of the system are described by the Lagrangian function using the generalized angular position coordinates (α and θ) [40]. The Lagrangian $L(\alpha,\theta)$ is defined as the difference between the kinetic energy (T) and potential energy (V) of the system, as expressed in Equation (1).

$$L(\alpha,\theta) = T - V \tag{1}$$

where,

$$T = \frac{1}{2}J_e(\dot{\alpha})^2 + \frac{1}{2}M_p\left(r\dot{\alpha} - l_p\dot{\theta}(cos\theta)\right)^2 + \frac{1}{2}M_p\left(-l_p\dot{\theta}(sin\theta)\right)^2 + \frac{1}{2}J_p\left(\dot{\theta}\right)^2,$$

and,

$$V = M_p l_p g(\cos\theta)$$

where $\dot{\alpha}$ is the arm's angular velocity, $\dot{\theta}$ is the pendulum link's angular velocity, J_e is the moment about the motor shaft, M_p is the mass of the pendulum, r is the length of the horizontal arm link, l_p is the fixed geometric distance from the pivot to the pendulum link's center of mass, and J_p is the moment about the pendulum. Table 1 quantifies the model parameters related to the Quanser rotary pendulum platform utilized in this study [41]. The formulation of $L(\alpha,\theta)$ is expressed as follows.

$$L(\alpha,\theta) = \frac{1}{2}\left(J_e + M_p r^2\right)\dot{\alpha}^2 + \left(\frac{2}{3}M_p l_p{}^2 + \frac{1}{2}J_p\right)\dot{\theta}^2 - M_p l_p r\dot{\alpha}\dot{\theta}\cos\theta - M_p l_p g\cos\theta \tag{2}$$

The formulation of the nonlinear equations of motion is presented in Equation (3) [40].

$$\frac{d}{dt}\left(\frac{dL}{d\dot{\alpha}}\right) - \frac{dL}{d\alpha} = \tau - b_v\dot{\alpha},\ \frac{d}{dt}\left(\frac{dL}{d\dot{\theta}}\right) - \frac{dL}{d\theta} = 0 \tag{3}$$

where τ denotes the DC motor's control torque, and b_v represents the motor's viscous friction coefficient. Due to its minimal influence, viscous friction is omitted from the model formulation. The torque produced by the DC motor is,

Table 1. Model specifications of Quanser SLRIP [41].

Parameter	Description	Value	Unit
M_p	Pendulum's mass	0.027	kg
l_p	Pivot to pendulum's center distance	0.153	m
L_p	Pendulum link length	0.191	M
r	Horizontal arm length	0.083	M
M_{arm}	Arm's mass	0.028	Kg
g	Acceleration due to gravity	9.810	m/s^2
J_e	Moment about motor shaft	1.23×10^{-4}	kgm^2
J_p	Moment about pendulum	1.10×10^{-4}	kgm^2
R_m	Motor's armature resistance	3.30	Ω
L_m	Motor's armature inductance	47.0	mH
K_t	Motor torque coefficient	0.028	Nm
K_m	Back-EMF coefficient	0.028	V/(rad/s)
T_m	Maximum torque	0.14	Nm

$$\tau = \frac{K_t(V_m - K_m\dot{\alpha})}{R_m} \tag{4}$$

where V_m represents the DC motor's input (control) voltage, K_t represents the motor's torque coefficient, K_m represents the back electromotive-force (EMF) coefficient, and R_m is the motor's internal resistance. By applying the Euler–Lagrange equations, the nonlinear equations of motion are derived as follows [39].

$$\ddot{\alpha} = \frac{-rM_p^2 l_p^2 g(\cos\theta)\theta - J_p M_p r^2 \cos\theta \sin\theta (\dot{\alpha})^2 - \left(J_p + M_p l_p^2\right)\tau}{(M_p r^2(\sin^2\theta) - J_e - M_p r^2)J_p - M_p l_p^2 J_e} \tag{5}$$

$$\ddot{\theta} = \frac{-M_p l_p \left((M_p r^2 g(\sin^2\theta) - J_e g - M_p r^2 g)\theta + rJ_e \sin\theta(\dot{\alpha})^2 - r\,\tau\cos\theta\right)}{(M_p r^2(\sin^2\theta) - J_e - M_p r^2)J_p - M_p l_p^2 J_e} \tag{6}$$

To facilitate controller design, the nonlinear system is linearized around the upright equilibrium position ($\alpha = 0,\ \theta = \pi$ rad., $\dot{\alpha} = 0,\ \dot{\theta} = 0$). Under the assumption of small angular displacements, the standard approximations $\sin\theta \approx \theta$ and $\cos\theta \approx 1$ are applied, leading to the following set of linearized equations.

$$\ddot{\alpha}(t) = \frac{1}{H}\left(rM_p^2 l_p^2 g\theta(t) - \frac{\left(J_p + M_p l_p^2\right)K_t K_m}{R_m}\dot{\alpha}(t) + \frac{\left(J_p + M_p l_p^2\right)K_t}{R_m}V_m\right) \tag{7}$$

$$\ddot{\theta}(t) = \frac{1}{H}\left(M_p l_p g\left(J_e + M_p r^2\right)\theta(t) - \frac{rM_p l_p K_t K_m}{R_m}\dot{\alpha}(t) + \frac{rM_p l_p K_t}{R_m}V_m\right) \tag{8}$$

$$\text{suchthat,}\quad H = J_e J_p + M_p r^2 J_p + M_p l_p^2 J_e$$

It should be noted that the derived dynamics correspond to the classical Furuta (rotary inverted) pendulum configuration, which is the actual physical setup employed in this study [2]. The Furuta pendulum model is well established in the literature [42]. Hence, it is directly applicable to the Quanser SLRIP (Markham, ON, Canada) used in our experiments.

The linearized state-space representation of the system is shown below.

$$\dot{x}(t) = \boldsymbol{A}x(t) + \boldsymbol{B}u(t),\ \ y(t) = \boldsymbol{C}x(t) + \boldsymbol{D}u(t) \tag{9}$$

where $x(t)$, $y(t)$, and $u(t)$ represent the state, output, and control input vectors, respectively. The matrices $\boldsymbol{A}$, $\boldsymbol{B}$, $\boldsymbol{C}$, $\boldsymbol{D}$ represent the system, input, output, and feedforward matrices, respectively. The state vector and input of the SLRIP are provided in Equation (10).

$$x(t) = \begin{bmatrix} \alpha(t) & \theta(t) & \dot{\alpha}(t) & \dot{\theta}(t) \end{bmatrix}^T, \ u(t) = V_m \tag{10}$$

The SLRIP system's nominal state-space model is provided as follows [39].

$$\boldsymbol{A} = \begin{bmatrix} 0 & 0 & 1 & 0 \\ 0 & 0 & 0 & 1 \\ 0 & a_1 & a_2 & 0 \\ 0 & a_3 & a_4 & 0 \end{bmatrix}, \boldsymbol{B} = \begin{bmatrix} 0 \\ 0 \\ b_1 \\ b_2 \end{bmatrix}, \boldsymbol{C} = \begin{bmatrix} 1 & 0 & 0 & 0 \\ 0 & 1 & 0 & 0 \\ 0 & 0 & 1 & 0 \\ 0 & 0 & 0 & 1 \end{bmatrix}, \boldsymbol{D} = \begin{bmatrix} 0 \\ 0 \\ 0 \\ 0 \end{bmatrix} \tag{11}$$

$$\text{where, } a_1 = \frac{rM_p^2 l_p^2 g}{J_p J_e + J_e l_p^2 M_p + J_p M_p r^2}, \ a_2 = \frac{-K_t K_m \left(J_p + M_p l_p^2\right)}{\left(J_p J_e + J_e l_p^2 M_p + J_p M_p r^2\right) R_m},$$

$$a_3 = \frac{M_p l_p g (J_e + M_p r^2)}{J_p J_e + J_e l_p^2 M_p + J_p M_p r^2}, \ a_4 = \frac{-r M_p l_p K_t K_m}{\left(J_p J_e + J_e l_p^2 M_p + J_p M_p r^2\right) R_m},$$

$$b_1 = \frac{K_t \left(J_p + M_p l_p^2\right)}{\left(J_p J_e + J_e l_p^2 M_p + J_p M_p r^2\right) R_m}, \ b_2 = \frac{r M_p l_p K_t}{\left(J_p J_e + J_e l_p^2 M_p + J_p M_p r^2\right) R_m}$$

Table 1 quantifies the system's model parameters. The reported value of the pendulum center of mass (identified in Table 1) of 0.153 m refers to the fixed geometric distance from the pivot to the pendulum link's center of mass, as provided in the Quanser datasheet [41]. This constant parameter is defined in the local body frame and is used in system modeling and controller design. It is to be noted that this link-level pendulum center of mass remains invariant and is the relevant quantity for linearization.

2.2. Baseline LQR Strategy

The LQR strategy is employed to stabilize and regulate the RIP system by utilizing full-state feedback [43]. The quadratic cost function, specified in (12), penalizes variations in the control input and state.

$$J_{lq} = \frac{1}{2} \int_0^\infty \left(x(t)^T \boldsymbol{Q} x(t) + u(t)^T \boldsymbol{R} u(t) \right) dt \tag{12}$$

where $\boldsymbol{Q} \in \mathbb{R}^{4 \times 4}$ is a positive semi-definite state weighting matrix and $\boldsymbol{R} \in \mathbb{R}$ is a positive control input weighting factor [39]. For the SLRIP system, the $\boldsymbol{Q}$ and $\boldsymbol{R}$ matrices are parameterized as follows.

$$\boldsymbol{Q} = \text{diag}\left(q_\alpha \quad q_\theta \quad q_{\dot{\alpha}} \quad q_{\dot{\theta}} \right), \ \boldsymbol{R} = \rho \tag{13}$$

where q_α, q_θ, $q_{\dot{\alpha}}$, $q_{\dot{\theta}}$, and ρ are tuning parameters that influence the control performance. These matrix parameters are selected such that $q_x \geq 0$ and $\rho > 0$. An offline tuning procedure (explained in Section 4) is used to select these parameter settings. The expression of the optimal control law is provided in Equation (14).

$$u(t) = -\boldsymbol{K} x(t) \tag{14}$$

where $K = \begin{bmatrix} k_\alpha & k_\theta & k_{\dot{\alpha}} & k_{\dot{\theta}} \end{bmatrix}$ is the state feedback gain matrix, given by Equation (15).

$$K = R^{-1}B^T P \tag{15}$$

where $P \in \mathbb{R}^{4\times4}$ is a symmetric positive definite matrix. The matrix P is acquired by solving the following Algebraic-Riccati-Equation (ARE).

$$A^T P + PA - PBR^{-1}B^T P + Q = 0 \tag{16}$$

The solution of the ARE ensures asymptotic stability, provided that the weighting matrices are chosen such that, $R = R^T > 0$ and $Q = Q^T \geq 0$. This stability criterion is discussed as follows.

***Stability Analysis*:** The asymptotic stability of the LQR is verified via the following Lyapunov function [21].

$$V(t) = x(t)^T P(t)x(t) > 0, \quad \text{for } x(t) \neq 0 \tag{17}$$

The time-derivative of $V(t)$ is presented below.

$$\begin{aligned} \dot{V}(t) &= 2x(t)^T P\dot{x}(t) \\ &= 2x(t)^T P(A - BK(t))x(t) \\ &= 2x(t)^T P(A - BR^{-1}B^T P)x(t) \\ &= x(t)^T (PA + A^T P)x(t) - 2x(t)^T (PBR^{-1}B^T P)x(t) \end{aligned} \tag{18}$$

By substituting the ARE, the expression above is simplified as shown in Equation (19).

$$\dot{V}(t) = -x(t)^T Qx(t) - x(t)^T \left(PBR^{-1}B^T P\right)x(t) \tag{19}$$

Since $Q \geq 0$ and $R > 0$, it follows that $\dot{V}(t) < 0$, ensuring the LQR law maintains asymptotic stability. The block diagram of the LQR law is depicted in Figure 2.

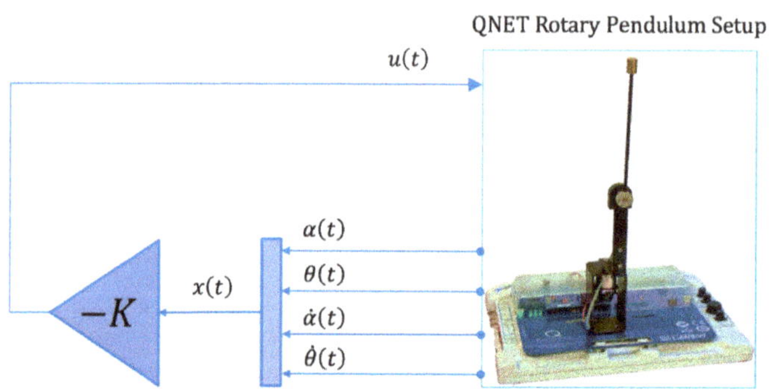

Figure 2. Block diagram of the baseline LQR law.

3. Proposed Control Methodology

The prescribed self-regulating FLQR method for the position regulation of the SLRIP systems is systematically formulated in this section.

3.1. Fuzzy-Based Linear Quadratic Regulator (FLQR)

The FLQR strategy is introduced to improve the resilience of the SLRIP system against parametric uncertainties and external disruptions. The proposed control procedure integrates the LQR's optimal control capabilities with the adaptive nature of fuzzy logic to adaptively reconfigure the control actions as the error dynamics vary [29]. The main goal of this approach is to enhance the system's transient behavior and strengthen its ability to damp oscillations while preserving its control input economy. The proposed FLQR is realized by designing a customized linear decomposition function (LDF) that decomposes the baseline LQR control law into two components: a CSE variable and a CSED variable. To acquire the said compounded error variables, the state feedback gain vector of the LQR law is rewritten as shown in Equation (20), [29].

$$\hat{K} = \begin{bmatrix} k_\alpha & k_\theta & 0 & 0 \\ 0 & 0 & k_{\dot{\alpha}} & k_{\dot{\theta}} \end{bmatrix} \quad (20)$$

The formulation of the LDF for the SLRIP is defined as follows [29].

$$\begin{bmatrix} \varepsilon(t) \\ \dot{\varepsilon}(t) \end{bmatrix} = -\hat{K}x(t) \quad (21)$$

such that,

$$\varepsilon(t) = -k_\alpha\, \alpha(t) - k_\theta\, \theta(t),$$

and,

$$\dot{\varepsilon}(t) = -k_{\dot{\alpha}}\, \dot{\alpha}(t) - k_{\dot{\theta}}\, \dot{\theta}(t).$$

where $\varepsilon(t)$ is the CSE variable and $\dot{\varepsilon}(t)$ is the CSED variable. The arrangement above consolidates the system's multiple state variables into the CSE and CSED. These aggregated variables encapsulate the system's dynamic behavior and simplify the implementation of the FIS to generate an adaptive and robust control input. The FIS leverages fuzzy logic principles to enhance the system's adaptability and robustness by dynamically re-adjusting control actions in response to state variations and disturbances. It employs a set of heuristically designed fuzzy logic (qualitative) rules and predefined membership functions (MFs) to infer precise control actions from input data. The fuzzy logical rules are formulated using the expert's knowledge to regulate the control effort as a function of the system's tracking error variations. Considering the operational requirements of the SLRIP's balancing control problem, the meta rules established to construct the fuzzy qualitative rules are listed below.

1. Amplify the control effort when the system deviates from the desired state.
2. Reduce the control effort as the system approaches the stable equilibrium.

The above rules are computationally implemented for the online refinement of control actions through a two-input Mamdani-type FIS, where the inputs are derived from the error dynamics of the system. The two compounded error variables are normalized (between -1 and +1) by processing them through a cubic HTF [21]. The cubic HTF is employed in this study due to its favorable combination of smooth saturation, odd symmetry, and adjustable steepness via tunable variation rate [21]. Its waveform ensures smooth normalization of the error signals within the bounded range of ± 1. Moreover, its cubic formulation introduces selective amplification and suppression regions, such that it significantly attenuates small error fluctuations, reducing unnecessary control effort and chattering, while sharply amplifying larger errors to ensure rapid corrective action [21]. The cubic HTF is preferred over other odd-symmetric functions, like the hard-tanh or soft-sign functions. The hard-tanh function renders abrupt transitions, which introduce discontinuities or chattering in control signals [44]. On the other hand, the soft-sign function offers gentler scaling but weaker nonlinearity near the origin. This cubic nonlinear scaling enhances the controller's sensitivity in critical conditions, improving control efficiency and robustness [45]. The normalized compounded error variables are expressed below.

$$\varepsilon_n(t) = \tanh\left(\gamma\ (\varepsilon(t))^3\right) \tag{22}$$

$$\dot{\varepsilon}_n(t) = \tanh\left(\delta\ (\dot{\varepsilon}(t))^3\right) \tag{23}$$

where tanh(.) represents the HTF, and γ and δ are the pre-determined variation rates of the HTFs. The parameters γ and δ are determined by using the offline tuning procedure (explained in Section 4). The cube of the CSE and CSED is fed the input to the HTFs. The waveform of the prescribed cubic HTF, formulated in Equations (22) and (23), with respect to an input error variable is depicted in Figure 3.

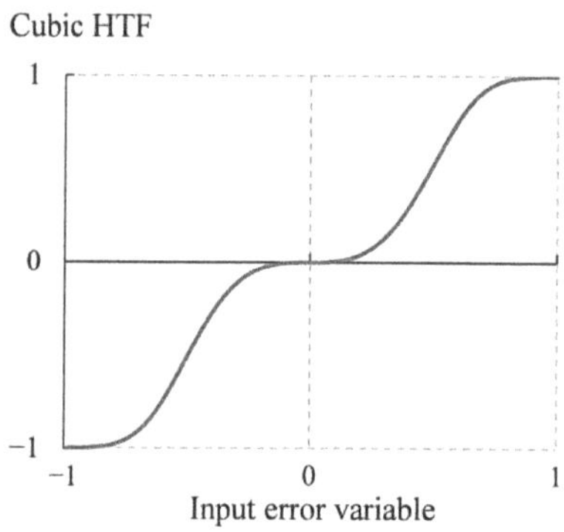

Figure 3. Waveform of the cubic HTF formulated for error normalization.

As discussed earlier, the formulation of the cubic HTF generates distinct amplified and suppressed error regions, as illustrated in Figure 3, leading to a sharp increase in the magnitude of $\varepsilon_n(t)$ and $\dot{\varepsilon}_n(t)$ under large error conditions, while significantly reducing it for smaller errors [46]. The preprocessing of CSE and CSED variables via the above-mentioned CHTFs not only normalizes them within ± 1 but also creates selective attenuation and amplification regions for improved control efficiency. The normalized error variables $\varepsilon_n(t)$ and $\dot{\varepsilon}_n(t)$ serve as inputs to the FIS, while the normalized control action $\lambda(\varepsilon_n, \dot{\varepsilon}_n)$ serves as the output of the FIS. The crisp input values are converted into fuzzy linguistic variables by the fuzzification module. The input and output variables are classified into seven linguistic categories: Negative Big (NB), Negative Medium (NM), Negative Small (NS), Zero (Z), Positive Small (PS), Positive Medium (PM), and Positive Big (PB). Table 2 presents the complete fuzzy rule-base, which consists of 49 qualitative rules. For fuzzy implication, the max-min inference method stated in Equation (24), is employed.

$$\mu_i = min(g_i(\varepsilon_n), g_i(\dot{\varepsilon}_n)) \tag{24}$$

where μ is the degree of the MF, i is the number of rule, and $g_i(.)$ is the triangular input MF of the following form [16].

$$g(f) = \begin{cases} 1 + \frac{f - c_i}{b_i^-}, & -b_i^- \le f - c_i \le 0 \\ 1 - \frac{f - c_i}{b_i^+}, & 0 \le f - c_i \le b_i^+ \\ 0, & \text{otherwise} \end{cases} \tag{25}$$

where f represents the normalized value of the input ε_n and $\dot{\varepsilon}_n$, while b_i^-, c_i, and b_i^+ correspond to the left-half width, centroid, and right-half width of the input MF, respectively. Symmetrical triangular MFs are adopted in this work to simplify computation during fuzzy implication and aggregation.

Table 2. Fuzzy inference rule-base used for online control adjustment.

$\varepsilon_n \downarrow / \dot{\varepsilon}_n \rightarrow$	NB	NM	NS	Z	PS	PM	PB
NB	NB	NB	NB	NB	NM	NS	Z
NM	NB	NB	NB	NM	NS	Z	PS
NS	NB	NB	NM	NS	Z	PS	PM
Z	NB	NM	NS	Z	PS	PM	PB
PS	NM	NS	Z	PS	PM	PB	PB
PM	NS	Z	PS	PM	PB	PB	PB
PB	Z	PS	PM	PB	PB	PB	PB

The waveforms of the input MFs as well as the output fuzzy MFs are depicted in Figure 4a and Figure4b, respectively. While the membership functions appear visually similar due to the use of identical triangular shapes and evenly spaced centroids for design uniformity, Figure 4a corresponds to the fuzzification of the compounded error variables, whereas Figure 4b represents the output control signal space. The centroid defuzzification method, defined in Equation (26), is used to acquire the crisp output $\lambda(\varepsilon_n, \dot{\varepsilon}_n)$ [16].

$$\lambda(\varepsilon_n, \dot{\varepsilon}_n) = \frac{\sum_{i=1}^{N} \mu_i \, w_i}{\sum_{i=1}^{N} \mu_{ij}} \tag{26}$$

where w represents the output MF's centroid, and $N = 49$ denotes the number of fuzzy rules. The resulting crisp output is then processed using the saturation function, formulated in Equation (27), to generate the refined control action.

$$u(t) = V_{sat} \tanh(\lambda(\varepsilon_n, \dot{\varepsilon}_n)) \tag{27}$$

where V_{sat} sets the lower and upper bounds of the applied control input, preventing the SLRIP's actuator (DC motor) from getting saturated. The value of V_{sat} is set to 15.0 V as per the guidelines provided in [32]. The schematic of the FLQR scheme is shown in Figure 5.

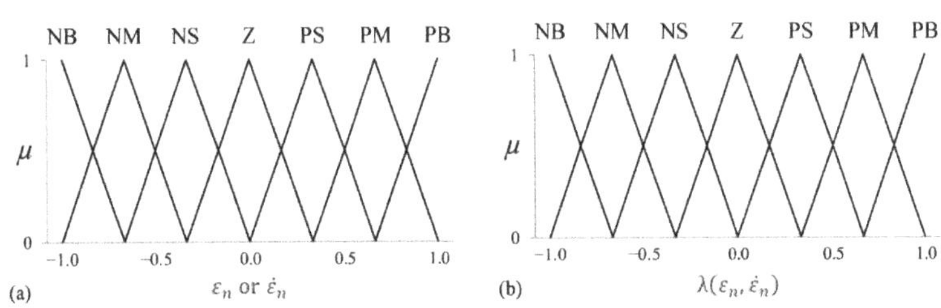

Figure 4. MF waveforms of (**a**) input variables, and (**b**) output variable.

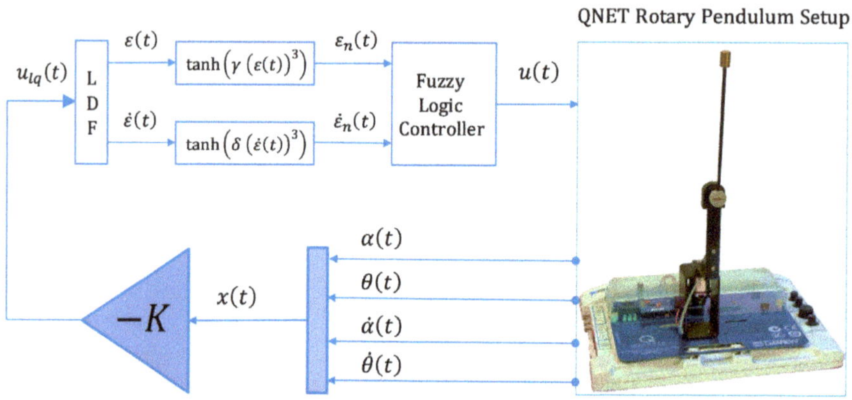

Figure 5. Block diagram of the FLQR law.

3.2. Self-Regulating FLQR Law

The selection of fixed variation rates for the CHTFs, in the FLQR law, formulated for the nonlinear preprocessing of the CSE and CSED variables, presents a fundamental limitation, as it constrains the controller's adaptability to reject parametric variations and environmental uncertainties. This issue is addressed by dynamically self-regulating the variation rates through a model-free online adaptive tuning law, which optimizes the inflation or depreciation of the applied control force in real time. This enhancement improves system sensitivity and responsiveness, allowing the controller to efficiently modulate control effort, minimize oscillations, and compensate for disturbances with reduced energy consumption.

The online self-tuning mechanism, proposed in [47], is employed due to its robust tracking performance. It is formulated by using pre-configured anti-dissipative and dissipative functions that adjust the hyperbolic function's variation rates after each sampling interval based on the system's CSE and CSED variables [39,45]. The formulations of the aforesaid Variance Adaptation Functions (VAFs) are expressed as follows [47].

$$\dot{\gamma}(t) = -\sigma_1\gamma(t) + \beta_1(\varepsilon(t))^2 \tag{28}$$

$$\dot{\delta}(t) = -\sigma_2\delta(t) + \beta_2\varepsilon(t)\dot{\varepsilon}(t) \tag{29}$$

where σ_1 and σ_2 represent the pre-configured positive decay rates, and β_1 and β_2 represent the pre-configured positive adaptation rates associated with the online adaptation functions. The parameters governing adaptation and damping rates are optimized as per the offline tuning method described in Section 4. The constructed VAFs dynamically regulate the stiffness or softness of control action depending on the state's deviations from the reference, ensuring adaptive performance. Each of these VAFs consists of a dissipative term and an anti-dissipative term, as demonstrated below.

Dissipative term: $\begin{cases} -\sigma_1\gamma(t) \\ -\sigma_2\delta(t) \end{cases}$ Anti-dissipative term: $\begin{cases} \beta_1(\varepsilon(t))^2 \\ \beta_2\varepsilon(t)\dot{\varepsilon}(t) \end{cases}$

Each of these terms serves a specific purpose while modifying the variation rate. The rationale is presented as follows [47].

1. Dissipative term: Reduces the variation rate under small error conditions, ensuring gentle settlement of the response as it approaches reference, reducing steady-state fluctuations, and preventing wind-up.

2. Anti-dissipative term: Amplifies the variation rate when the state error increases, promoting an aggressive control action to suppress overshoots, while speeding up the transient recovery response.

The VAFs formulated in Equations (28) and (29) are unified as a first-order differential equation, expressed in Equation (30).

$$\dot{v}(t) = \boldsymbol{H}v(t) + \boldsymbol{M}\epsilon(t) \tag{30}$$

such that,

$$v(t) = \begin{bmatrix} \gamma(t) \\ \delta(t) \end{bmatrix}, \boldsymbol{H} = \begin{bmatrix} -\sigma_1 & 0 \\ 0 & -\sigma_2 \end{bmatrix}, \boldsymbol{M} = \begin{bmatrix} \beta_1 & 0 \\ 0 & \beta_2 \end{bmatrix}, \epsilon(t) = \begin{bmatrix} (\varepsilon(t))^2 \\ \varepsilon(t)\dot{\varepsilon}(t) \end{bmatrix}$$

where $v(t)$ represents a vector of time-varying variation rate gains, while $\epsilon(t)$ denotes a vector containing state-error-dependent terms. The $\boldsymbol{H}$ and $\boldsymbol{M}$ matrices are positive-definite, where $\boldsymbol{H}$ contains the decay rates σ_1 and σ_2, and $\boldsymbol{M}$ includes the adaptation rates β_1 and β_2. At each sampling interval, the ordinary differential equation in Equation (30) is integrated numerically to determine $v(t)$, thereby updating the variation-rates. The variation rate adaptation process is computationally realized by solving the ordinary differential equation in Equation (30), as expressed in Equation (31).

$$v(t) = exp(-\boldsymbol{H}t)v(0) + \int_0^t (exp(-\boldsymbol{H}(t-p))\boldsymbol{M}\epsilon(p))dp \tag{31}$$

where $exp(.)$ denotes the exponential function. The initial variation rate vector $v(0)$ contains the offline optimized values of γ and δ (refer to Section 4). The coefficients of the vector $v(t)$ are adaptively modified online after each sampling interval based on the system's state and error variations. The rate adaptation process initiates with the initial variation rates, $v(0)$, prescribed for FLQR in Section 4. As long as the exponent of the exponential term $-\boldsymbol{H}$ remains negative definite, $exp(-\boldsymbol{H}t)$ decays to zero over time. This behavior ensures that the proposed adaptation law exhibits exponential stability, as the output progressively diminishes. These VAFs deliver time-varying variation rates $\gamma(t)$ and $\delta(t)$, which are then incorporated into the cubic HTFs. The expressions of the self-regulating CSE and CSED variables are shown in Equations (32) and (33).

$$\varepsilon_{n,sr}(t) = \tanh\left(\gamma(t)\ (\varepsilon(t))^3\right) \tag{32}$$

$$\dot{\varepsilon}_{n,\ sr}(t) = \tanh\left(\delta(t)\ (\dot{\varepsilon}(t))^3\right) \tag{33}$$

where $\varepsilon_{n,sr}(t)$ and $\dot{\varepsilon}_{n,\ sr}(t)$ represent the self-regulating compounded error variables. After adaptive preprocessing via VAFs, the modified compounded error variables are fed directly to the FIS as input variables for the computation of the final control action. This arrangement dynamically reconfigures the shape and form of the cubic HTFs as per the state deviations, influencing the normalized CSE and CSED variables to manipulate the strength of the damping control action yielded by the FIS. Thus, apart from normalizing the input variables within ± 1 for improved fuzzy reasoning and introducing selective attenuation and amplification regions to enhance control input efficiency, the designed VAFs adaptively reconfigure the waveform of CHTF to improve the FLQR's flexibility and responsivity to effectively reject external disturbances. Consequently, the system efficiently mitigates overshoots, steady-state fluctuations, and disruptive control actions. The proposed control procedure is referred to as the Self-Regulating FLQR (SR-FLQR) law. The block diagram of SR-FLQR law is illustrated in Figure 6.

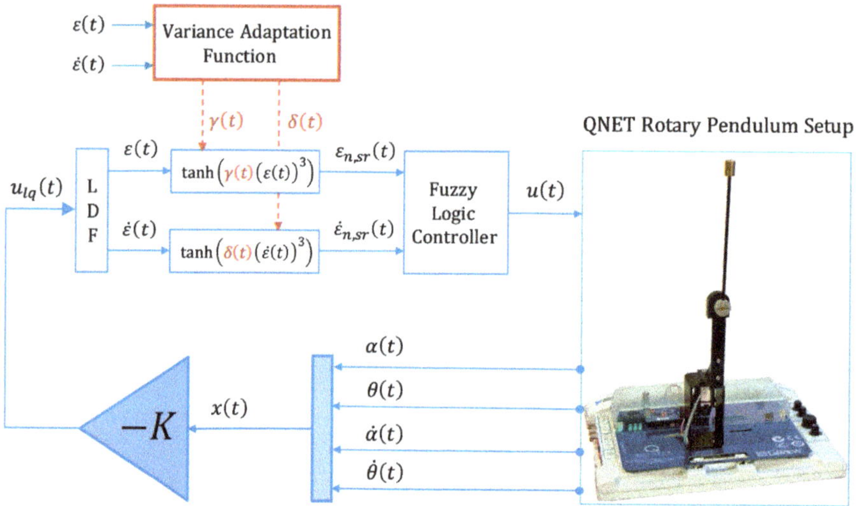

Figure 6. Block diagram of the proposed SR-FLQR law.

4. Parameter Optimization Method

The effectiveness of the LQR depends on both the system's state deviations and the adjustments in the control signal. The optimal control performance is achieved by appropriately selecting the weighting factors for the state matrix $\boldsymbol{Q} = \mathrm{diag}\begin{pmatrix} q_\alpha & q_\theta & q_{\dot{\alpha}} & q_{\dot{\theta}} \end{pmatrix}$ and control input factor $\boldsymbol{R} = \rho$. The operation of the FLQR scheme is dictated by the optimal selection of the pre-determined variation rates, γ and δ, of the HTFs. Similarly, the performance of the proposed self-regulating FLQR scheme, formulated in Section 3.2, is influenced by the predefined selection of the learning gains (σ_1, σ_2, β_1, and β_2) associated with the VAFs that dynamically reconfigure the waveform of the HTFs.

4.1. Tuning Algorithm

To optimize the aforementioned controller parameters offline, the cost function in Equation (34) is minimized [21].

$$J_c = \int_0^{\mathrm{T}} \left[(e_\alpha(t))^2 + (e_\theta(t))^2 + (u(t))^2 \right] dt \tag{34}$$

where T represents the duration of an experimental trial, $e_\alpha(t) = \alpha(t) - \alpha(0)$ and $e_\theta(t) = \theta(t) - \pi$, denote the deviation (or error) in the horizontal arm and pendulum link from their respective reference positions. The pendulum link's reference position is set at π radians, while the arm's reference is set at its initial position, $\alpha(0)$. Equal weights are applied to the variables in the cost function J_c to ensure a balanced compromise between tracking accuracy and control efficiency. This approach avoids biasing the controller toward any single objective. This setting ensures that neither the position errors nor the control effort are overly prioritized, leading to a well-rounded controller that maintains accurate reference tracking while minimizing excessive control actions. For the baseline LQR design, the weighting coefficients for the state error and control cost terms are selected within the range [0, 100] to balance performance and energy efficiency [1]. For the FLQR design, the variation rates, γ and δ, of the HTFs are selected within the range [0, 10]. For the self-regulating FLQR design, the decay rates (σ_1 and σ_2) and adaptation rates (β_1 and β_2) associated with the VAFs are selected within the range [0, 1] and [0, 10], respectively. The parameter selection and tuning process is depicted in Figure 7 [21]. The experimental implementation details are provided in Section 5.

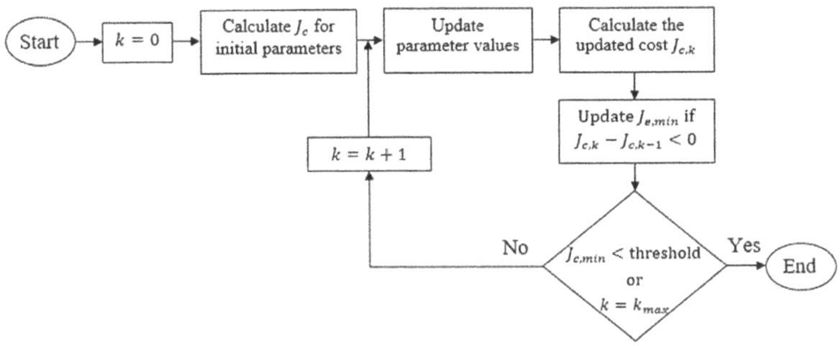

Figure 7. Parameter selection process [21].

The tuning procedure begins by setting all initial controller parameters to unity. Each iteration involves modifying the designated control parameters, manually positioning the pendulum rod in an upright state, and balancing it for a duration of 10 s to evaluate the corresponding cost $J_{c,k}$, where k denotes the trial number. The optimization follows a gradient-based approach, iteratively refining the parameters to minimize the cost function [21]. If the cost associated with the current trial, $J_{c,k}$, is less than that of the preceding trial, $J_{c,k-1}$, the local minimum cost, $J_{c,min}$, is revised accordingly.

The tuning process terminates when either the predefined maximum number of trials, k_{max}, is reached or $J_{c,min}$ falls below a specified threshold. The termination threshold is empirically determined through preliminary algorithmic trials, which assess different threshold values to achieve the best trade-off between solution accuracy and computational efficiency while avoiding premature convergence. In this study, the initial minimum cost is recorded as $J^0_{c,min} \approx 0.89 \times 10^6$ [21]. A scaling factor of 0.01 is applied to establish the stopping criterion, ensuring efficient convergence while mitigating unnecessary computational complexity. A larger scaling factor would impose an excessive computational cost, whereas a smaller factor risks premature termination. Consequently, the algorithm concludes when $J_{c,min}$ reaches 0.01 $J^0_{c,min}$. For this study, the final threshold values for $J_{c,min}$ and k_{max} are set at 1×10^4 and 30, respectively.

4.2. Controller Parameterization

Upon completion of the optimization process, the following set of parameters is acquired. For the LQR, the weighting coefficients of the state and control matrices are obtained as $Q = \mathrm{diag}\begin{pmatrix}32.82 & 52.23 & 6.12 & 2.51\end{pmatrix}$ and $R = 1.02$. The corresponding state feedback gain matrix, computed using (15), is $K = \begin{bmatrix}-6.21 & 130.56 & -4.22 & 17.83\end{bmatrix}$. For the FLQR design, the variation rates of the cubic HTF used to normalize the compounded error and its derivative are obtained as $\gamma = 1.59$ and $\delta = 0.87$. For the SR-FLQR design, the learning gains associated with the VAFs are selected as $\sigma_1 = 0.082$, $\sigma_2 = 0.065$, $\beta_1 = 3.14$, and $\beta_2 = 2.61$.

5. Experimental Results and Discussions

This section details the hardware-in-loop (HIL) implementation of the designed control procedures on Quanser's SLRIP platform.

5.1. HIL Implementation

The resilience of the designed control procedures is evaluated through customized experiments conducted on the Quanser rotary pendulum platform, depicted in Figure 8.

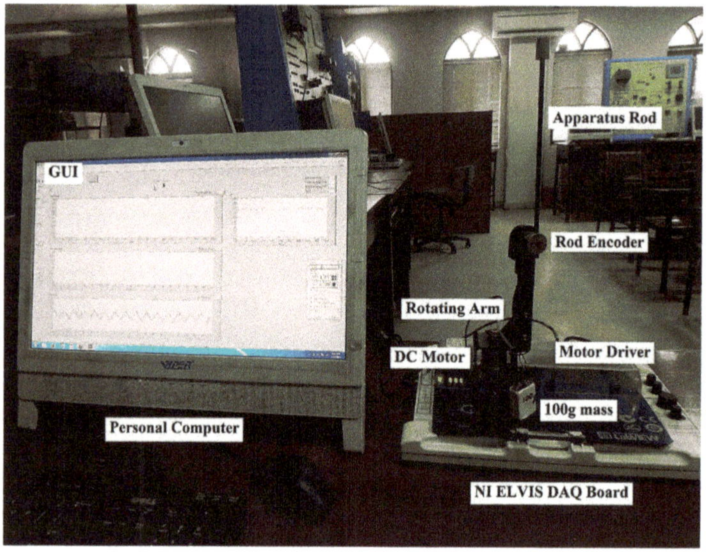

Figure 8. Quanser rotary pendulum platform used for HIL experiments [21].

The angular positions of the SLRIP's arm and rod are measured using high-resolution encoders at a 1.0 kHz sampling frequency, with data acquisition performed via an NI-DAQ Board. The encoder resolution was explicitly accounted for in the digital implementation. The measured angles are obtained from raw encoder counts as shown in Equation (35).

$$\theta = \frac{2\pi}{N} C \tag{35}$$

where N is the encoder counts and C is the encoder resolution in counts per revolution. This conversion ensures that the control algorithm operates on precise angular states rather than raw counts, thereby maintaining accuracy in the feedback loop. The received signals are filtered and then communicated to the control application over a 9600-bps serial link. A customized LabVIEW-based software application is developed using LabVIEW 2018's Block Diagram tool. The control software runs on a 64-bit, 2.1 GHz Intel Core i7 processor with 16 GB RAM, which efficiently handles the fuzzy logic computations as well as the recursive computations associated with the online variance adaptation functions. These functions are implemented using C-language programming within LabVIEW's Math Script tool, while additional required functions are selected from the function palette. The customized graphical user interface of the control application provides real-time monitoring and data logging of the system's control input and state variations. The angular velocities of both the pendulum link and the rotary arm were computed from encoder position measurements by numerical differentiation, followed by low-pass filtering to suppress measurement noise. These filtered velocity estimates were then used in the control computations. The control algorithm continuously performs the fuzzy logic computations and updates the control signal at each sampling interval, leveraging the real-time clock of the embedded processor. The SLRIP's onboard motor driver modulates these updated signals to actuate the DC servomotor, ensuring safe operation under discontinuous control demands. Each experiment begins with the manual stabilization of the pendulum link, ensuring consistent initial conditions for a fair performance comparison.

5.2. Experimental Validation

To assess the performance and robustness against disturbances, each prescribed control procedure is tested to maintain the pendulum link in the upright position and the horizontal arm at its initial (reference) position. The following HIL experiments are used to validate system resilience.

A. Position Regulation: This baseline experiment evaluates the system's ability to maintain its reference positions without external disturbances. It replicates real-world scenarios where a robotic system or an industrial manipulator must remain stable under normal operating conditions. The time-domain profiles of $\theta(t)$, $\alpha(t)$, and $V_m(t)$ are manifested in Figure 9.

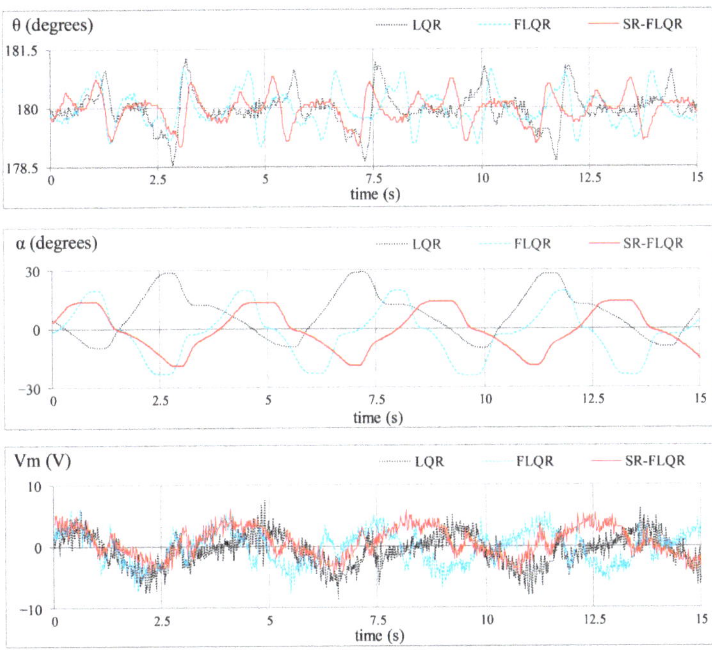

Figure 9. Experiment A: Normal position regulation behavior of SLRIP.

B. Impulse Disturbance Rejection: This experiment is conducted to evaluate the system's response to transient impacts. An impulse disturbance is injected into the control input to assess the system's capability to recover from sudden external shocks, such as power transients, abrupt mechanical impacts, sudden hardware failures, or seismic forces in engineering applications. A simulated 5.0 V pulse, having a span of 100 ms, is applied whenever the horizontal arm reaches its maximum position. The corresponding profiles of $\theta(t)$, $\alpha(t)$, and $V_m(t)$ are illustrated in Figure 10.

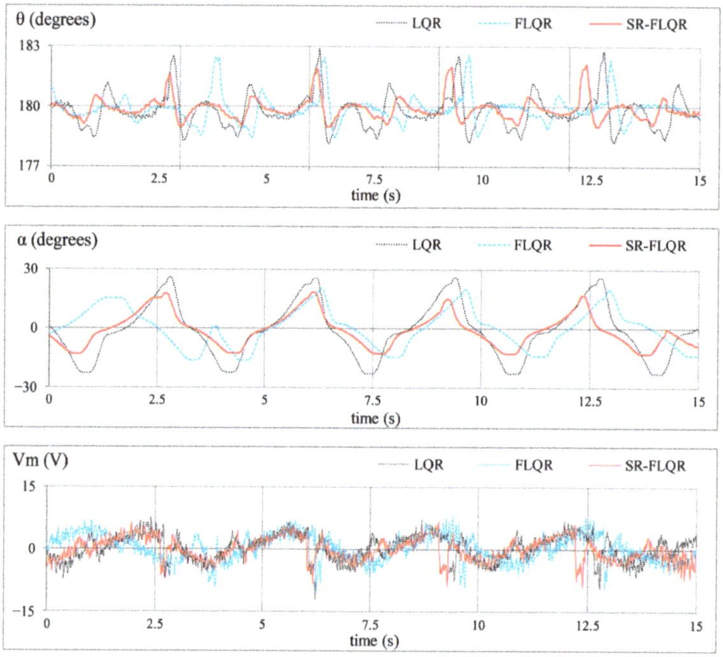

Figure 10. Experiment B: Impulsive disturbance compensation behavior of SLRIP.

C. Step Disturbance Compensation: This experiment evaluates the effect of a sudden and sustained external force, similar to wind gusts on aircraft, ocean currents affecting marine vessels, or sustained mechanical loads on industrial robots. A −5.0 V step input is introduced in the control signal at t = 10 s to examine the control scheme's ability to manage abrupt but constant disturbances. The resulting variations in $\theta(t)$, $\alpha(t)$, and $V_m(t)$ are depicted in Figure 11.

D. Sinusoidal Disturbance Attenuation: This test examines the control system's resilience against periodic disturbances and mechanical vibrational effects, which are very common in robotics, aerospace, and automotive control systems. Although real disturbances such as sensor noise, friction, and backlash are non-periodic in nature, a sinusoidal disturbance is employed for this test case as a structured input to systematically test the controller's robustness. The sinusoidal disturbance provides a repeatable and well-defined disturbance with continuous excitation across time, which makes it suitable for stressing both the tracking and regulation capabilities of the controller [48]. Moreover, stochastic disturbances are inherently present during the hardware experiments, meaning the proposed controller was simultaneously exposed to realistic non-periodic perturbations as well. A sinusoidal disturbance signal of the form, $\mathrm{d}(\mathrm{t}) = \sin(20\pi \mathrm{t})$, is injected into the control signal to evaluate robustness. The corresponding responses of $\theta(t)$, $\alpha(t)$, and $V_m(t)$ are shown in Figure 12.

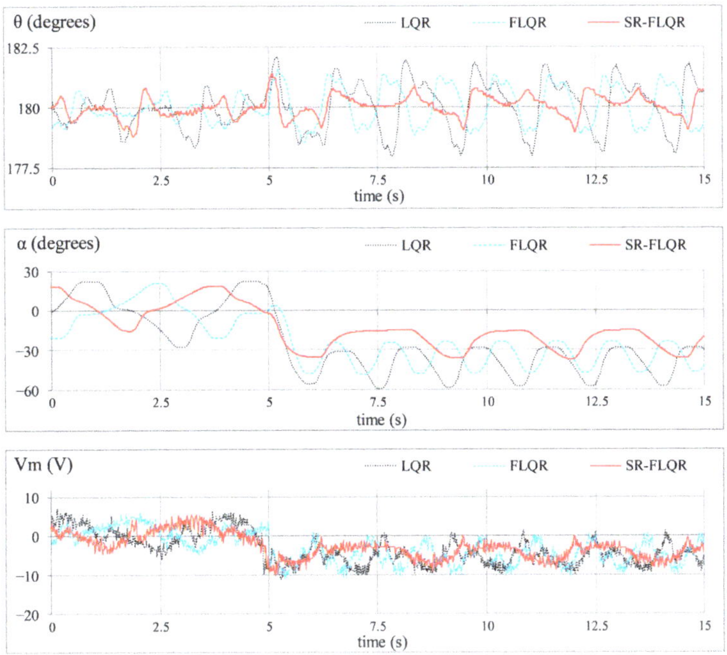

Figure 11. Experiment C: Step disturbance compensation behavior of SLRIP.

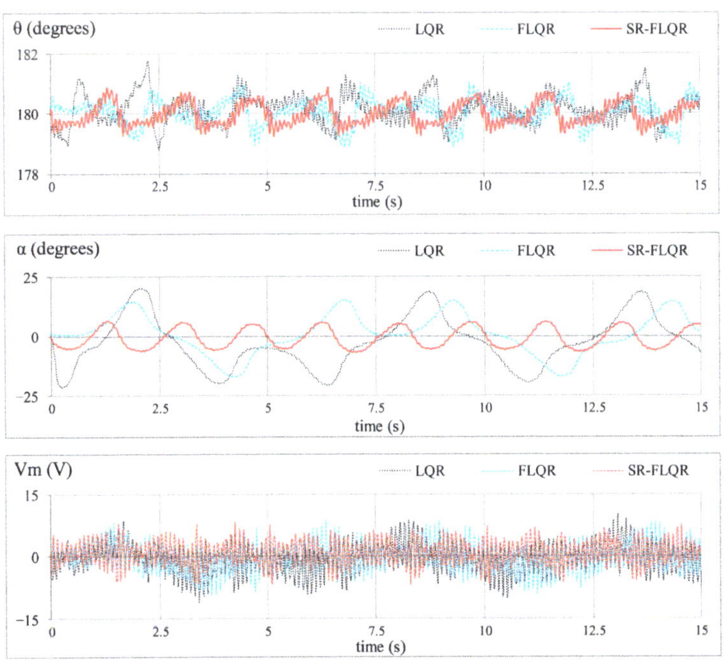

Figure 12. Experiment D: Sinusoidal disturbance attenuation behavior of SLRIP.

E. Model Error Rejection: This scenario assesses the controller's capacity to compensate for model inaccuracies and parametric variations, which occur in real-world applications where the system dynamics evolve/alter over time. To simulate this effect, a 0.1 kg mass is attached beneath the pendulum link, creating a mismatch between the real and modeled system dynamics. The system's adaptability to this perturbation is investigated via the time domain profiles of $\theta(t)$, $\alpha(t)$, and $V_m(t)$ shown in Figure 13.

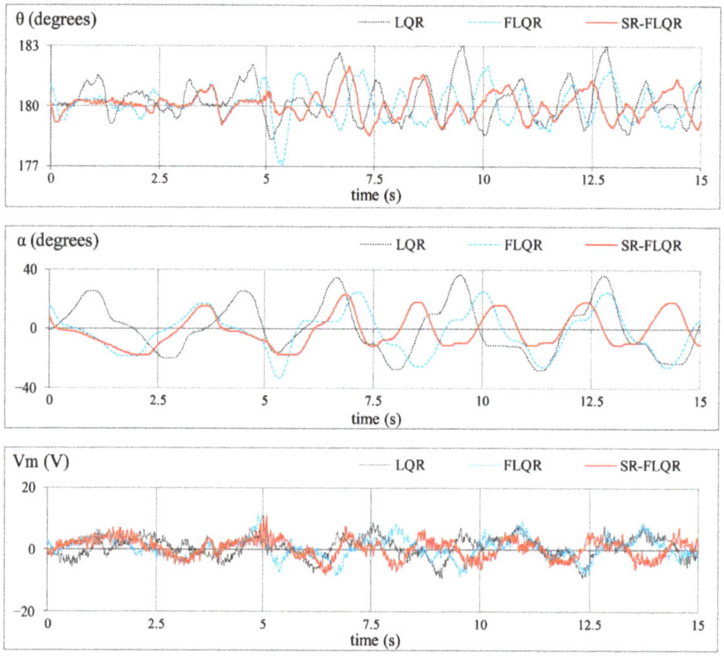

Figure 13. Experiment E: Model error rejection behavior of SLRIP.

5.3. Analysis and Discussions

The following seven Quantitative Performance Measures (QPMs) are used to evaluate each controller's performance under the given test scenarios.

1. $RMSE_z$—Root mean square (RMS) error in the pendulum link and arm positions.
2. $M_{p,\theta}$—Peak overshoot magnitude of the rod following a disturbance.
3. $T_{s,\theta}$—Time required for the pendulum link to settle within ±5% of π rad. reference of the pendulum (~0.157 rad.) after a transient disturbance.
4. α_{off}—Offset in the horizontal arm's position after disturbance application.
5. $\alpha_{p\text{-}p}$—Peak-to-peak magnitude of oscillations in the arm after disturbance application.
6. MSV_m—Mean squared value of the motor control voltage.
7. V_p—Maximum control voltage after disturbance application.

The RMS error is evaluated as shown in Equation (36).

$$\mathrm{RMSE_z} = \sqrt{\frac{1}{M}\sum_{k=1}^{M}(e_z(t))^2} \tag{36}$$

where $z = \alpha$ or θ, and M is the number of samples. The settling time $T_{s,\theta}$ is evaluated from the pendulum's response as expressed in Equation (37).

$$\mathrm{T_{s,\theta}} = min\{t \mid |e_z(t)| \leq 0.157\} \tag{37}$$

The mean-squared value of motor control voltage is evaluated as shown in (38).

$$\mathrm{MSV_m} = \frac{1}{M}\sum_{k=1}^{M}(u(t))^2 \tag{38}$$

The HIL experimental results confirm that the proposed SR-FLQR scheme exhibits superior position regulation, robust disturbance compensation, rapid response speed, and minimal control energy consumption. A quantitative analysis of the experimental results is presented in Table 3, with a qualitative discussion presented below.

Table 3. Quantitative evaluation of experimental results.

Experiment	QPM		Control Scheme		
	Symbol	**Unit**	**LQR**	**FLQR**	**SR-FLQR**
A	RMSE_θ	deg.	0.49	0.43	0.38
	RMSE_α	deg.	14.64	13.38	10.81
	$\mathrm{MSV_m}$	$\mathrm{V^2}$	7.53	7.09	5.87
B	RMSE_θ	deg.	0.82	0.66	0.52
	$\mathrm{M}_{p,\theta}$	deg.	2.87	2.58	2.15
	$\mathrm{T}_{s,\theta}$	s.	0.78	0.61	0.52
	RMSE_α	deg.	13.74	9.82	8.42
	$\mathrm{MSV_m}$	$\mathrm{V^2}$	10.06	10.19	7.52
	$\mathrm{V_p}$	V	−11.94	−10.05	−8.99
C	RMSE_θ	deg.	0.96	0.68	0.45
	RMSE_α	deg.	34.81	26.98	21.31
	α_{off}	deg.	−42.12	−29.92	−25.40
	$\alpha_{\text{p-p}}$	deg.	30.78	33.88	21.37
	$\mathrm{MSV_m}$	$\mathrm{V^2}$	27.70	21.01	17.98
D	RMSE_θ	deg.	0.51	0.47	0.38
	RMSE_α	deg.	10.93	8.21	4.29
	$\mathrm{MSV_m}$	$\mathrm{V^2}$	13.21	11.34	8.57
E	RMSE_θ	deg.	1.01	0.83	0.62
	RMSE_α	deg.	16.84	14.22	11.18
	$\mathrm{MSV_m}$	$\mathrm{V^2}$	11.29	12.42	10.03

In Experiment A, the fixed-gain LQR exhibits the weakest performance among the tested controllers, exhibiting the highest position-regulation errors and excessive control energy usage. The FLQR improves accuracy and reduces control costs, while the SR-FLQR further enhances tracking precision and energy efficiency. In Experiment B, LQR demonstrates weak disturbance rejection, leading to slow recovery, large overshoots, and high control demands. FLQR mitigates overshoots and speeds up response, but at the cost of disrupted control activity. SR-FLQR ensures faster recovery, better damping, and lower peak control input demands, optimizing efficiency. In Experiment C, LQR fails to compensate for step disturbances, resulting in large angular offsets. FLQR reduces fluctuations, but with erratic control energy usage. SR-FLQR provides faster recovery, better disturbance rejection, and smoother control with lower peak torque requirements. In Experiment D, LQR struggles with sinusoidal disturbances, causing large state fluctuations and poor control efficiency. The FLQR offers better immunity, but SR-FLQR outperforms both by minimizing fluctuations and reducing chattering while maintaining smooth control. In Experiment E, LQR inadequately handles modeling errors, leading to large perturbations. The FLQR improves position regulation and control efficiency, while SR-FLQR delivers the best error compensation with strong damping and optimal energy usage.

Quantitatively, SR-FLQR improves position regulation, energy efficiency, transient recovery, and overshoot reduction by significant margins over both LQR and FLQR. As compared to the baseline LQR, SR-FLQR exhibits a 22.4% improvement in $RMSE_\theta$, a 26.2% reduction in $RMSE_\alpha$, and a 22.0% reduction in MSV_m under nominal conditions. In the disturbed state, as compared to LQR, the SR-FLQR demonstrates a 25.1% reduction in $M_{p,\theta}$, a 33.3% reduction in $T_{s,\theta}$, a 24.7% reduction in the peak servo control demands, a 39.7% improvement in α_{off}, and a 30.6% enhancement in $\alpha_{p\text{-}p}$. As compared to the baseline FLQR, SR-FLQR exhibits an 11.6% improvement in $RMSE_\theta$, a 19.2% improvement in $RMSE_\alpha$, and a 17.2% improvement in MSV_m under nominal conditions. In the disturbed state, as compared to FLQR, the SR-FLQR demonstrates a 16.7% reduction in $M_{p,\theta}$, a 14.8% reduction in $T_{s,\theta}$, a 10.5% reduction in the peak servo control demands, a 15.1% improvement in α_{off}, and a 36.9% enhancement in $\alpha_{p\text{-}p}$.

These enhancements stem from its self-regulating weight-adjusting function, which increases adaptability and enables dynamic restructuring of the control scheme for optimal performance. The superior responsiveness of SR-FLQR is evident in its rapid yet precise variance adjustments, allowing it to effectively counteract disturbances.

In practical digital implementations, the proposed SR-FLQR offers strengths such as online self-tuning without requiring explicit system models, robustness against parameter variations, and ease of integration with existing LQR frameworks. However, its limitations include increased computational demand due to the fuzzy inference and variance adaptation layers, potential sensitivity to discretization delays at low sampling rates, and the need for careful tuning of adaptation gains to avoid excessive actuator effort. These aspects should be considered when deploying the controller in embedded hardware with limited resources.

5.4. Comparison with a Robust SMC Controller

To benchmark the proposed control strategy further, a comparison was performed against Gao's power-rate SMC law, a well-known robust method designed to handle model uncertainties and external disturbances. The SMC was implemented using the same system model and experimental setup. The SMC formulation adopted for comparison in this study follows the control law design proposed in [7], and is expressed in Equation (39).

$$u(t) = -\left(\boldsymbol{F}^T\boldsymbol{B}\right)^{-1}\left(\boldsymbol{F}^T\boldsymbol{A}x(t) + m|s(t)|^{\gamma} sgn(s(t))\right) \quad (39)$$

where $s(t)$ is the sliding surface, $\boldsymbol{F}^T$ is a state-error weighting vector, m is a positive scaling factor, γ is the positive fractional power of $s(t)$, and $sgn(.)$ is the "signum" function. The sliding surface $s(t)$ is computed as shown in Equation (40).

$$s(t) = \boldsymbol{F}^T p(t) \quad (40)$$

such that,

$$p(t) = x(t) - x_{ref}$$

where $x_{ref} = \begin{bmatrix} 0 & \pi & 0 & 0 \end{bmatrix}^T$ is the reference state vector. The tuned parameter values used for controlling the Quanser SLRIP system were adopted from the study in [7], with $m = 4.07$, $\gamma = 0.45$, and $\boldsymbol{F}^T = \begin{bmatrix} -3.16 & 73.48 & -3.02 & 9.25 \end{bmatrix}$. The SMC-regulated SLRIP system is subjected to the same tests as discussed in Section 5.2. The graphical results yielded by Experiments A to E, conducted to compare the performance of the SR-FLQR against the SMC, are shown in Figures 14–18.

The experimental results are quantitatively presented in Table 4. The experimental results show that while the SMC law achieves acceptable stabilization of the rotary inverted pendulum, it exhibits chattering in the control input and larger steady-state fluctuations in both the pendulum angle and the rotary arm position. In contrast, the proposed SR-FLQR significantly reduces oscillations, improves tracking precision, and requires lower control effort. Quantitatively, SR-FLQR yielded smaller variance and RMSE values across all state variables compared to SMC, thereby demonstrating superior disturbance rejection and smoother control action.

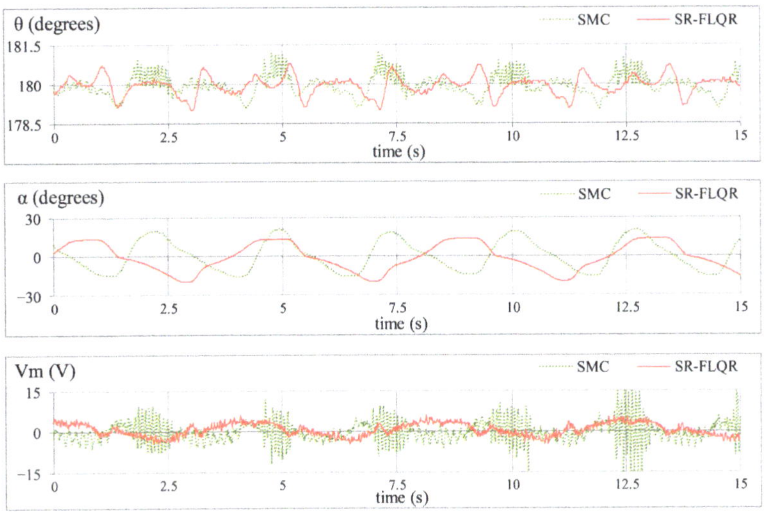

Figure 14. Experiment A: Normal position regulation behavior of SLRIP.

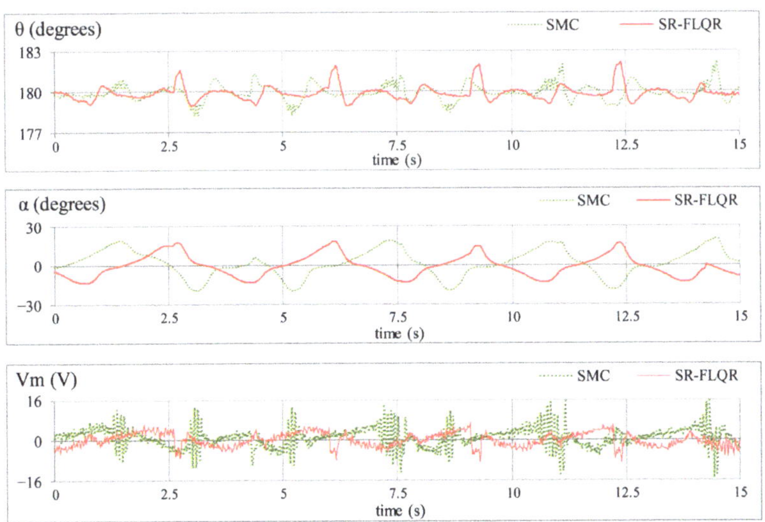

Figure 15. Experiment B: Impulsive disturbance compensation behavior of SLRIP.

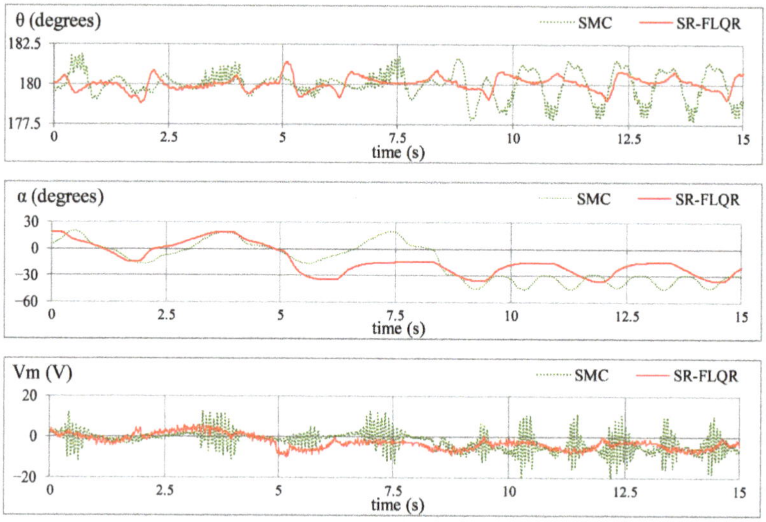

Figure 16. Experiment C: Step disturbance compensation behavior of SLRIP.

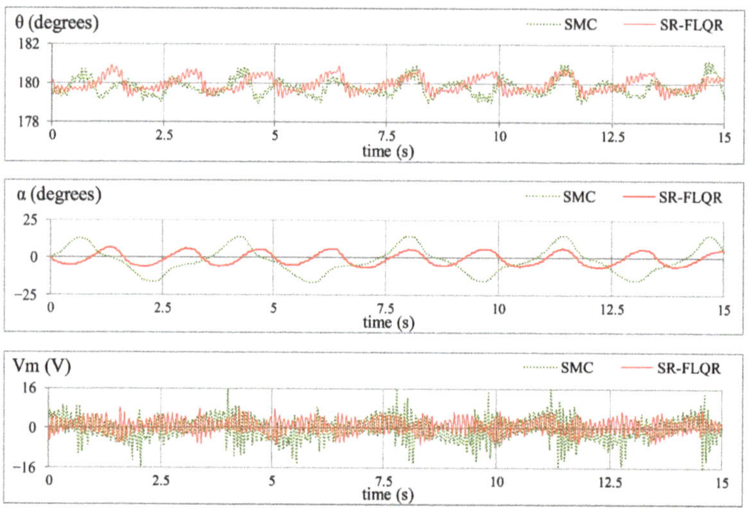

Figure 17. Experiment D: Sinusoidal disturbance attenuation behavior of SLRIP.

These findings confirm that the proposed SR-FLQR framework not only enhances performance over conventional LQR and FLQR but also outperforms an established robust control method such as SMC, highlighting its practical effectiveness for inverted pendulum stabilization.

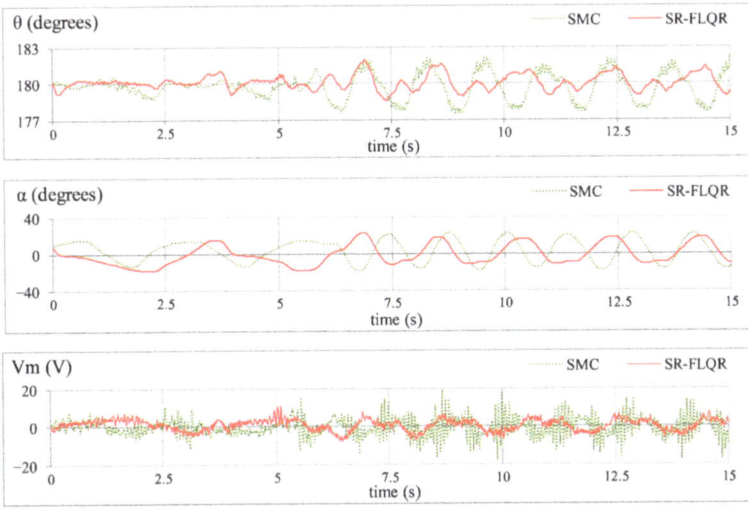

Figure 18. Experiment E: Model error rejection behavior of SLRIP.

Table 4. Comparative analysis of SMC with SR-FLQR.

Experiment	QPM		Control Scheme		Improvement in SR-FLQR (%)
	Symbol	Unit	SMC [7]	SR-FLQR	
A	$RMSE_\theta$	deg.	0.40	0.38	5.0
	$RMSE_\alpha$	deg.	11.85	10.81	8.8
	MSV_m	V^2	19.71	5.87	70.2
B	$RMSE_\theta$	deg.	0.61	0.52	14.8
	$M_{p,\theta}$	deg.	2.20	2.15	2.3
	$T_{s,\theta}$	s.	0.72	0.52	27.8
	$RMSE_\alpha$	deg.	10.34	8.42	18.6
	MSV_m	V^2	14.71	7.52	48.9
	V_p	V	−16.02	−8.99	43.9
C	$RMSE_\theta$	deg.	0.91	0.45	50.5
	$RMSE_\alpha$	deg.	27.70	21.31	23.1
	α_{off}	deg.	−35.91	−25.40	29.3
	MSV_m	V^2	30.33	17.98	40.7
D	$RMSE_\theta$	deg.	0.44	0.38	13.6
	$RMSE_\alpha$	deg.	8.69	4.29	50.6
	MSV_m	V^2	19.94	8.57	57.0
E	$RMSE_\theta$	deg.	1.06	0.62	41.5
	$RMSE_\alpha$	deg.	12.42	11.18	10.0
	MSV_m	V^2	22.54	10.03	55.5

6. Conclusions

This study validates the effectiveness of a self-regulating fuzzy-LQR control approach incorporating reconfigurable hyperbolic functions for intermediate error modulation in SLRIP systems. The proposed controller leverages self-regulating nonlinear scaling techniques to enhance adaptability, resilience, and transient response against bounded external disturbances. By dynamically modifying the variances of cubic HTFs, the scheme eliminates inaccuracies in heuristic calibration and optimally reconfigures its control structure in real time. The experimental results demonstrate that the adaptive variance adjustments significantly improve transient behavior, reinforce control action against tracking fluctuations, and reduce the actuator's control demands while preserving the system's closed-loop stability under diverse operating conditions. The proposed methodology effectively mitigates nonlinear complexities and parametric uncertainties, offering a more flexible and resilient control strategy. The proposed SR-FLQR technique is notably scalable and can be extended to numerous underactuated electromechanical systems, including flexible joint manipulators, bipedal robots, quadrotors, or VTOL platforms. In all these cases, the LQR component of the control law would necessitate an accurate linearized state-space model around a predefined equilibrium point. However, as system dynamics become more complex, the selection of fuzzy rules, membership functions, and scaling parameters would also become nontrivial and computationally demanding.

Future research may explore several directions to strengthen the robustness of the proposed control framework further. First, the LQR component may be replaced with state-dependent Riccati equation control or nonlinear MPC to address the challenges associated with the system's nonlinearities. Second, the proposed control scheme can be validated on a broader class of underactuated electromechanical systems, such as flexible manipulators, quadrotors, and humanoid robots. Third, the manual formulation and tuning of fuzzy rules and membership functions can be streamlined through data-driven techniques, including adaptive neuro-fuzzy inference systems (ANFISs), evolutionary algorithms, or reinforcement learning. Fourth, online self-organizing algorithms can be integrated with the proposed control framework for real-time shape adaptation of scaling functions. Fifth, the integration of a fuzzy adaptive system with fractional-order PID control to adaptively regulate the control input can also be investigated in future work. Finally, the influence of viscous friction in DC motors must be explicitly incorporated into the system model in future work. Accounting for this effect is expected to improve the accuracy of the system's dynamic model representation and provide deeper insight into the controller's robustness under more realistic actuator conditions.

Author Contributions: Conceptualization, O.S.; methodology, O.S.; software, O.S. and S.A.; validation, J.I.; formal analysis, S.A.; investigation, O.S. and S.A.; writing—original draft preparation, O.S. and S.A.; writing—review and editing, J.I.; visualization, J.I.; supervision, J.I.; validation, J.I. All authors have read and agreed to the published version of the manuscript.

Funding: This research received no external funding.

Data Availability Statement: The original contributions presented in this study are included in the article. Further inquiries can be directed to the corresponding author.

Conflicts of Interest: The authors declare no conflicts of interest.

References

1. Aguilar-Avelar, C.; Moreno-Valenzuela, J. A composite controller for trajectory tracking applied to the Furuta pendulum. *ISA Trans.* **2015**, *57*, 286–294. [CrossRef] [PubMed]
2. Costanzo, M.; Mazza, R.; Natale, C. Digital Control of an Inverted Pendulum Using a Velocity-Controlled Robot. *Machines* **2025**, *13*, 528. [CrossRef]
3. Ramírez-Neria, M.; Sira-Ramírez, H.; Garrido-Moctezuma, R.; Luviano-Juarez, A. Linear active disturbance rejection control of underactuated systems: The case of the Furuta pendulum. *ISA Trans.* **2014**, *53*, 920–928. [CrossRef]
4. Gritli, H.; Belghit, S. Robust feedback control of the underactuated Inertia Wheel Inverted Pendulum under parametric uncertainties and subject to external disturbances: LMI formulation. *J. Frankl. Inst.* **2018**, *355*, 9150–9191. [CrossRef]
5. Kim, S.; Kwon, S. Nonlinear optimal control design for underactuated two-wheeled inverted pendulum mobile platform. *IEEE/ASME Trans. Mechatron.* **2017**, *22*, 2803–2808. [CrossRef]
6. Wang, F.-C.; Chen, Y.-H.; Wang, Z.-J.; Liu, C.-H.; Lin, P.-C.; Yen, J.-Y. Decoupled Multi-Loop Robust Control for a Walk-Assistance Robot Employing a Two-Wheeled Inverted Pendulum. *Machines* **2021**, *9*, 205. [CrossRef]
7. Saleem, O.; Mahmood-ul-Hasan, K. Adaptive State-space Control of Under-actuated Systems Using Error-magnitude Dependent Self-tuning of Cost Weighting-factors. *Int. J. Control Autom. Syst.* **2021**, *19*, 931–941. [CrossRef]

8. Prasad, L.B.; Tyagi, B.; Gupta, H.O. Optimal control of nonlinear inverted pendulum system using PID controller and LQR: Performance analysis without and with disturbance input. *Int. J. Autom. Comput.* **2014**, *11*, 661–670. [CrossRef]
9. Boubaker, O. The Inverted Pendulum Benchmark in Nonlinear Control Theory: A Survey. *Int. J. Adv. Robot. Syst.* **2013**, *10*, 233. [CrossRef]
10. Wang, J.J. Simulation studies of inverted pendulum based on PID controllers. *Sim. Mod. Pract. Theory* **2011**, *19*, 440–449. [CrossRef]
11. Wang, J.J.; Kumbasar, T. Optimal PID control of spatial inverted pendulum with big bang–big crunch optimization. *IEEE/CAA J. Autom. Sin.* **2018**, *7*, 822–832. [CrossRef]
12. Saleem, O.; Mahmood-ul-Hasan, K. Robust stabilisation of rotary inverted pendulum using intelligently optimised nonlinear self-adaptive dual fractional-order PD controllers. *Int. J. Syst. Sci.* **2019**, *50*, 1399–1414. [CrossRef]
13. Batiha, I.M.; Momani, S.; Batyha, R.M.; Jebril, I.H.; Judeh, D.A.; Oudetallah, J. DC motor speed control via fractional-order PID controllers. *Int. J. Fuzzy Log. Intell. Syst.* **2024**, *24*, 74–82. [CrossRef]
14. Huang, Y.; Wang, D.; Zhang, J.; Guo, F. Controlling and synchronizing a fractional-order chaotic system using stability theory of a time-varying fractional-order system. *PLoS ONE* **2018**, *13*, e0194112. [CrossRef]
15. Nguyen-Vinh, Q.; Pham-Tran-Bich, T. Sliding mode control of induction motor with fuzzy logic observer. *Electr. Eng.* **2023**, *105*, 2769–2780. [CrossRef]
16. Saleem, O.; Alsuwian, T.; Amin, A.A.; Ali, S.; Alqarni, Z.A. Stabilization control of rotary inverted pendulum using a novel EKF-based fuzzy adaptive sliding-mode controller: Design and experimental validation. *Automatika* **2024**, *65*, 538–558. [CrossRef]
17. Chu, T.D.; Chen, C.K. Design and Implementation of Model Predictive Control for a Gyroscopic Inverted Pendulum. *Appl. Sci.* **2017**, *7*, 1272. [CrossRef]
18. Waheed, F.; Yousufzai, I.K.; Valášek, M. A TV-MPC Methodology for Uncertain Under-Actuated Systems: A Rotary Inverted Pendulum Case Study. *IEEE Access* **2023**, *11*, 103636–103649. [CrossRef]
19. Le, H.D.; Nestorović, T. Integral Linear Quadratic Regulator Sliding Mode Control for Inverted Pendulum Actuated by Stepper Motor. *Machines* **2025**, *13*, 405. [CrossRef]
20. Xue, D.; Chen, Y.Q.; Atherton, D.P. *Linear Feedback Control: Analysis and Design with MATLAB*; SIAM: Philadelphia, PA, USA, 2007.
21. Saleem, O.; Iqbal, J. Phase-Based Adaptive Fractional LQR for Inverted-Pendulum-Type Robots: Formulation and Verification. *IEEE Access* **2024**, *12*, 93185–93196. [CrossRef]
22. Saleem, O.; Abbas, F.; Iqbal, J. Complex Fractional-Order LQIR for Inverted-Pendulum-Type Robotic Mechanisms: Design and Experimental Validation. *Mathematics* **2023**, *11*, 913. [CrossRef]
23. Szuster, M.; Hendzel, Z. *Intelligent Optimal Adaptive Control for Mechatronic Systems*; Springer: Berlin/Heidelberg, Germany, 2017.
24. Zabihifar, S.H.; Yushchenko, A.S.; Navvabi, H. Robust control based on adaptive neural network for Rotary inverted pendulum with oscillation compensation. *Neural Comput. Appl.* **2020**, *32*, 14667–14679. [CrossRef]
25. Ochoa, P.; Peraza, C.; Melin, P.; Castillo, O.; Geem, Z.W. Type-3 Fuzzy Dynamic Adaptation of the Crossover Parameter in Differential Evolution for the Optimal Design of Type-3 Fuzzy Controllers for the Inverted Pendulum. *Machines* **2025**, *13*, 450. [CrossRef]
26. Moness, M.; Mahmoud, D.; Hussein, A. Real-time Mamdani-like fuzzy and fusion-based fuzzy controllers for balancing two-wheeled inverted pendulum. *J. Ambient Intell. Humaniz. Comput.* **2022**, *13*, 3577–3593. [CrossRef]
27. Nguyen, T.V.A.; Dao, Q.T.; Bui, N.T. Optimized fuzzy logic and sliding mode control for stability and disturbance rejection in rotary inverted pendulum. *Sci. Rep.* **2024**, *14*, 31116. [CrossRef]
28. Hazem, B.Z.; Bingül, Z. A comparative study of anti-swing radial basis neural-fuzzy LQR controller for multi-degree-of-freedom rotary pendulum systems. *Neural Comput. Appl.* **2023**, *35*, 17397–17413. [CrossRef]
29. Hazem, B.Z.; Fotuhi, M.J.; Bingül, Z. Development of a Fuzzy-LQR and Fuzzy-LQG stability control for a double link rotary inverted pendulum. *J. Frankl. Inst.* **2020**, *357*, 10529–10556. [CrossRef]
30. Shang, W.W.; Cong, S.; Li, Z.X.; Jiang, S.L. Augmented Nonlinear PD Controller for a Redundantly Actuated Parallel Manipulator. *Adv. Robot.* **2009**, *23*, 1725–1742. [CrossRef]
31. Li, Z.; Zhang, X.; Huang, J. Tank-level robust control of LNG carrier based on hyperbolic tangent function nonlinear modification. *Transp. Saf. Environ.* **2024**, *6*, tdae010. [CrossRef]
32. Qasem, O.; Gutierrez, H.; Gao, W. Experimental validation of data-driven adaptive optimal control for continuous-time systems via hybrid iteration: An application to rotary inverted pendulum. *IEEE Trans. Ind. Electron.* **2023**, *71*, 6210–6220. [CrossRef]
33. Unluturk, A.; Aydogdu, O. Machine learning based self-balancing and motion control of the underactuated mobile inverted pendulum with variable load. *IEEE Access* **2022**, *10*, 104706–104718. [CrossRef]
34. Lee, T.; Ju, D.; Lee, Y.S. Transition Control of a Double-Inverted Pendulum System Using Sim2Real Reinforcement Learning. *Machines* **2025**, *13*, 186. [CrossRef]
35. Oh, Y.; Lee, T.; Ryoo, S.; Koh, K.C.; Han, S.; Lee, Y.S. Reinforcement Learning to Achieve Real-time Control of a Quadruple Inverted Pendulum. *Int. J. Control Autom. Syst.* **2025**, *23*, 2797–2806. [CrossRef]

36. Israilov, S.; Fu, L.; Sánchez-Rodríguez, J.; Fusco, F.; Allibert, G.; Raufaste, C.; Argentina, M. Reinforcement learning approach to control an inverted pendulum: A general framework for educational purposes. *PLoS ONE* **2023**, *18*, e0280071. [CrossRef]
37. Hernandez, R.; Garcia-Hernandez, R.; Jurado, F. Modeling, simulation, and control of a rotary inverted pendulum: A reinforcement learning-based control approach. *Modelling* **2024**, *5*, 1824–1852. [CrossRef]
38. Razmjooei, H.; Palli, G.; Abdi, E.; Strano, S.; Terzo, M. Finite-time continuous extended state observers: Design and experimental validation on electro-hydraulic systems. *Mechatronics* **2022**, *85*, 102812. [CrossRef]
39. Saleem, O.; Iqbal, J.; Afzal, M.S. A robust variable-structure LQI controller for under-actuated systems via flexible online adaptation of performance-index weights. *PLoS ONE* **2023**, *18*, e0283079. [CrossRef] [PubMed]
40. Balamurugan, S.; Venkatesh, P. Fuzzy sliding-mode control with low pass filter to reduce chattering effect: An experimental validation on Quanser SRP. *Sadhana* **2017**, *42*, 1693–1703. [CrossRef]
41. Astom, K.J.; Apkarian, J.; Karam, P.; Levis, M.; Falcon, J. *Student Workbook: QNET Rotary Inverted Pendulum Trainer for NI ELVIS*; Quanser Inc.: Waterloo, ON, Canada, 2011.
42. Peláez, G.; Izquierdo, P.; Peláez, G.; Rubio, H. A Multibody-Based Benchmarking Framework for the Control of the Furuta Pendulum. *Actuators* **2025**, *14*, 377. [CrossRef]
43. Lewis, F.L.; Vrabie, D.; Syrmos, V.L. *Optimal Control*, 3rd ed.; John Wiley & Sons: Hoboken, NJ, USA, 2012.
44. Lupu, D.; Necoara, I. Exact representation and efficient approximations of linear model predictive control laws via HardTanh type deep neural networks. *Syst. Control Lett.* **2024**, *186*, 105742. [CrossRef]
45. Wang, Y.; Li, Y.; Song, Y.; Rong, X. The Influence of the Activation Function in a Convolution Neural Network Model of Facial Expression Recognition. *Appl. Sci.* **2020**, *10*, 1897. [CrossRef]
46. Alagoz, B.B.; Ates, A.; Yeroglu, C.; Senol, B. An experimental investigation for error-cube PID control. *Trans. Inst. Meas. Control* **2015**, *37*, 652–660. [CrossRef]
47. Saleem, O.; Rizwan, M.; Iqbal, J. Adaptive optimal control of under-actuated robotic systems using a self-regulating nonlinear weight-adjustment scheme: Formulation and experimental verification. *PLoS ONE* **2023**, *18*, e0295153. [CrossRef] [PubMed]
48. Kumar, S.; Ajmeri, M. Robust stabilization and tracking control of spatial inverted pendulum using super-twisting sliding mode control. *Int. J. Dyn. Control* **2023**, *11*, 1178–1189. [CrossRef]

machines

MDPI

Article

Autonomous Sewer Robot: A Laser Marker-Based Detection System

Vygantas Ušinskis, Andrius Dzedzickis, Justas Nekrašas and Vytautas Bučinskas *

Department of Mechatronics, Robotics and Digital Manufacturing, Vilnius Gediminas Technical University (VilniusTech), Plytinės 25, LT10505 Vilnius, Lithuania; vygantas.usinskis@vilniustech.lt (V.U.); andrius.dzedzickis@vilniustech.lt (A.D.); justas.nekrasas@gmail.com (J.N.)
* Correspondence: vytautas.bucinskas@vilniustech.lt; Tel.: +370-5-27447

Abstract: Navigation technologies are becoming more advanced, helping to solve complicated problems in various fields. Navigation can be classified as global, in which predefined data and reference points from the working environment are used to generate a path, and local, where the map is generated momentarily by acquiring data from outside using sensors. As navigation tasks become more demanding, working environments can become very complicated with an increasing number of dynamic obstacles or, in some cases, a lack of global references, which may have particularly notable impacts on communication. Inspection robots that are required to work in underground sewers are often required to work completely locally, relying on sensor data. For this reason, in this study, a cost-efficient laser marker-based obstacle detection and measurement system is designed and tested for future use in autonomous sewer robot local path generation. Our experiments show the convenience of applying four linear laser markers with an RGB camera to fully inspect upcoming obstacles in the region covering the front robot dimensions. The results show distance measurement accuracy of up to ±1.33 mm and obstacle width accuracy of up to ±0.28 mm measured from a 130 mm range. Nevertheless, accuracy is strongly dependent on the segmentation of image resolution, lighting, and reflection of the inspected surfaces. Also, it depends on the configuration of laser markers and the RGB camera position.

Keywords: dynamic navigation; mobile robot; laser scanner; autonomous robot; obstacle detection

1. Introduction

Navigation technologies are advancing rapidly with the adoption of sensor fusion and artificial intelligence (AI). Recent advances in AI are paving the way for autonomous systems. However, as technologies advance, tasks are also becoming increasingly complex.

Navigation can be categorized as global or local. Global navigation involves relying on predefined information about the working environment, which can contain marked tracks, labels, and global communication systems. In this case, the optimal path to the target is generated in advance [1,2]. On the other hand, local navigation relies on sensor data to acquire information from the environment and momentarily make a decision about which direction to move in [3,4]. Nowadays, even systems that are integrated in a well-prepared environment, like automated warehouses, cannot fully rely on global information. One of the reasons for this is that various dynamic obstacles can introduce unpredictable obstructions [5,6]. Furthermore, for inspection-type robots that work in places that are difficult for humans to reach, the robots often need to completely rely on local sensor data [7,8]. Such environments also involve underground sewers, which require regular checkups, and performing such checks physically by hand can cost a lot or may be completely impossible. For this reason, a fully autonomous mobile sewer robot is being designed.

An essential step for navigation system development is localization. Local navigation sensors for measuring distance are of utmost importance for mobile robots. Optical sensors are very popular due to their high accuracy and fast response time, including time-of-flight (ToF) sensors, which evaluate distance by measuring the amount of time it takes for light to travel from the source and return from a reflective surface [9]. However, they are usually designed for one-point measurements and require a multiple-sensor solution for wider field-of-view capability. LiDAR sensors are another alternative that often have integrated drives for rotation and point cloud measurement in the 2D plane or even 3D space [10–12]. These sensors provide a wide measurement range and accuracy. On the other hand, optical sensors are sensitive to lighting and weather conditions. In this case, Radar sensors have an advantage, but at the cost of precise point cloud mapping [13–15].

To maximize localization, different sensors are fused together to compensate for their limitations. Distance sensors are often mapped with cameras to expand the amount of information received from the environment for better interpretation of detected objects [16]. Some of the cameras that are widely used for mobile robot navigation include RGB cameras used for color recognition. By mapping color data with distance sensor data, accurate estimation can be made about detected objects' shape, geometry, type, and other characteristics [17,18]. Cameras can also be categorized into 2D cameras, which convert data to a plane from a specific field of view varying from 40° to 360°, including fish-eye cameras, or 3D cameras, which also include depth data. These cameras work by integrating ToF or LiDAR sensors or using several integrated cameras, like stereo cameras [19–21]. Another alternative, which is more robust to environmental conditions, is ultrasonic sensors, which are cost-efficient and accurate for short-range measurement [22,23].

An RGB camera was selected for designing an autonomous sewer robot localization. In combination with four linear laser pointers, depth information can be received from the environment, thus creating an accurate, robust, and cost-efficient object detection solution, which is the main criterion for this system. In the literature, similar methods were experimentally analyzed. In [24,25], one linear laser marker is applied in combination with an RGB camera to detect and measure the distance from an obstacle. On the other hand, this combination only allows inspection of one surface or obstacles attached to the ground. Using two linear laser markers, as described in [26], allows for better surface scanning, minimizing blind spots. Nevertheless, in a sewer-like environment, obstacles can hang from the ceiling or walls. Four laser pointers covering the main planes allow for an optimal field of view, saving the computational resources required for image processing. The proposed laser scanner configuration is shown in the principal scheme of the developed sewer robot system presented in Figure 1.

The red lines represent the possible detection field of view marked by linear laser pointers. To further extend the robot's field of view, 2D LiDAR is added to better recognize the sides of the robot, which comes in handy in close gaps. In addition to object recognition, odometry data are also important for local navigation systems. Some of the sensors most frequently used for this purpose are accelerometers and gyroscopes. An accelerometer is very useful for measuring tilt, which is important in challenging terrain scenarios, while a gyroscope helps to check the robot's rotation [21,27,28]. Both of these sensors are integrated into the designed robot system for better orientation. Magnetometers are also very useful for estimating heading, which is commonly combined with accelerometers and gyroscopes, but due to their dependency on the magnetic field, it is not reliable for underground sewer conditions [29].

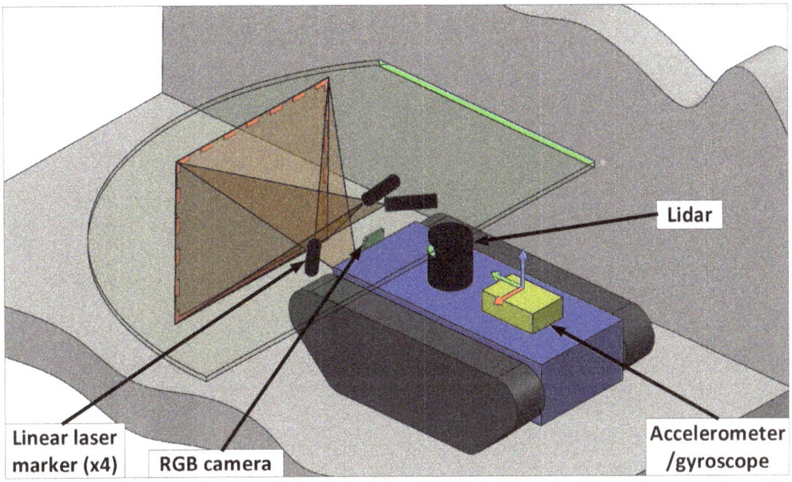

Figure 1. Principal scheme of an autonomous sewer robot.

In this work, the effectiveness of the designed laser scanner is analyzed. In the first part, the methodology of laser triangulation principles and adaptation for the designed laser scanner is discussed. Then, the working method of object detection is analyzed, and is further used for experimentation. Moreover, the experimental setup for determining object recognition accuracy is explained, including the chosen equipment, calibration of the system, and image processing. After that, the results of the experiments are presented, and the standard deviations of measuring different obstacles with different image segmentations are evaluated. Finally, conclusions are drawn and future works are discussed.

2. Materials and Methods

As previously mentioned, an autonomous robot for underground sewers requires a reliable and computationally cost-efficient solution for object detection. One of the most versatile devices for object recognition is an RGB camera. Nevertheless, to simplify the amount of information gathered from images, it is important to simplify the task as much as possible and search for specific patterns. In this case, laser pointers are used to provide linear contours that can be recognized from an image and interpreted for obstacle detection purposes. This reduces the amount of information required to process compared to depth camera point cloud data. Further experiments are performed to estimate the accuracy of the designed laser scanner distance measurement system.

2.1. Laser Scanner Obstacle Detection System Configuration

As previously mentioned, a combination of an RGB camera and a laser pointer has been researched in the literature as a practical obstacle detection, distance, and geometry measurement solution. The main idea behind this method is laser triangulation, as shown in Figure 2.

The obstacle obstructs laser light, which is reflected from the surface of an obstacle, resulting in displacement h_1 from the laser reflection on the ground. Displacement h_1 is then projected onto the CCD sensor, resulting in displacement $h_{2,}$ which is processed and scaled on the image by the RGB camera. To calculate the distance from the obstacle $d_{2,}$ additional parameters are needed, which are configured initially for the designed system.

For this project, the system is configured by fixing the focus distance—d_1, the linear laser tilt angle—θ_1, and the camera perspective to the focus point tilt angle—ω_g. The following formulas are used to detect the actual distance:

$$d_1 - d_2 = \cos\omega_g \cdot \frac{h_1 \cdot L_2}{L_1}, \tag{1}$$

$$d_2 = d_1 - \left(\cos\omega_g \cdot \frac{h_1 \cdot L_2}{L_1}\right), \tag{2}$$

Then, the obstacle is present along the measuring range equal to d_1. The distance from an object can also be defined by two resulting angles tilted from laser and camera projections, respectively—θ_2 and ω_c. Tilt angles for a specific distance can be found using the following formulas:

$$\tan\omega_c = \frac{d_2}{\left(\tan(\omega_g - \theta_1)\cdot d_1\right) - \left(\sin\omega_g \cdot \frac{h_1 \cdot L_2}{L_1}\right)}, \tag{3}$$

$$\sin\theta_2 = \frac{L_3 \cdot \sin\omega_c}{d_2}, \tag{4}$$

These formulas are further used for camera input data post-processing to create a map of occupied regions, later used for path planning. Nevertheless, one linear laser pointer is not enough to recognize all obstacles in front of the robot covering d_1 range. In Figure 3a, it can be seen that the upper linear laser is able to mark only the obstacle from the left, which is fixed to the ground. This could lead the robot to move right and collide with an obstacle fixed to the wall.

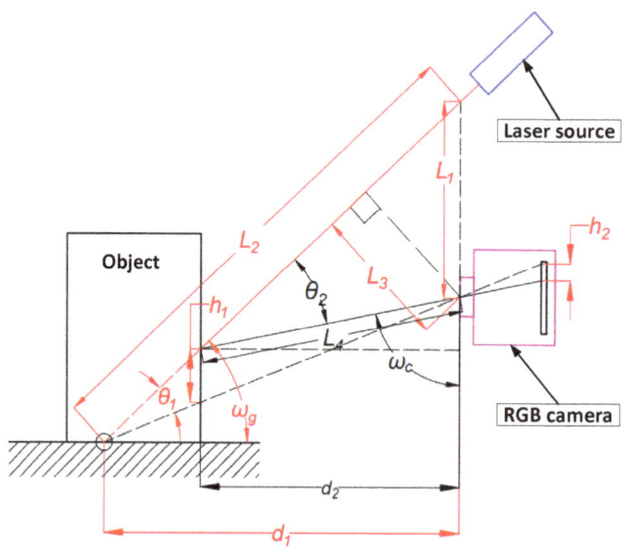

Figure 2. Configuration of one linear laser pointer based on triangulation principles.

Adding another laser marker pointing from the left side allows us to detect the obstacles fixed to the right wall, as shown in Figure 3b. Moreover, adding laser markers from the bottom and right sides allows for the detection of obstacles fixed to all surfaces, creating a field of view whose size depends on the system configuration, linking each laser marker with the camera, as shown in Figure 2. Thus, the rectangular field of view is created by adding additional tolerance around the dimensions of the robot, as shown in Figure 4b.

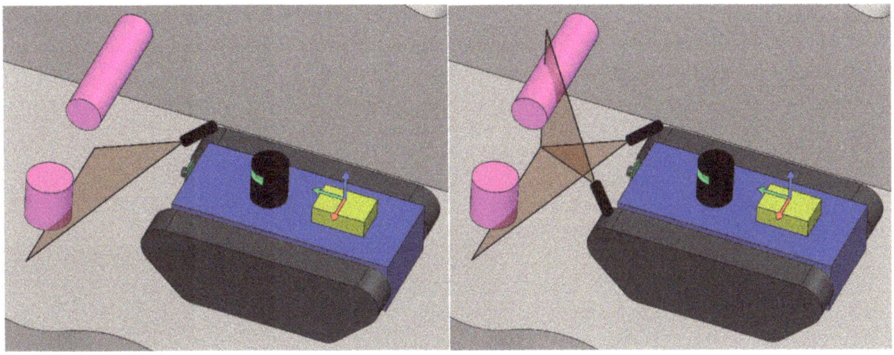

(a) (b)

Figure 3. Obstacle recognition with mobile robot (obstacles marked in purple and laser linear projections in red): (**a**) obstacle detection with one laser pointer; (**b**) obstacle detection with two laser pointers.

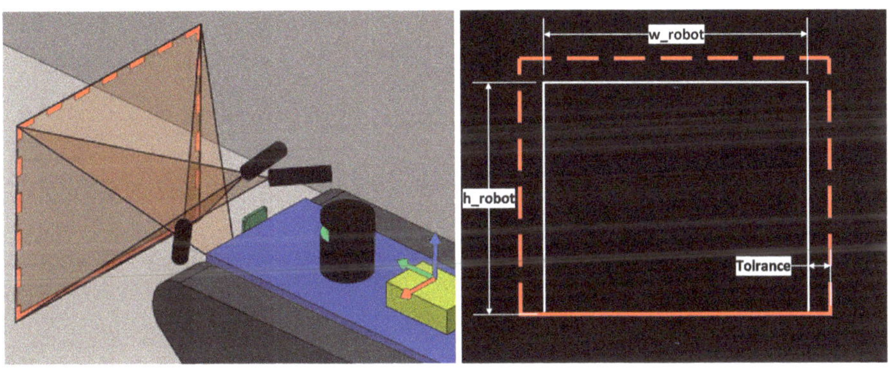

(a) (b)

Figure 4. Laser scanner field of view using four linear laser markers (laser linear projections marked in red): (**a**) side view of laser scanner field of view; (**b**) front plane of laser scanner field of view.

In Figure 4b, it can be seen that tolerance is added around the dimensions of the front robot projection. As a laser scanner is used as the main detection method for nearest obstacles, additional tolerance is necessary for detecting incoming obstacles. In Figure 4a, a laser scanner view from the side can also be seen, each projection marking a different plane in the rectangular tunnel.

2.2. Laser Scanner Experimental Setup

To test the designed scanner, the first experiment will be performed with one laser pointer. The goal of this experiment is to check the accuracy of distance measurement to an obstacle using an RGB camera and a linear laser marker. Accuracy is estimated by comparing the measured value with a laser scanner and the real dimensions. The laser that projects the light to the ground is used as shown in the simplified configuration of the system in Figure 4a.

In Figure 5b, a camera view example is shown corresponding to the position of the measurement system in Figure 5a. From an image, not only can displacement—h_1—be measured, as explained earlier, but width—w—can also be measured, and the result can be directly converted to the real width of an obstacle. During the experiment, only cuboid forms are used as obstacles for testing the system. To be able to evaluate the accuracy of the designed measurement system, it is important to gather relevant data. For this reason, obstacles of different widths are used, including obstacles that are 5, 10, 30, and 50 mm wide. During the experiment, an obstacle is put in front of the laser scanner in a tunnel-like environment, as shown in Figure 6.

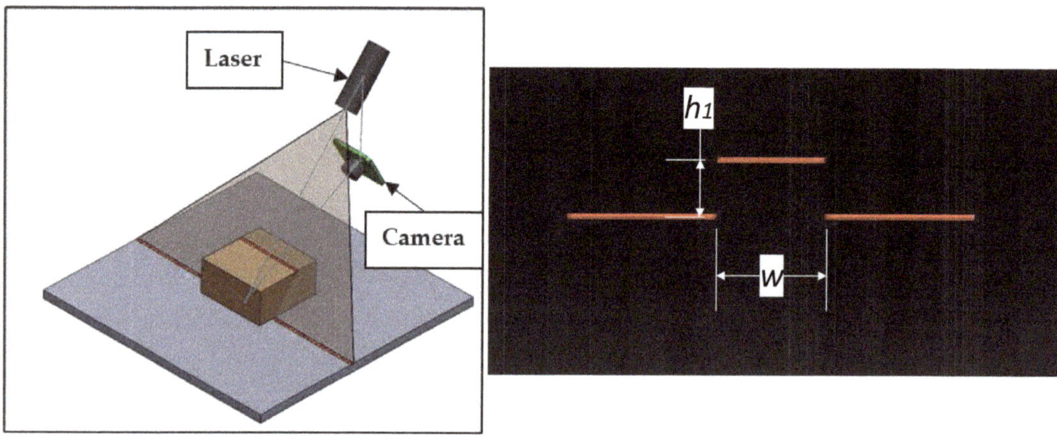

(a) (b)

Figure 5. Experiment principal scheme—laser scanner with one laser marker (laser marker linear projection marked in red): (**a**) 3D view of the system; (**b**) camera view of laser scanner when an obstacle is present.

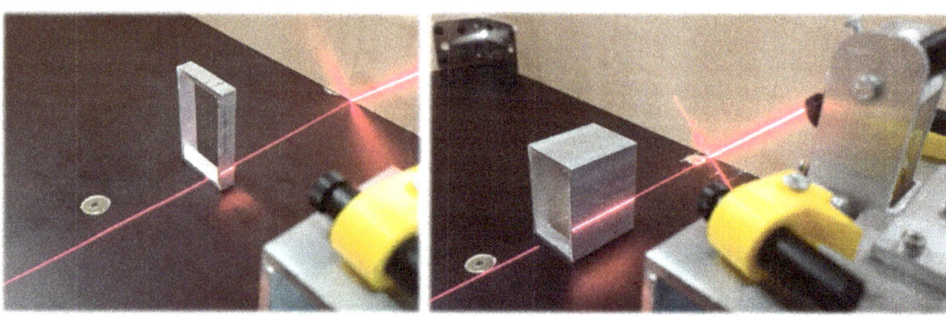

Figure 6. Distance and obstacle width measurement using laser scanner in a tunnel-like environment.

The object is also centered on the camera. Initially, it is placed 130 mm from the scanner, which is the measuring range of the camera—d_1. After that, the laser scanner is moved to an obstacle in increments of 5 mm up to 50 mm. The image is taken and processed at each increment. This procedure is performed with all four obstacles of different widths and additional changes in the image processing segmentation parameter, which will be discussed further.

For this experiment, a measuring system prototype was created according to the geometrical specifications of the mobile sewer robot, which is in development. The prototype was mounted on the "Yaskawa MOTOMAN HC10DTB" industrial robot, as shown in Figure 7. An industrial robot was chosen to ensure positioning accuracy, and it was used to move and measure dimensions with the scanner during experimentation. The robot ensured positioning accuracy of up to ±0.1 mm.

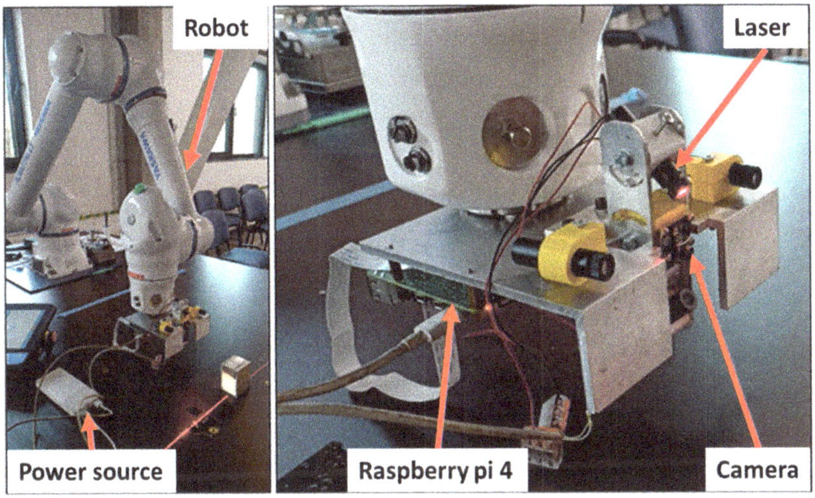

Figure 7. Laser scanner experiment hardware setup.

A camera with a wide field-of-view angle was required for this project because a laser scanner was used for detecting obstacles at close distances while covering an area no smaller than the robot's front dimensions. In this case, the "Raspberry Pi 3 wide Noir" RGB camera was integrated. This camera provides a 120° angle field of view with a maximum resolution of 4608 × 2592 pixels. Because the camera is programmed to search for linear laser projections in the captured image, it is also important to have laser markers with a wide angle. This allows for marking obstacles in close range, covering the whole width and height of the robot. The selected laser angle is 130°. The Raspberry Pi 4 microcontroller is used to control the camera due to its direct connection. An external power bank is used as a power source for the "Raspberry Pi 4". The Scanner program is developed with C++, which was chosen because of the higher computational speed compared to Python 3.1. This is essential because this program will be integrated for mobile robot purposes in the future, which will have a demanding control system.

After the laser scanner prototype is assembled with the robot, the next step is to calibrate the system according to the designed mobile sewer robot specifications. For this purpose, an additional calibration plate is designed, as shown in Figure 8. The plate is placed 130 mm from the laser scanner, which is a distance of d_1, according to Figure 2. This is achieved by positioning an industrial robot with a simple motion program.

From Figure 8, it can be seen that the calibration plate has linear markings indicating the robot's dimensions and a tolerance of 10 mm. The camera is positioned straight, while the upper laser pointer, which will be used in the experiment, is tilted at an angle, lining up with the laser projection at the bottom of the calibration plate. According to Figure 2, angle ω_g is equal to 40.5°. Other dimensions marked in red can also be estimated by using a 3D modeling program and a 3D model of the system. The following dimensions are determined in advance:

- $\omega_g = 40.5°$;
- d_1 – 130 mm;
- $\theta_1 = 6.5°$;
- $\omega_g = 40.5°$;
- $L_1 = 24.43$ mm;
- $L_2 = 180.12$ mm;

- L_3 = 18.58 mm.

Figure 8. Calibration plate for laser scanner.

With these parameters, the distance from an object can be calculated, and then h_1 is detected with the laser scanner. But to achieve this, the first camera has to be calibrated by estimating the image pixel relation to the real length of the captured projection. To achieve this, an image is taken with a laser scanner in the initial position by putting a calibration plate in the center 130 mm away from the scanner, as d_1 is a measuring range. An image of the highest field of view and resolution is taken—4608 × 2592 pixels. The resulting image is shown in Figure 9.

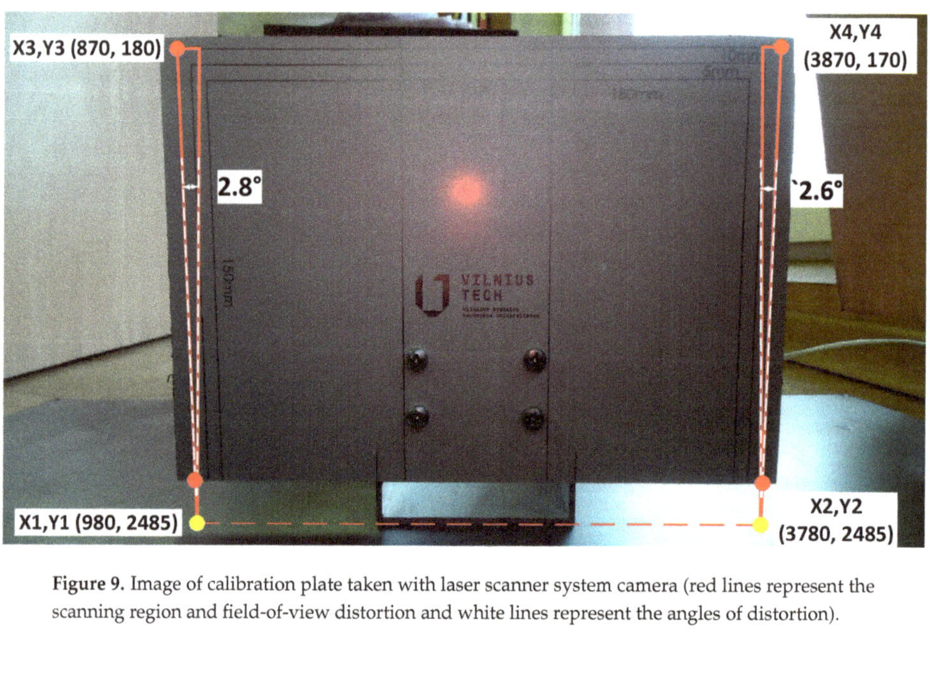

Figure 9. Image of calibration plate taken with laser scanner system camera (red lines represent the scanning region and field-of-view distortion and white lines represent the angles of distortion).

The calibration plate is marked with lines representing the scanning region with specified values—width w_r = 200 mm and height h_r = 160 mm. To convert pixel size to the real length, firstly, the corners of the calibration plate are marked in the captured image as shown in Figure 8. Red corners represent the corners of the plate, and yellow corners are projections of the bottom corners of the scanning region, which are estimated geometrically via solidworks 2024 (Student's version) modeling software. To find pixel real dimensions as width w_{pxi} and height h_{pix}, the following proportions are used, as defined in Equations (5) and (6):

$$h_{pi} = \frac{h}{y_1 - y_3} = \frac{160}{2485 - 180} = 0.0694\ \text{mm}, \tag{5}$$

$$\text{w}_{pi} = \frac{w}{x_2 - x_1} = \frac{200}{3780 - 980} = 0.0714\ \text{mm}, \tag{6}$$

By calculating the image pixel width w_{pix} and height h_{pix} of the calibrated region, it is possible to directly measure the width and height of an obstacle based on the displacement and length of linear laser projections. In Figure 10, a simplified view of the pixel grid and the laser line projected on the obstacle is shown as an estimated data processing model.

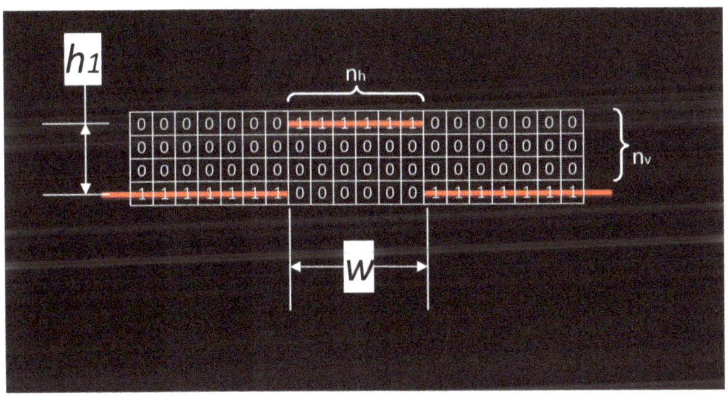

Figure 10. Simplified laser scanner camera view with present obstacle and pixel grid (laser linear projection marked in red).

Pixels marked with 1 represent the detected color of the laser, while 0 s represent the filtered zone of a different color. The obstacle height can be calculated by the following formula:

$$h_1 = h_{pix} \times n_v \tag{7}$$

where n_v is the number of pixels shown vertically. Nevertheless, as seen in Figure 9, the vertical view is stretched, and the width at the bottom of the calibration plate and the top is different. This happens because the camera used has a very wide field of view, which results in a slightly distorted view on the sides. Because of this, width has to be compensated depending on the distance from the obstacle, which is determined by displacement h_1. Another factor that significantly affects width calculation is the camera perspective angle from different distances, which is visually presented in Figure 11.

If an obstacle is closer to the camera from the calibration plane, the width of the obstacle is perceived as bigger because d_1 divided by d_2 is proportional to b_1 divided by b_2. Because of this, the calculated width must be additionally multiplied by the ratio of d_1/d_2. Taking these aspects into account, the obstacle width can be calculated with the following formula:

$$\mathrm{w} = \left((w_{pix} \cdot n_h) - 2 \cdot \left(\tan\left(\frac{w_{pix} \cdot n_h}{w_r} \cdot \theta_k \right) \cdot h_1 \right) \right) \cdot \frac{d_2}{d_1}, \tag{8}$$

where n_h is the number of pixels shown horizontally and θ_k is the compensation angle, and the width $\mathrm{w_r}$ is 180 mm, which can be seen in Figure 9. The compensation angle increases proportionally from the center to the left or right corner of the picture because of the wide-angle camera view distortion. The compensation angle is slightly different on each side because of system deviations. For this reason, the average value is taken—θ_k = 2.7°.

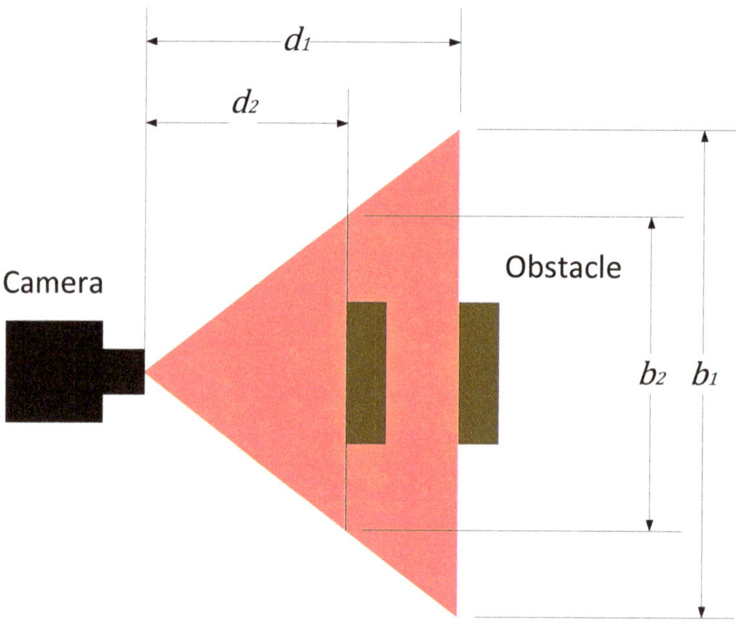

Figure 11. Camera object perspective from different distances (red color represents laser linear projection and brown color represents obstacle positions).

2.3. *Image Processing*

As mentioned before, the camera takes 4608 × 2592-pixel size images. To be able to recognize obstacles which are marked with linear laser projections, it is necessary to filter unnecessary colors from the raw camera picture. There are many models which are used for image processing. For this project, the hue, saturation, value (HSV) model is used, programmed with the C++ language, integrating OpenCV open-source libraries for faster computational processing. The HSV method is more advantageous than regular RGB color processing, most importantly because it is more robust in terms of external lighting changes. It is able to evaluate color intensity [30–32]. For example, an unusual color that is being searched for can have different intensities in the photo because of external light and shadows, and the HSV method allows the color to be distinguished. HSV color space is shown in Figure 12, defined by three parameters according to the model's name, as stated previously.

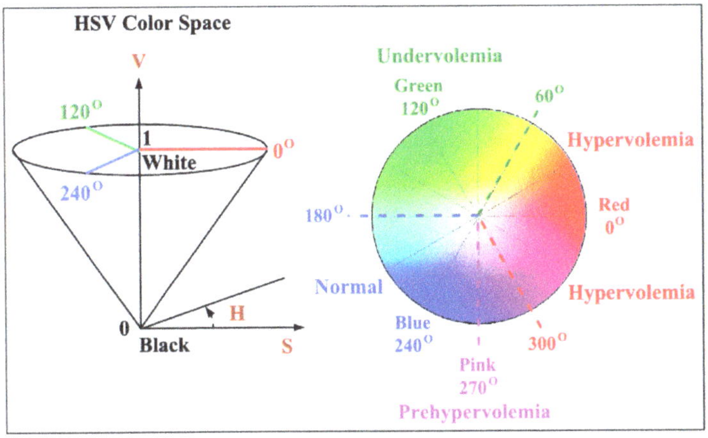

Figure 12. HSV color space graph and parameters [33].

Hue defines the different color regions and has a value from 0° to 360°. When using OpenCV libraries, hue values are scaled between 0 and 180. Value defines brightness, and saturation defines color depth. Both of these values range between 0 and 255.

For this experiment, two red color ranges were chosen for the upper and lower thresholds. Also, a saturation value of 10,000 and a gain value of 2 were chosen to allow the camera to take darker images and mimic a sewer-like view without outside lighting. For the lower range, values from [0, 50, 50] to [10, 255, 255] were selected, while for the upper range, values from [170, 50, 50] to [180, 255, 255] were selected. This range allows for the detection of shades of red of varying intensity. After that, the image is transformed into a grayscale, as shown in Figure 13.

Figure 13. Original image of the present obstacle taken with an RGB camera and transformed to grayscale.

Filtered pixels are changed to a white color with a value of 255, and background pixels are changed to black with a value of 0. This simplifies the calculations used to find regions of interest. Another factor that greatly affects image recognition accuracy and processing speed is segmentation. Reading the value of every pixel and saving it in the memory is computationally demanding. It is more convenient to separate the image into larger square segments, as shown in Figure 14.

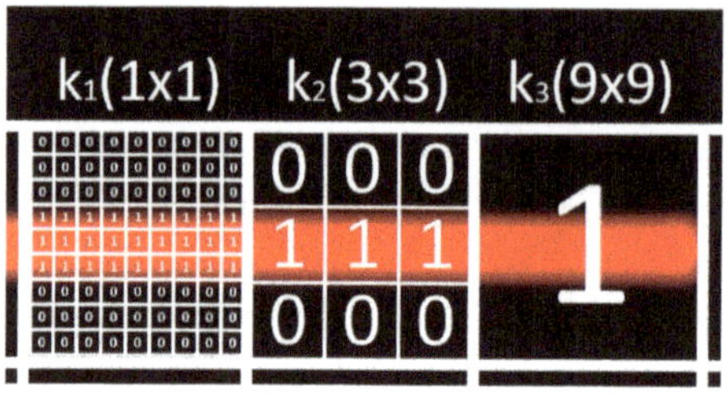

Figure 14. Simplified view of the original image with different segmentation values (red color represents laser linear projection).

The described procedure allows for a reduction in memory load. Segmentation value will be defined as $k\ (n_{h2}, n_{v2})$, where n_{h2} represents how many pixels are stored horizontally in the analyzed squared region, and n_{v2} represents the number of pixels stored vertically. Looking at Figure 11, it can be seen that depending on the segmentation value, the detected laser line can be divided into several pixels. For this reason, the calculation of pixels horizontally and vertically is performed as shown in Figure 15.

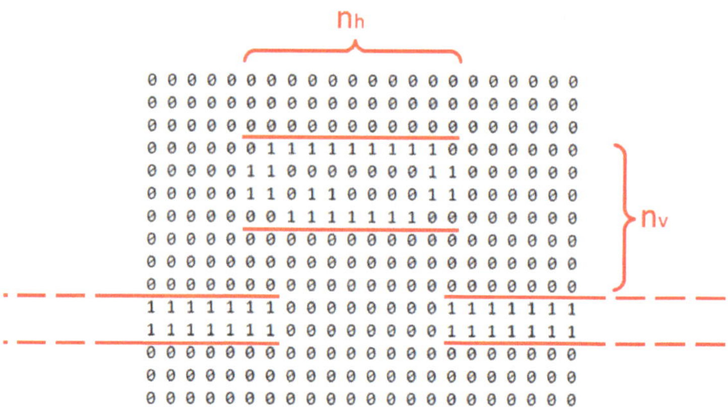

Figure 15. Obstacle width and height of pixel selection from the processed image (red lines represent laser linear projection width).

The figure above shows the segmentation of a real camera image taken from the experiment. The number of pixels stored vertically is calculated from the top of the straight-line projection of the laser to the top of the highest projection disturbed by the obstacle. This logic is applied because the object observed during the experiment is of a square shape, so the top of the object can be assumed as a straight line in this case. For this experiment, segmentation values of 32 × 32, 16 × 16, 9 × 9, and 4 × 4 are used for each of the different measurements.

3. Results

According to the explained methodology, an experiment was performed and data were gathered. In total, 160 measurements of the distance from and the width of obstacles were taken, including four different obstacles that were 5, 10, 30, and 50 mm wide. The deviation between real and measured dimensions was calculated, and the following formulas are used for statistical analysis:

$$\overline{x} = \frac{\sum_{i=1}^{n} x_i}{n}, \tag{9}$$

$$s_i = x_i - \overline{x}, \tag{10}$$

$$s = \sqrt{\frac{\sum(x_i - \overline{x})^2}{n-1}}, \tag{11}$$

where x_i is the measurement deviation, $\overline{x}$ is the mean, n—number of measurements with corresponding obstacle and segmentation value, s_i—deviation from the mean value, and s—standard deviation. During each distance increment from 5 to 50 mm, an image was taken, and each step was increased by 5 mm, moving closer to an obstacle. All images were processed using four different segmentation values of 32 × 32, 16 × 16, 8 × 8, and 4 × 4. The collected data and analysis are presented further.

3.1. Distance Measurements

Measurements were carried out starting from the thinnest obstacle, which was 5 mm wide. The results of different segmentation values are compared in a column-type diagram for a clearer view. The diagram is presented in Figure 16.

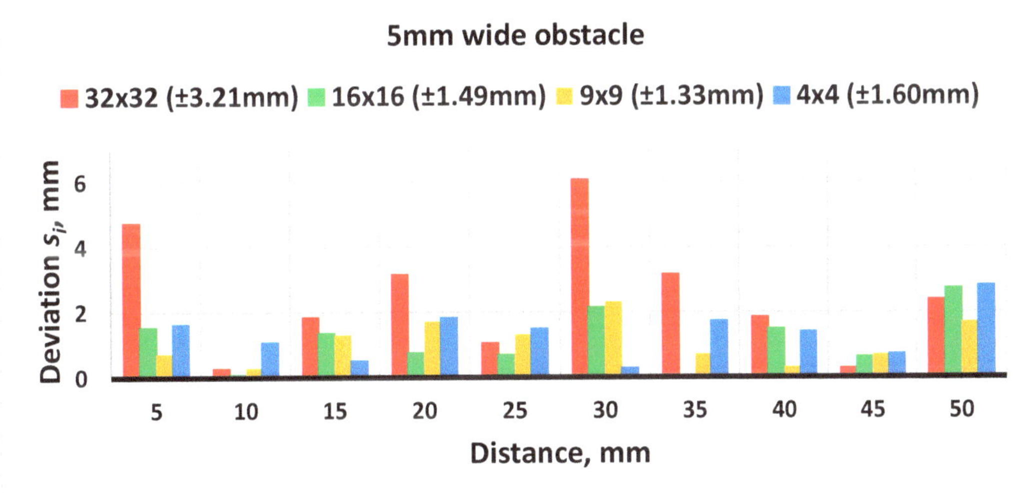

Figure 16. Results of distance measurements with a 5 mm wide obstacle.

The horizontal axis represents distance increments, and the laser scanner is moving closer to an obstacle from an initial position, which is 130 mm away from the obstacle. The vertical axis represents deviation from the mean of the measured distance and the real distance estimated by the industrial robot movement, which ensures ±0.1 mm accuracy. At the top of the table, different segmentation values are shown. Each segmentation value has a corresponding standard deviation value. Looking at Figure 15, it is noticeable that the standard deviation is highest with 32 × 32 segmentation and gradually reduces as the segmentation value is lowered. Nevertheless, the lowest deviation is achieved with a 9 × 9 segmentation value. This would indicate that the 9 × 9 value is more optimal compared to the 4 × 4, as it generates smaller deviation and requires fewer computational resources for processing optical information.

The experiment is further proceeded by measuring a 10 mm wide obstacle. The results are shown in Figure 17.

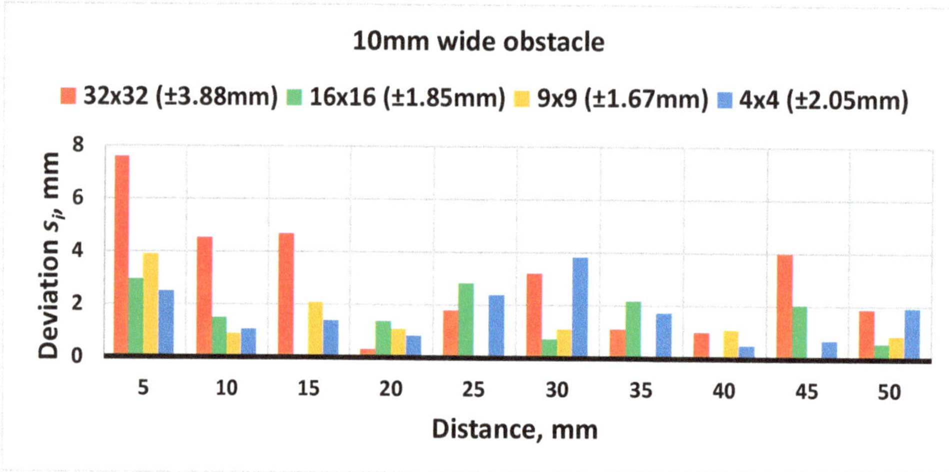

Figure 17. Results of distance measurements with a 10 mm wide obstacle.

Looking at the figure above, it can be noticed that the standard deviation has a similar tendency when measuring a 10 mm wide obstacle. It is highest at a segmentation value of 32 × 32. Then, increasing the number of pixels processed, the lowest deviation is reached with a 9 × 9 value. Decreasing the value further increases the deviation.

The experiment is further proceeded by measuring a 30 mm wide obstacle. The obtained results are shown in Figure 18.

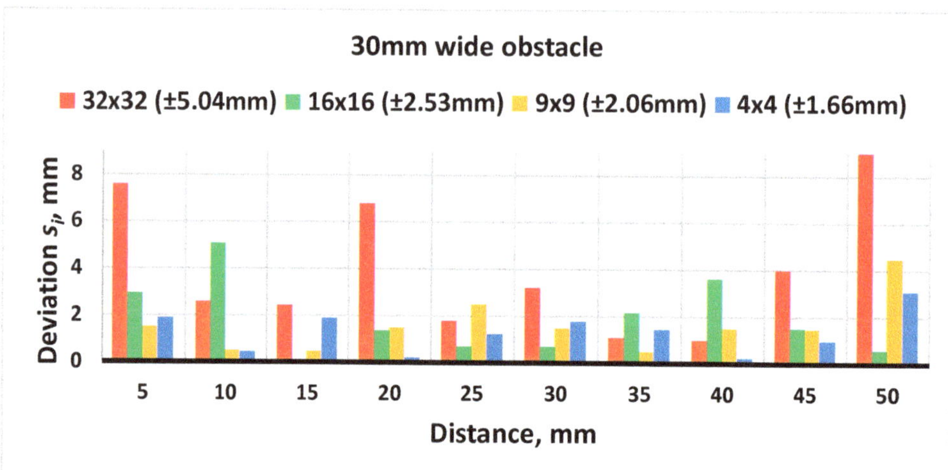

Figure 18. Results of distance measurements with a 30 mm wide obstacle.

Looking at the figure above, it can be seen that, unlike the first two obstacles, lowering the segmentation value up to 4 × 4 gradually reduces deviation, although the most significant difference still remains between 32 × 32 and 16 × 16. Also, the largest error of ±5.04 was measured with a 32 × 32 segmentation value during this experiment.

The experiment proceeded with the measurement of a 30 mm wide obstacle. The results are shown in Figure 19.

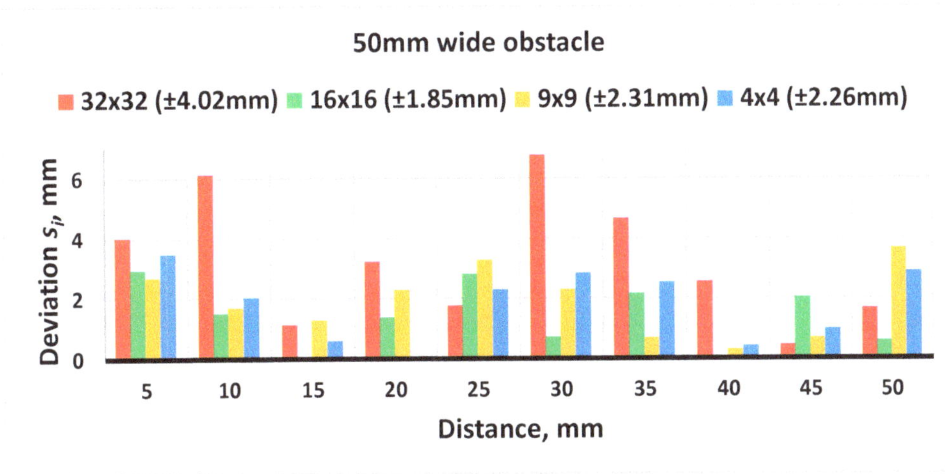

Figure 19. Results of distance measurements with a 50 mm wide obstacle.

Looking into the results acquired from measuring a 50 mm wide obstacle, it is noticeable that the highest deviation remains when the segmentation value is 32 × 32. Nevertheless, the lowest resolution is reached with a segmentation value of 16 × 16, and this is increased by further lowering the segmentation to 9 × 9 and 4 × 4.

Based on an analysis of distance measurements acquired during an experiment, it was observed that changing the segmentation value greatly affects distance measurement accuracy. Nevertheless, the relation is not linear, as seen in Figure 20.

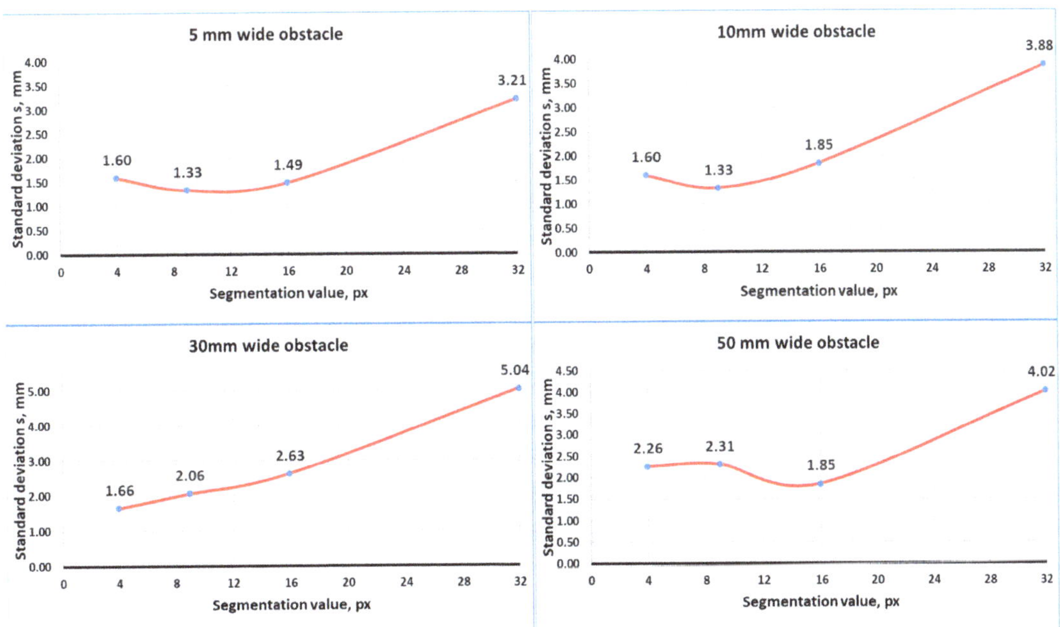

Figure 20. Standard deviation values of distance measurements.

The lowest deviation was reached using a segmentation value of 9 × 9. However, it is noticeable that while measuring wider obstacles (30 and 50 mm), the lowest deviation varies between 4 × 4 and 16 × 16. This can be affected by the reflection of laser light, which is more noticeable on wider areas of surfaces. Reflection introduces noise in the image, which makes the laser beam projection appear thicker and more distorted, as shown in Figure 21.

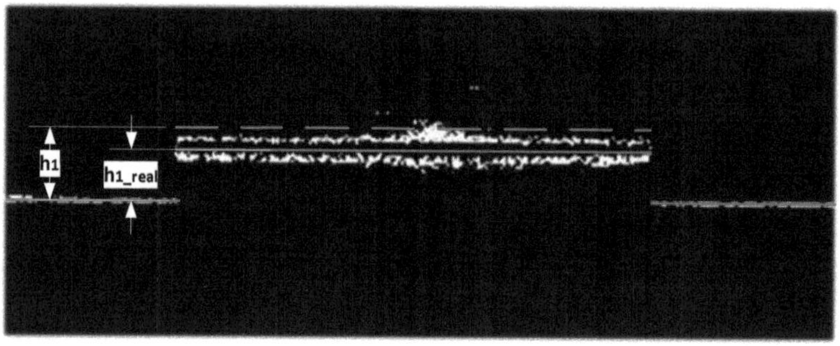

Figure 21. Filtered image processing (red lines represent center position of laser linear projection and dashed line estimated upper position).

It is noticeable that the laser line projected on the obstacle is thicker than the line projected on the ground due to geometrical conditions. Because of this measured displacement, h_1 is between the estimated lines by the algorithm, which is larger than the displacement between the line projected on the ground and the nominal real line corresponding to $h_{\_real}$ shown in the figure above. Moreover, additional peaks of reflected light also increase h_1 and influence measurement deviation. To compensate for the reflection influence coefficient—a is adjusted, which is expressed in Formula (2):

$$d_1 - d_2 = \cos\omega_g \cdot \frac{h_1 \cdot L_2}{L_1} = h_1 \cdot \frac{\cos\omega_g \cdot L_2}{L_1} => h_1 \cdot a, \tag{12}$$

Using known values from the calibrated system, the geometric value of coefficient a is equal to 5.6. Nevertheless, experimentally, the lowest deviations were achieved with the value of 3.2. A smaller value of a compensates for the higher displacement estimated by the calculation algorithm.

It is also possible to reduce picture noise with saturation and gain values of the camera, which reduces the received light, resulting in a darker picture with fewer reflections. Nevertheless, these values have already sufficiently decreased according to the experimental environment; therefore, further reducing these values makes laser projection discontinuous, leading to calculation errors.

3.2. Width Measurements

The image processing algorithm also calculates the width, which is necessary for determining the occupancy grid of the scanned environment. Width is calculated using formula 8, where h_2 is the measured laser projection deviation and d_2 is the calculated distance from an obstacle. The camera's field of view develops slight distortion in the corners, as previously explained in Figure 9. For this reason, h_2 has a slight influence on width calculation accuracy. The results of the experimental width measurements for each obstacle are shown in Figure 22.

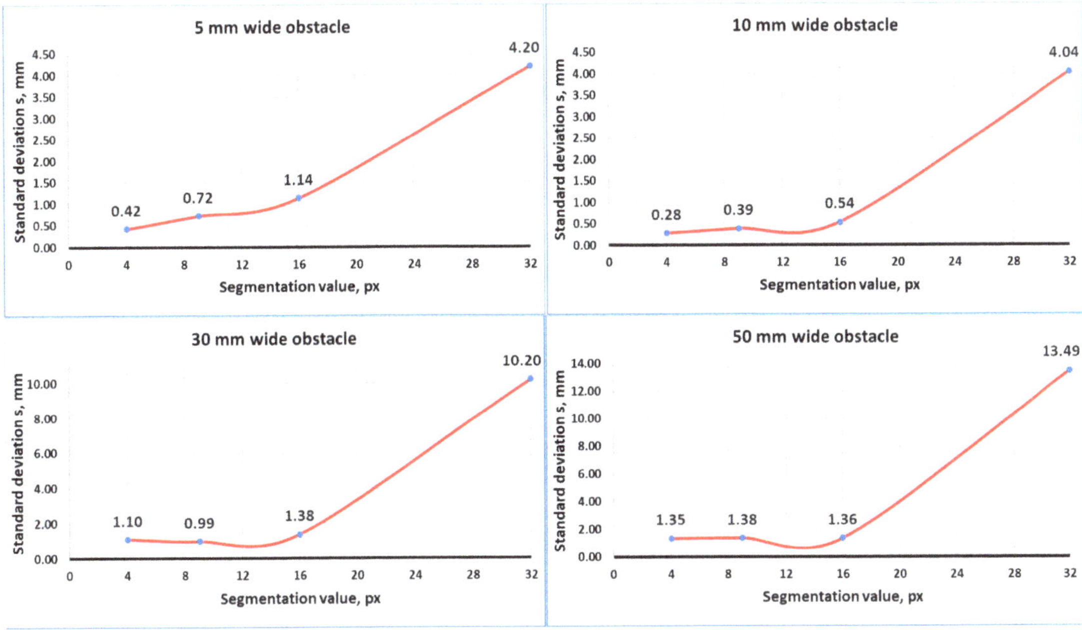

Figure 22. Standard deviation values of obstacle width measurements.

From the figure above, it is noticeable that the standard deviation pattern of width measurements is similar to that of distance measurements. Nevertheless, with a segmentation value of 32 × 32, deviations are significantly higher, especially with wider obstacles of 30 and 50 mm. Additional noise at the corners of processed obstacle laser projection taken in the image influences the measured width more significantly, the larger the segmentation value. Also, the perspective of the camera from different distances increases the error when the obstacle width is higher, which can be seen from gathered data, as the wider the obstacle, the more significant the increase in deviations.

Similarly, as with distance measurement, it is not worth going lower than a 9 × 9 segmentation value, as deviations do not fall or decrease by a very small margin, and such distances are not significant for navigation purposes. Moreover, computational resources can be saved considerably by choosing 9 × 9 or even 16 × 16 segmentation values.

3.3. *Integration into the Mobile Robot*

After calibrating and testing the laser marker and the RGB camera-based measurement system, it can be applied to the intended autonomous sewer robot. Calibration of the measurement system with the industrial robot was performed according to the designed mobile robot prototype geometry. Laser makers and cameras could then be assembled directly on the mobile robot platform shown in Figure 23.

There are marked lines for adjusting the angles of each laser, or the same calibration plate as used in the experiment can be used to calibrate the system. The prototype also includes previously mentioned sensors, including LiDAR and a gyroscope/accelerometer module. The prototype is made of lightweight aluminum and has rubber belts for better friction. Also, all parts are either made from aluminum, 3D-printed, or zinc-coated to prevent corrosion. The tank-type chassis allows the robot to pass through obstacles that are 4 cm high, but for more accurate navigation, only obstacles up to 2 cm are considered acceptable. This platform is suitable for further experiments in harsher sewer-like environments.

Figure 23. Autonomous sewer robot prototype.

4. Discussion and Conclusions

A laser scanner consisting of four linear laser markers and an RGB camera provides a versatile obstacle detection and measurement system. Four laser pointers allow for the detection of obstacles present in a rectangular field of view, which can easily cover the mobile robot's dimensions.

For image processing, we decided to filter red laser light color from the background using low camera saturation and gain values. This created a darker image, reducing light reflections. For distance and width measurement, filtered image data were transformed into a one-dimensional array storing relevant data for both dimensions. An array could then be used for direct fusion with LiDAR data, expanding the field of the mobile sewer robot system.

During experiments, the accuracy of distance measurement was determined using one laser marker, as a similar procedure was performed with all four laser markers. The lowest distance measurement deviation was ±1.33 mm, and this was reached using a 9 × 9 segmentation value with 4608 × 2592 pixel resolution. This deviation was reached measuring a 10 mm wide obstacle with a range of 130 mm. It was observed that the deviation was increased by increasing the obstacle width, which was influenced by the surface reflection of aluminum material, which was more reflective compared to the ground. Nevertheless, 1.33 mm is sufficient for navigation purposes. Moreover, reducing the segmentation value to lower than 9 × 9 increased accuracy very minimally, or in some cases, higher deviation was encountered. Also, lowering the segmentation value increased computational resources.

During the experiment, the obstacle width was also tested, reaching the lowest deviation of ±0.28 mm, then measuring a 10 mm wide obstacle with a 4 × 4 segmentation value. However, the 9 × 9 segmentation value was more efficient, as the increase in deviation was minimal. Similarly, as with distance measurement, an increase in obstacle width increases deviation, which is influenced by surface reflection and also by camera perspective.

The tested optical obstacle detection and distance measurement system allows for accurate detection of obstacles in a short range of 130 mm. Further experiments are needed to combine all four laser markers to detect obstacles attached to the walls and the ceiling. Also, further experimentation is needed to fuse laser scanner data with LiDAR data.

Moreover, mobile robot rotation and swinging from uneven ground surface can also influence laser scanner measurements, as the ground or wall can also deform the laser projection. For this reason, an accelerometer and a gyroscope signal should be added for relative and absolute measurement of mobile robot rotation in each axis for laser beam projection compensation.

Author Contributions: Conceptualization, V.U. and V.B.; methodology, V.B.; software, J.N.; validation, V.U., A.D. and J.N.; formal analysis, A.D.; investigation, V.U.; resources, V.B.; data curation, J.N.; writing—original draft preparation, V.U.; writing—review and editing, V.B. and A.D.; visualization, V.U.; supervision, V.B.; project administration, A.D.; funding acquisition, V.B. All authors have read and agreed to the published version of the manuscript.

Funding: The research was supported by Joint Research Collaborative Seed Grant Program between Taiwan Tech and Vilnius Gediminas Technical University (Grant No Taiwan Tech—VILNIUS TECH-2024-03).

Data Availability Statement: The original contributions presented in this study are included in the article. Further inquiries can be directed to the corresponding author(s).

Conflicts of Interest: The authors declare no conflicts of interest.

References

1. Akhshirsh, G.S.; Al-Salihi, N.K.; Hamid, O.H. A Cost-Effective GPS-Aided Autonomous Guided Vehicle for Global Path Planning. *Bull. Electr. Eng. Inform.* **2021**, *10*, 650–657. [CrossRef]
2. Qin, H.; Shao, S.; Wang, T.; Yu, X.; Jiang, Y.; Cao, Z. Review of Autonomous Path Planning Algorithms for Mobile Robots. *Drones* **2023**, *7*, 211. [CrossRef]
3. Belkin, I.; Abramenko, A.; Yudin, D. Real-Time Lidar-Based Localization of Mobile Ground Robot. *Procedia Comput. Sci.* **2021**, *186*, 440–448. [CrossRef]
4. Ou, X.; You, Z.; He, X. Local Path Planner for Mobile Robot Considering Future Positions of Obstacles. *Processes* **2024**, *12*, 984. [CrossRef]
5. Li, K.; Gong, X.; Tahir, M.; Wang, T.; Kumar, R. Towards Path Planning Algorithm Combining with A-Star Algorithm and Dynamic Window Approach Algorithm. *Int. J. Adv. Comput. Sci. Appl.* **2023**, *14*, 511–519. [CrossRef]
6. Alshammrei, S.; Boubaker, S.; Kolsi, L. Improved Dijkstra Algorithm for Mobile Robot Path Planning and Obstacle Avoidance. *Comput. Mater. Contin.* **2022**, *72*, 5939–5954. [CrossRef]
7. Tatsch, C.; Bredu, J.A.; Covell, D.; Tulu, I.B.; Gu, Y. Rhino: An Autonomous Robot for Mapping Underground Mine Environments. In Proceedings of the 2023 IEEE/ASME International Conference on Advanced Intelligent Mechatronics (AIM), Seattle, WA, USA, 28–30 June 2023; pp. 1166–1173. [CrossRef]
8. Assaf, E.H.; von Einem, C.; Cadena, C.; Siegwart, R.; Tschopp, F. High-Precision Low-Cost Gimballing Platform for Long-Range Railway Obstacle Detection. *Sensors* **2022**, *22*, 474. [CrossRef]
9. Shamshiri, R.R.; Navas, E.; Dworak, V.; Auat Cheein, F.A.; Weltzien, C. A Modular Sensing System with CANBUS Communication for Assisted Navigation of an Agricultural Mobile Robot. *Comput. Electron. Agric.* **2024**, *223*, 109112. [CrossRef]
10. You, H.; Xu, F.; Ye, Y.; Xia, P.; Du, J. Adaptive LiDAR Scanning Based on RGB Information. *Autom. Constr.* **2024**, *160*, 105337. [CrossRef]
11. Weinmann, K.; Simske, S. Design of Bluetooth 5.1 Angle of Arrival Homing Controller for Autonomous Mobile Robot. *Robotics* **2023**, *12*, 115. [CrossRef]
12. Wang, H.; Yin, Y.; Jing, Q. Comparative Analysis of 3D LiDAR Scan-Matching Methods for State Estimation of Autonomous Surface Vessel. *J. Mar. Sci. Eng.* **2023**, *11*, 840. [CrossRef]
13. Yao, S.; Guan, R.; Huang, X.; Li, Z.; Sha, X.; Yue, Y.; Lim, E.G.; Seo, H.; Man, K.L.; Zhu, X.; et al. Radar-Camera Fusion for Object Detection and Semantic Segmentation in Autonomous Driving: A Comprehensive Review. *IEEE Trans. Intell. Veh.* **2024**, *9*, 2094–2128. [CrossRef]
14. Toderean, B.; Rusu-Both, R.; Stan, O. Guidance and Safety Systems for Mobile Robots. In Proceedings of the 2020 IEEE International Conference on Automation, Quality and Testing, Robotics (AQTR), Cluj-Napoca, Romania, 21–23 May 2020. [CrossRef]
15. Wang, T.; Guan, X. *Research on Obstacle Avoidance of Mobile Robot Based on Multi-Sensor Fusion*; Springer International Publishing: Berlin/Heidelberg, Germany, 2020; Volume 928, ISBN 9783030152345.
16. Sarmento, J.; Neves dos Santos, F.; Silva Aguiar, A.; Filipe, V.; Valente, A. Fusion of Time-of-Flight Based Sensors with Monocular Cameras for a Robotic Person Follower. *J. Intell. Robot. Syst. Theory Appl.* **2024**, *110*, 30. [CrossRef]

17. Lin, Z.; Gao, Z.; Chen, B.M.; Chen, J.; Li, C. Accurate LiDAR-Camera Fused Odometry and RGB-Colored Mapping. *IEEE Robot. Autom. Lett.* **2024**, *9*, 2495–2502. [CrossRef]
18. Gatesichapakorn, S.; Takamatsu, J.; Ruchanurucks, M. ROS Based Autonomous Mobile Robot Navigation Using 2D LiDAR and RGB-D Camera. In Proceedings of the 2019 First International Symposium on Instrumentation, Control, Artificial Intelligence, and Robotics (ICA-SYMP), Bangkok, Thailand, 16–18 January 2019; pp. 151–154. [CrossRef]
19. Liu, Y.; Zhang, W.; Li, F.; Zuo, Z.; Huang, Q. Real-Time Lidar Odometry and Mapping with Loop Closure. *Sensors* **2022**, *22*, 4373. [CrossRef] [PubMed]
20. Miranda, J.C.; Arnó, J.; Gené-Mola, J.; Lordan, J.; Asín, L.; Gregorio, E. Assessing Automatic Data Processing Algorithms for RGB-D Cameras to Predict Fruit Size and Weight in Apples. *Comput. Electron. Agric.* **2023**, *214*, 108302. [CrossRef]
21. De Silva, V.; Roche, J.; Kondoz, A. Robust Fusion of LiDAR and Wide-Angle Camera Data for Autonomous Mobile Robots. *Sensors* **2018**, *18*, 2730. [CrossRef]
22. Naga, P.S.B.; Hari, P.J.; Sinduja, R.; Prathap, S.; Ganesan, M. Realization of SLAM and Object Detection Using Ultrasonic Sensor and RGB-HD Camera. In Proceedings of the 2022 International Conference on Wireless Communications Signal Processing and Networking (WiSPNET), Chennai, India, 24–26 March 2022; pp. 167–171. [CrossRef]
23. Yang, L.; Li, P.; Qian, S.; Quan, H.; Miao, J.; Liu, M.; Hu, Y.; Memetimin, E. Path Planning Technique for Mobile Robots: A Review. *Machines* **2023**, *11*, 980. [CrossRef]
24. Fu, G.; Corradi, P.; Menciassi, A.; Dario, P. An Integrated Triangulation Laser Scanner for Obstacle Detection of Miniature Mobile Robots in Indoor Environment. *IEEE/ASME Trans. Mechatron.* **2011**, *16*, 778–783. [CrossRef]
25. Ding, D.; Ding, W.; Huang, R.; Fu, Y.; Xu, F. Research Progress of Laser Triangulation On-Machine Measurement Technology for Complex Surface: A Review. *Meas. J. Int. Meas. Confed.* **2023**, *216*, 113001. [CrossRef]
26. So, E.W.Y.; Munaro, M.; Michieletto, S.; Antonello, M.; Menegatti, E. Real-Time 3D Model Reconstruction with a Dual-Laser Triangulation System for Assembly Line Completeness Inspection. In *Advances in Intelligent Systems and Computing*; Springer: Berlin/Heidelberg, Germany, 2013; Volume 194, pp. 707–716. [CrossRef]
27. Ai, C.; Qi, Z.; Zheng, L.; Geng, D.; Feng, Z.; Sun, X. Research on Mapping Method Based on Data Fusion of Lidar and Depth Camera. In Proceedings of the 2021 4th International Conference on Advanced Electronic Materials, Computers and Software Engineering (AEMCSE), Changsha, China, 26–28 March 2021; pp. 360–365. [CrossRef]
28. Kolar, P.; Benavidez, P.; Jamshidi, M. Survey of Datafusion Techniques for Laser and Vision Based Sensor Integration for Autonomous Navigation. *Sensors* **2020**, *20*, 2180. [CrossRef]
29. Hassani, S.; Dackermann, U.; Mousavi, M.; Li, J. A Systematic Review of Data Fusion Techniques for Optimized Structural Health Monitoring. *Inf. Fusion* **2024**, *103*, 102136. [CrossRef]
30. Yang, K.; Wang, K.; Bergasa, L.M.; Romera, E.; Hu, W.; Sun, D.; Sun, J.; Cheng, R.; Chen, T.; López, E. Unifying Terrain Awareness for the Visually Impaired through Real-Time Semantic Segmentation. *Sensors* **2018**, *18*, 1506. [CrossRef] [PubMed]
31. Le, N.M.D.; Nguyen, N.H.; Nguyen, D.A.; Ngo, T.D.; Ho, V.A. ViART: Vision-Based Soft Tactile Sensing for Autonomous Robotic Vehicles. *IEEE/ASME Trans. Mechatron.* **2023**, *29*, 1420–1430. [CrossRef]
32. Maitlo, N.; Noonari, N.; Arshid, K.; Ahmed, N.; Duraisamy, S. AINS: Affordable Indoor Navigation Solution via Line Color Identification Using Mono-Camera for Autonomous Vehicles. In Proceedings of the 2024 IEEE 9th International Conference for Convergence in Technology (I2CT), Pune, India, 5–7 April 2024. [CrossRef]
33. Chen, W.-L.; Kan, C.-D.; Lin, C.-H.; Chen, Y.-S.; Mai, Y.-C. Hypervolemia Screening in Predialysis Healthcare for Hemodialysis Patients Using Fuzzy Color Reason Analysis. *Int. J. Distrib. Sens. Netw.* **2017**, *13*, 155014771668509. [CrossRef]

machines

MDPI

Article

Automatic Jig-Assisted Battery Exchange for Lightweight Drones

Aleksi Vilkki [1,*], Antti Tikanmäki [2] and Juha Röning [2]

[1] Biomimetics and Intelligent Systems Group (BISG), University of Oulu, 90570 Oulu, Finland
[2] Biomimetics and Intelligent Systems Group, Faculty of Information Technology and Electrical Engineering, University of Oulu, 90014 Oulu, Finland; antti.tikanmaki@oulu.fi (A.T.); juha.roning@oulu.fi (J.R.)
* Correspondence: aleksi.vilkki@oulu.fi

Abstract: Drones utilize rechargeable batteries as a power source. Operating a drone requires human interaction with the exchange or recharge process of these batteries. This can provide limits for drones, which is why some use cases for drones could benefit from automated battery exchange. The purpose of this work was to research how a robotic arm could be used to automatically exchange flight batteries in flying drones without interference from operators and create a proof-of-concept system for evaluation. The devised method was based around a separate jig, which would hold the drone in place during the exchange operation, where battery exchange itself was handled by the use of a robotic arm. The constructed prototype could exchange flight batteries in drones in under two minutes with reasonable perpetuity and reliability.

Keywords: drone; battery replacement; robotic arm; machine vision; flight battery

1. Introduction

Unmanned aerial vehicles (UAVs) are types of flying drones that do not have a human operator on board. Such drones have found various possible use cases in multiple fields, from aerial photography to the delivery industry and even medical services [1,2]. Drones require a power source to operate. A variety of different methods of powering drones have been studied extensively regarding aspects of endurance and efficiency. While drones with tethered power sources do exist, drones often rely on portable power due to their remote controlled nature. Current drones often utilize removable batteries as a portable power source and can be equipped with additional payloads, both of which are usually manually installed by the pilot or other human operator. As drones can, by design, operate far away from their operators, the manual replacement of a battery by an operator can be a limiting and ineffective method that sets limits to the drone and its operational purpose. Therefore, a system designed to automatically and remotely handle these tasks could potentially benefit existing and emerging use cases for drones.

Research on different autonomous flight-time-extending solutions for drones is ongoing, as drones remain a popular topic across different industries. Examples of methods for automatic flight battery exchange include designs such as that of Lee et al., where the battery can be exchanged with a mechanical carriage, if the drone has successfully landed in the correct position [3]. Their system could exchange a flight battery in around one minute. However, it relies on the landing accuracy of the drone, which could be problematic when wind or other natural factors disturb the landing process. In their work, Fujii et al. [4] present a system where a landed drone could be mechanically guided in place before swapping the battery to compensate for inaccurate landings. Alternatively, Zhao et al. [5] disclose the use of a funnel and its counterpart attached to a drone for guidance before battery exchange. Other methods which could be more tolerant for inaccuracies in landing include methods with wireless or contact-based charging areas with guidance, but these can usually accommodate a singular drone for the duration of charging time [6–8].

Some solutions have emerged that have attempted to address both issues with mechanical containers for drones, often dubbed drone-in-a-box solutions, which can store a drone and provide either charging capabilities or a battery exchange for a singular drone, where they often rely on mechanical positioning of the drone in the landing area [9,10]. Similarly to those, some recent methods have utilized machine vision locating of drones and an extendable arm for exchanging batteries [?]. Using such an open-chained solution has created a possibility to house multiple batteries and possibly even service multiple drones while not relying on mechanical positioning systems or the pinpoint landing accuracy of the drone. Table 1 summarizes the advantages and disadvantages that could be identified from methods to extend the flight time of drones.

Table 1. Comparison of range-extending methods.

Method	Advantages	Disadvantages
Wireless charging	Lightweight Simple	Services a singular drone while charging Relies on charging time
Contact-based charging	Lightweight Simple	Services a singular drone while charging Relies on charging time Requires a guidance solution
Tethered drone	Does not rely on charging time	Limits the range of drone
Battery exchange, guidance by a drone	Does not rely on charging time Less reliant on charging time	Relies on landing accuracy Complex
Battery exchange, guidance by an external solution	Does not rely on charging time Does not rely on landing accuracy Less reliant on charging time	Complex

This work presents a prototype system built around a robotic arm and a simple machine vision system to recognize and move the drones through a separate battery exchange jig, which can be used to store drones when not needed and hold them in place during the exchange process. Our system's novelty and contribution can be summed up as follows: instead of using a more complex mechanisms that hold the drone and manipulate the battery simultaneously, we divide those into separate functionalities resembling how you could exchange the battery while holding it by hand by using separate jigs as a tool.

Background for this work begun as a master's thesis of one of the authors, which is publicly available online [12]. In an effort to keep this paper short, some of the more trivial technical aspects that are shared between these works are not covered in this paper.

2. Methodology and Construction

One of the common challenges in flight-time-extending solutions for drones is the problem of accurately positioning the drone for a recharge or battery exchange. Our attempt to overcome this challenge was to locate the drone with machine vision and physically move the drone from the landing area to a separate battery exchange location with a robotic arm, which would also handle the battery exchange operation. Our presumption was that this would not require pinpoint accurate landings from the drone itself as the drone would only have to land in the general field of vision of the robot's camera for the localization to work. Therefore, this concept would be central to the built system's design.

The proposed system (Figure 1) was designed and constructed from sub-assemblies attached to an aluminum support frame. The Kinova Gen3 robotic arm was placed in position where it could reach the drones in the entirety of the landing area while maintaining a capability to operate within preset static points of interest (POIs). These POIs consisted of positions which would not move during operation and included the battery exchange jig, the battery chargers and a separate storage position for the drone. The construction also included a peripheral monitoring system for keeping track of occupied drone positions, a landing area to function as a good background for machine vision operation and for drones to land, and a support frame constructed from aluminum profile to fasten everything down. A list of the main parts in the final construction and their availability can be seen in Table 2.

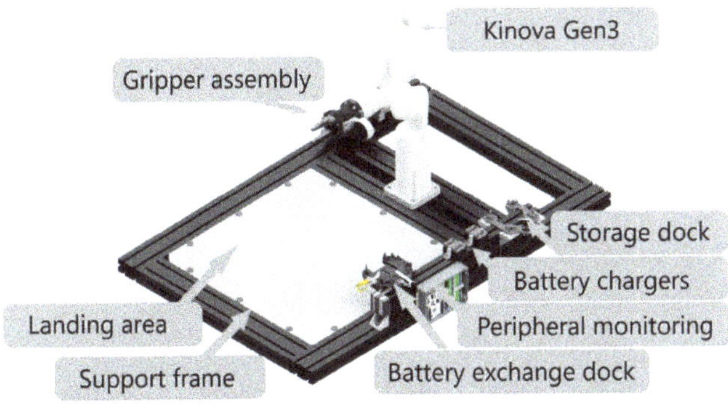

Figure 1. Three-dimensional model of the entire assembly.

Table 2. List of main parts.

Part	Description	Availability
Kinova Gen3 7DOF	Robotic arm	Commercial
Robotiq Hand-E	Mechanical gripper interface	Commercial
RealSense L515	Camera	Commercial
Minitec 45 × 90 UL 4800 mm	Support frame material	Commercial
Corrugated plastic panel 598 × 598 mm	Landing area material	Commercial
DJI Tello drones products	Drones, batteries and chargers	Commercial
Gripper fingers	Gripper fingers fitted for Tello	3D printed
Exchange & storage jigs	Holding drones	3D printed
Miscellaneous parts	Funnels, custom fasteners, supports	3D printed
Actuonix PQ12-P	Linear actuator in exchange dock	Commercial
Microswitch D2VW-5	Used in battery exchange & storage dock	Commercial
STM32 Nucleo L433RC-P	Controller for microswitches	Commercial

The drones utilized in this project were DJI Tello drones equipped with a gripping surface and augmented reality (AR) marker. During a normal, manual battery exchange, the Tello drone is held between a person's fingers (Figure 2, left) while the flight battery is removed by pulling it from the rear of the drone, and it is reinstalled in the opposite direction. A similar approach was used as a starting point for our automatic exchange, where the drone would be mounted on a jig instead of a hand.

A 3D-printed jig (Figure 2, right) was designed to attach directly to the feet of the Tello drone through the holes in the jig. These holes function similarly to a funnel and position the feet of the drone accurately when they are correctly pushed in. Once pushed in, movement of the drone is restricted, except for the same direction in which it was pushed in. In order to make sure that the feet, and by extension the drone, are pushed in the correct position and orientation, chamfered poles were designed next to each hole. These poles would act as preliminary guides and would also provide additional support during the manipulation of the drone. The jig also provided attachment points for a microswitch, which was used to detect the correct installation of the drone, and a small linear actuator, which could be used to toggle the power on the drone by pressing the power button of the drone.

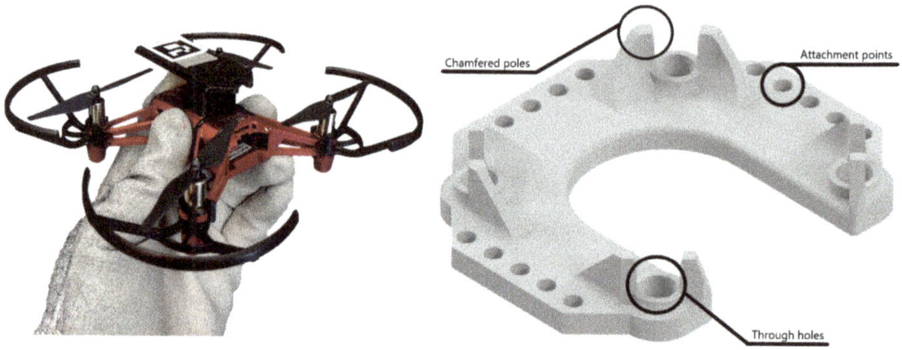

Figure 2. (**Left**) Tello drone held in hand. (**Right**) Three-dimensional model of the jig designed for a Tello drone.

The battery exchange jig was mounted in such a way that the robotic arm could reach partially underneath it in order to grasp the flight battery from the rear of the drone. As the movement of the drone is bound in this direction by the jig, the battery could simply be removed by pulling it away from the drone with the robot. For safety and monitoring purposes, a microswitch was placed on the jig as a fail-safe to detect the correct installation of the drones. This microswitch would not trigger unless the drone was installed correctly. The storage position was a simplified version of the exchange jig assembly (Figure 3), but without the linear actuator, as it was unnecessary.

Moving drones to a separate jig for battery exchange required a method of locating drones accurately from the landing area and a method to hold both the batteries and the drones. We used a Robotiq Hand-E gripper, which was fitted with a photogrammetrically calibrated RealSense L515 camera and proprietary gripper fingers with notches suitable for grasping both the batteries and the drones (Figure 4).

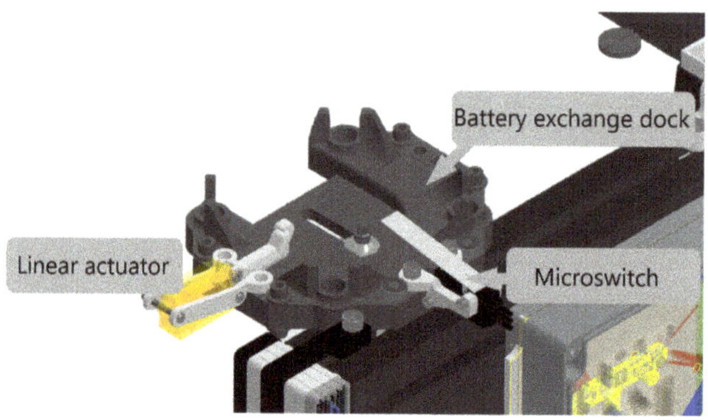

Figure 3. Three-dimensional model of the battery exchange jig assembly.

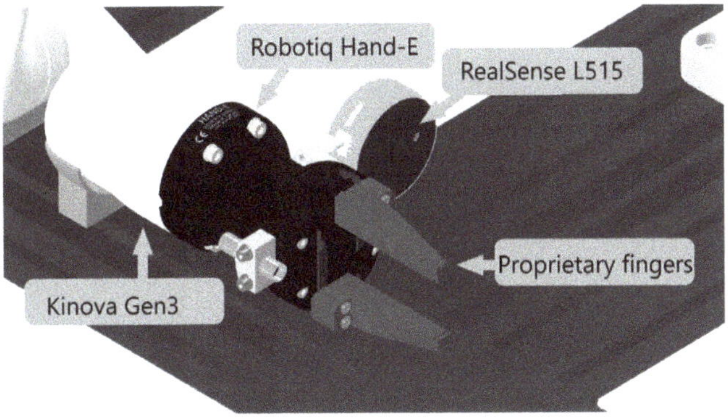

Figure 4. Three-dimensional model of the gripper assembly.

The machine vision system for locating the drone from the landing area used AR markers attached to a visible location on the drones. From these markers, the position and orientation of the drone in the landing area could be calculated by measuring the difference between the center point of the AR marker on the drone and the center point of the visual frame of the camera. This solution relied on the method introduced in the OpenCV library for Python [13]. The localization worked by driving the robot in a position where the arm of the robot would be perpendicular to the planar landing area and the attached camera would have a good field of view (FOV) of the landing area. This is illustrated in Figure 5, where FOV X and FOV Y resemble the edges of the FOV of the attached camera, and distance D resembles the height from the arm to the landing area. This height was assumed to be static as long as positions of the robot or landing area did not change, which eliminated the requirement of measuring it every time a drone was picked up.

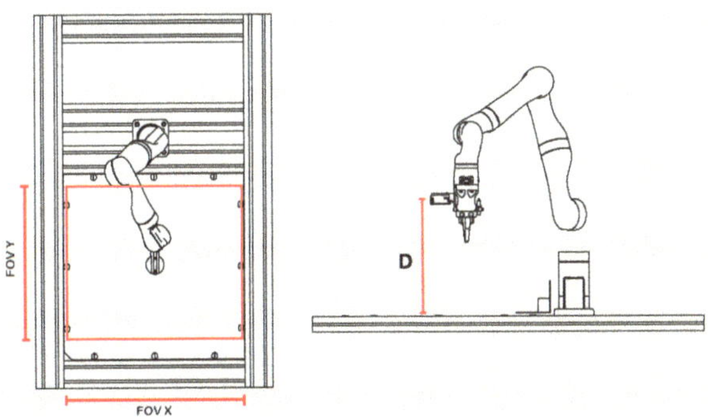

Figure 5. Illustration of localization position.

The location of the AR marker in the frame of the FOV could be obtained using the detector feature built into OpenCV 4.2, yielding the positions of corners of the detected marker, which could further be converted to Euler angles with the pose estimation tools built into OpenCV. This method provided distances from the AR marker to the corner of the visual frame and rotation along the Z-axis of the marker. The difference between the AR marker and the center point of the camera could then be converted to Cartesian coordinates and angular values, which could further be passed to the robotic arm as corrections toward the pick-up point of the drone (Figure 6). Detection in the planar field yielding values for X and Y corrections and the angle of orientation was deemed sufficient, as the distance between the base of the robot and the top surface of the landing area, and by extension the height of the landed drone, was known.

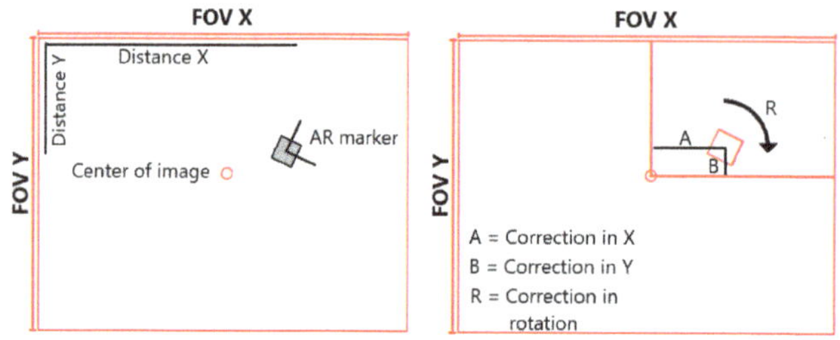

Figure 6. Visualization of correction calculation.

Since the camera would already be used to locate the drones, it was also used for monitoring the status of the battery chargers. In this case, the chargers were a set of two chargers made for DJI Tello batteries, each capable of charging three batteries. The status of each charger position was determined by monitoring the position and the LED next to it with the camera, indicating the presence of a battery and readiness of the charge. The middle position in the battery chargers was too narrow to access it with the gripper and was therefore utilized as an attachment point to fasten the chargers to the support frame. This resulted in the system having a total of four possible charging positions. The unavailable middle position would also be used for attaching funnels directly above the remaining charger positions, which would guide batteries to the correct positions. As the system could identify empty slots and readiness of a charge in occupied positions, a simple pick-and-place action was deemed to be sufficient when returning or fetching batteries from the chargers.

As the points of interest were static, apart from the action to pick up the drone, movements of the robotic arm were created with positional sets containing Cartesian coordinates with angle/axis representations and angular position values. Cartesian coordinates were used in movements requiring finesse, such as the insertion of the drone to the exchange dock and the manipulation of the flight batteries within the drone, as Cartesian coordinates were easier to use in this application. Broader, more general movements, such as movements from one safe point above the exchange dock to another safe point above the drone storage dock, were made with angular position values to avoid unreachable intermediate points and high-joint-rate near-joint singularity in Cartesian paths.

The correct operation of the system was monitored with a separate peripheral monitoring system. This peripheral monitoring system consisted of an STM32 micro-controller that controlled an array of multiplexers that were connected to the microswitches in the storage and battery exchange jigs. The STM32 transmitted changes in the statuses of the microswitches to a main computer responsible for running the system. This main computer was a laptop computer responsible for monitoring data and controlling actions using Kinova's Kortex application interface version 2.6.0 for robotic movements and the aforementioned OpenCV 4.2 for machine vision purposes.

Three modes of operation were programmed into the system (Figure 7). These modes were essentially looping sets of commands and checks based on parameters, such as the statuses of the battery chargers or the occupation of the drone docks. These modes were run in a loop until certain conditions were no longer met or the operator interrupted the program. The modes were called 'Catch', 'Release', and 'Catch and release'. 'Catch' was used to find the drone in the landing area, remove its flight battery, and move the drone to storage. 'Release' was the opposite mode, where the drone was moved from the storage to the landing area with a charged battery. 'Catch and release' was used to exchange the flight battery for a charged battery and immediately release the drone back to the landing area.

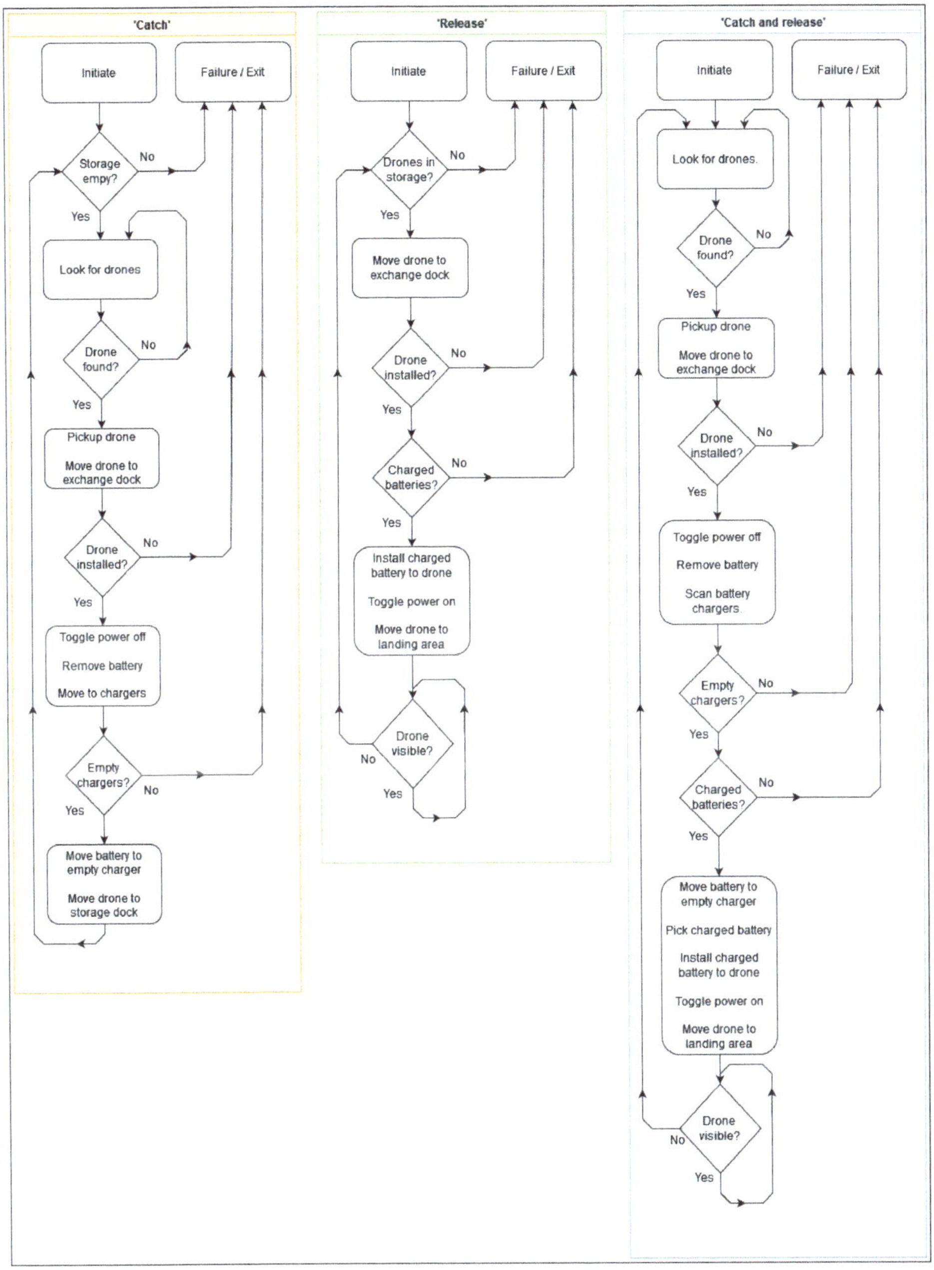

Figure 7. Simplified state machine diagrams of all programmed modes.

In 'Catch' mode, the system would check if any storage positions were available, and if they were, it would begin to look for AR markers from the landing area using the configured detection model (Figure 8, left). If any markers would be detected, the system would move to a better position for scanning the landing area for the position of those markers and, by extension, the drones (Figure 8, right).

If a marker was found and correction coordinates were successfully extracted, the arm would move to pick up the drone from the preset pick-up height (Figure 9 left) and transport the drone on top of the battery exchange dock (Figure 9, right).

The drone could then be inserted into the dock by pushing it in from above (Figure 10, left). If a correct installation was detected via the activation of the microswitch, the robot would move into position to remove the battery from the rear of the drone (Figure 10, right).

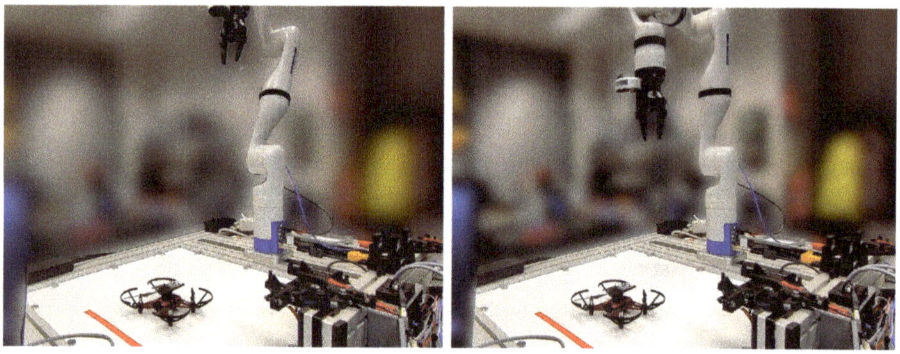

Figure 8. (**Left**) 'Overlooking' position. (**Right**) 'Scanning' position.

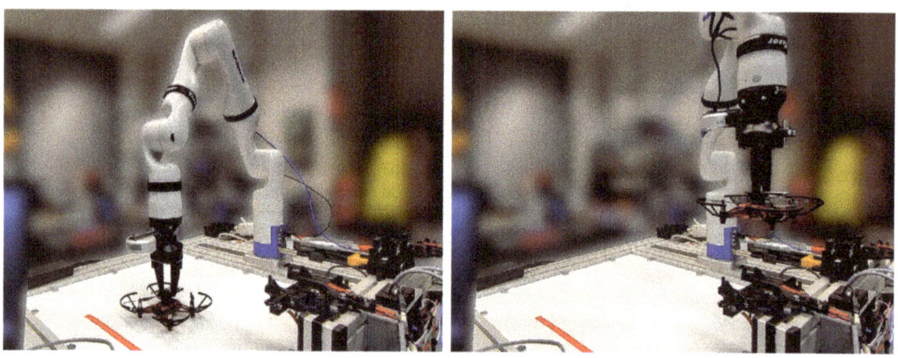

Figure 9. (**Left**) Picking up the drone. (**Right**) Taking the drone to the battery exchange dock.

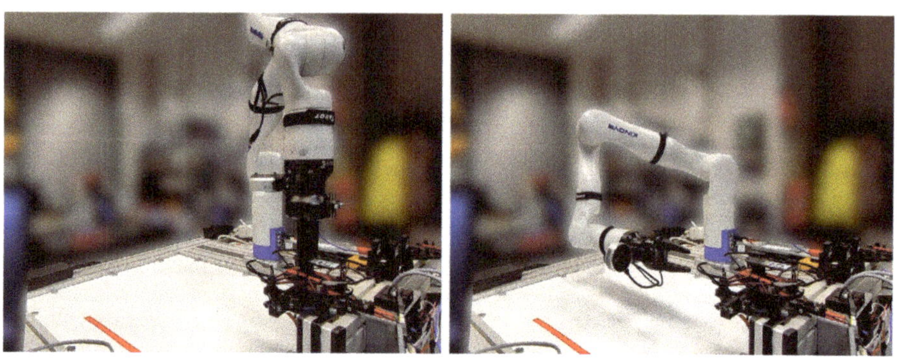

Figure 10. (**Left**) Inserting the drone to the exchange dock. (**Right**) Moving in position to remove the battery from the drone.

Once the battery was removed (Figure 11), the robot would then move to a position where the camera could see the battery chargers. The system would then attempt to detect the statuses of all battery chargers and insert the battery to the first empty charger position for charging. After leaving the empty battery at the charger, the robotic arm would move to pick up the empty drone from the exchange dock. The empty drone would then be transported to the storage position that was deemed empty in the first phase, and the loop began over again until all storage positions, which in our case comprised of just one actual storage position, were full. The 'Release' mode was configured to be the opposite of 'Catch', releasing drones from storage. The 'Catch and release' mode would insert a fully charged battery into the drone and then move it back to the landing area. By using static coordinates for the POIs, positions of the POIs did not change between different modes.

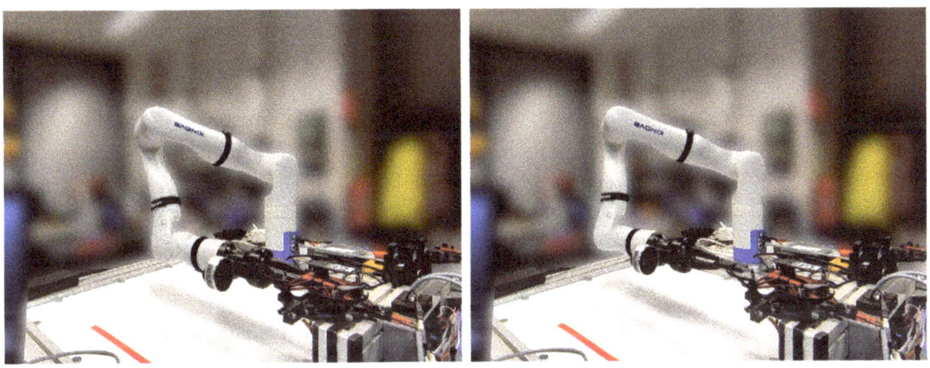

Figure 11. (**Left**) Gripping the battery in a drone. (**Right**) Pulling the battery out from a drone.

3. Tests

The system was tested by running the longest of the three modes, the 'Catch and release' mode, for 200 consecutive timed test runs. Tests were performed in an indoor laboratory with good lighting conditions, at around 20 degrees Celsius, and without wind or external vibrations. Between each run, the drone was manually placed by hand in various positions in the landing area. In every position, the orientation of the drone was set to be roughly between −90 and 90 degrees and generally facing the robot. The system also had three out of four possible battery charger locations occupied, leaving room for one battery from the drone. With this configuration, the location of the available empty battery position and the available fully charged batteries would rotate as the robot exchanged batteries between the drone and the chargers. Each test run was deemed successful if the drone was picked up from the landing area with an empty battery and released back to the landing area in correct orientation, facing away from the robot, with a charged battery and powered on, i.e., in a condition where it would realistically be ready for take-off.

During these 200 tests, the system failed this task five times. Two failures resulted from inaccurate readings of the battery charger LEDs, which caused the system to think that all battery charger positions were occupied, which in turn triggered an error state and halted operation. Three failures occurred from a poor pick-up accuracy of the drone, which led to a misalignment when inserting the drone to the battery exchange dock. However, it was also observed that the chamfered surfaces in the exchange dock managed to correct minor misalignments from poor pick-up action six times during all test runs by guiding the drone in the correct position. Therefore, even if the chamfered surfaces functioned as they were designed and provided correction when they were supposed to, the accuracy for the pick-up action would have been insufficient in a total of 9 times out of 200.

Durations of steps from all successful test runs were recorded with a Python 3.8.2 script, keeping log of when different steps were reached. The average time from the first detection of the drone to the robot returning the drone back to the landing area and moving back into the ready position was recorded to be 118.27 s, with most runs ranging roughly from 117 to 120 s. Average values of recorded times between steps are shown in Table 3. Test times were recorded with the highest available speed and acceleration settings for the Kinova Gen3 robot, which were 80 degrees/s and 297.9 degrees/s^2 for joints 1–4 and 70 degrees/s and 350 degrees/s^2 for joints 5–7, respectively.

Table 3. Average times between steps.

Step	Average Time in Step (s)	Average Time from Start to End of Step (s)
Time to detect the drone	8.94	8.94
Time to locate the drone	5.75	14.69
Time to pick up the drone	25.96	40.65
Time to remove the battery	4.48	45.13
Time to move to the battery chargers	16.41	61.54
Time to deposit the empty battery & pick up the charged battery	11.43	72.97
Time to release the drone back to the landing area	39.73	112.7
Time to completion	118.27	118.27

Recorded times show that, on average, most of the time was spent on three steps. These steps were 'Pickup the drone' (25.96 s), 'Move to the battery chargers' (16.41 s), and 'Release the drone back to the landing area' (39.73 s).

The 'Pickup the drone' step incorporates the actions of picking up the drone from the landing area and moving it to the battery exchange dock, installing the drone in the dock, turning the power off in the drone using the small linear actuator in the exchange dock, and verifying correct installation of the drone with the peripheral system. As such, it is a fairly large step that could be considered movement-intensive. Therefore, the length of this step could likely be shortened by making movements of the robot either shorter or faster through optimization of the movements or via a different placement of the battery exchange dock. However, in the case of this system, these gains in performance would likely be close to negligible, as the movements were reasonably short in comparison to the speed settings of the robot.

The 'Moving to the battery chargers' step includes the short movement from the top of the landing area to above the battery chargers with the battery that was removed from the drone, as well as the measurement of the charger LEDs in each charger position. The latter spends five seconds measuring the frequency of the blinking LEDs in each occupied battery charger position to determine whether the battery in the charger position is charged or not. Therefore, with three batteries available in the chargers, this process takes at least 15 s in every test iteration, which is less than ideal. However, this was a placeholder solution, and while implementing a better solution could provide major gains in performance, it was considered to be outside the scope of this work.

The 'Release the drone back to the landing area' was another large step. It included transporting the charged battery to the drone, installing it in the drone, toggling the drone's power on, removing the drone from the exchange dock, releasing the drone in the landing area, and finally looking for the presence of the marker to confirm that the drone has left the landing area. Similarly to the 'Pickup the drone' step, this is a movement-intensive step with multiple short movements. However, as the timing stops when the system can no longer detect the drone, with the drone was removed from the landing area by the operator in these tests, this step also relies on the speed of the operator removing the drone from the landing area. While the performance of the system in this paper could be increased by stopping the timing once the system considers the drone to be ready for take-off, the authors felt that this approach provided a more realistic scenario for testing.

Overall, for a system with similar functionalities, the speed of the system could potentially be increased by configuring it to require as few different steps of movement as possible and by conducting peripheral functions, such as battery monitoring, as parallel operations to the main system.

4. Conclusions

This article describes how an independent mechanical jig could be used to hold a drone in place during an automated battery replacement. As the drones were located with machine vision and handled by a robotic arm, the placement of a drone in the landing area could be flexible, allowing drones to land in the landing area with relatively loose precision. This was also tested during an evaluation by placing a drone in various positions and orientations in the landing area. Additionally, our system could be used to store and release drones using the same method. During the evaluation, the constructed prototype failed to exchange the battery 5 times out of 200 iterations, with the average time for a complete battery replacement being under two minutes. Compared to existing similar solutions, our method was neither the fastest nor the most reliable, as other comparable solutions could complete the flight battery exchange in even under one minute with near perfect reliability. Additionally, our setup was relatively complex and showed susceptibility to malfunctions during testing. However, our system offers the possibility for drones to land with relatively loose precision, and it could also be used as a storage for drones, which adds utility beyond battery exchange.

Therefore, the method of using independent jigs for handling drones during an automatic flight battery exchange shows promise as a concept but could benefit from additional research. The jigs could possibly be scaled up and redesigned for larger drones, as real-world use cases for DJI Tello drones are limited outside of research and hobbyist applications. Based on evaluations, other possible areas for future improvements include drones developed with such battery replacement methods in consideration, which could be beneficial for long-term operational reliability. It was observed that the localization of the drone in our system resulted in inaccurate pick-up actions, beyond what could be corrected three times, and pick-up actions within the margin of correction six times during testing. While the localization itself was likely the main cause, the mechanical guidance could possibly also be improved by design features such as centering gripping surfaces in drones for better pick up or by broader margins of guidance in the mechanical jig. Our system also took advantage of using static coordinates for non-moving POIs, which likely contributed to higher reliability once properly adjusted. However, using static coordinates may not be feasible in all use cases, as even a slight offset between the true POI location and its set coordinates can cause additional wear or even malfunctions due to the poor alignment of the robotic actions and the POIs.

Future development plans include designing scalable and modular features that could support various drone types to improve the overall usefulness of the system. The Tello drones win this study ere attached to jigs from their feet since they provided a sufficient support for the removal of batteries. However, using the same exact design may not feasible for larger drones that can require a greater surface area and a wider stance between their feet for stability. Scaling the jig to accommodate larger drones may require the development of separate attachment features for these drones, i.e., points of contact between a drone and a jig. These attachment features could possibly be designed to be universal between drones to increase compatibility. Similarly, batteries and battery housings could likely be designed in a way that would provide compatibility with different types of drones. Operational reliability could also benefit from having a method to recognize and correct for the misalignment of different POIs, i.e., a method of error correction instead of static coordinates, allowing for dynamic error correction rather that relying on static coordinates.

Author Contributions: Conceptualization, A.V. and A.T.; methodology, A.V.; software, A.V.; resources, J.R. and A.T.; writing—original draft preparation, A.V.; writing—review and editing, J.R.

and A.T.; visualization, A.V.; supervision, J.R.; project administration, A.T. All authors have read and agreed to the published version of the manuscript.

Funding: This research was funded by the Academy of Finland's support via Aeropolis (grant number: 348479).

Institutional Review Board Statement: Not applicable.

Informed Consent Statement: Not applicable.

Data Availability Statement: Data are contained within this article.

Conflicts of Interest: The authors declare no conflicts of interest.

References

1. Bitar, A.; Jamal, A.; Sultan, H.; Alkandari, N.; El-Abd, M. Medical Drones System for Amusement Parks. In Proceedings of the 2017 IEEE/ACS 14th International Conference on Computer Systems and Applications (AICCSA), Hammamet, Tunisia, 30 October–3 November 2017; Volume 14, pp. 19–20.
2. Alkouz, B.; Shahzaad, B.; Bouguettaya, A. Service-Based Drone Delivery. In Proceedings of the 2021 IEEE 7th International Conference on Collaboration and Internet Computing (CIC), Atlanta, GA, USA, 13–15 December 2021; Volume 7, pp. 68–76.
3. Lee, D.; Zhou, J.; Tze Lin, W. Autonomous Battery Swapping System for Quadcopter. In Proceedings of the 2015 International Conference on Unmanned Aircraft Systems (ICUAS), Denver, CO, USA, 9–12 June 2015; pp. 118–124.
4. Fujii, K.; Higuchi, K.; Rekimoto, J. Endless Flyer: A Continuous Flying Drone with Automatic Battery Replacement. In Proceedings of the 2013 IEEE 10th International Conference on Ubiquitous Intelligence and Computing and 2013 IEEE 10th International Conference on Autonomic and Trusted Computing, Sorrento Peninsula, Italy, 18–21 December 2013; Volume 10, pp. 216–223.
5. Zhao, T.; Currier C.; Bonnin, A.; Mellos, G.; Martinez N.; Chen, Y. Low Cost Autonomous Battery Replacement System for Quadrotor Small Unmanned Aerial Systems (sUAS) using 3D Printing Components. In Proceedings of the 2018 International Conference on Unmanned Aircraft Systems (ICUAS), Dallas, TX, USA, 12–15 June 2018; pp. 103–107.
6. Costea, I.M.; Plesca, V. Automatic battery charging system for electric powered drones. In Proceedings of the 2018 IEEE 24th International Symposium for Design and Technology in Electronic Packaging (SIITME), Sibiu, Romania, 25–28 October 2018; Volume 24, pp. 377–381.
7. Hoon Choi, C.; Jun Jang, H.; Guy Lim, S.; Chul Lim, S.; Ho Cho, S.; Gaponov, I. Automatic wireless drone charging station creating essential environment for continuous drone operation. In Proceedings of the 2016 International Conference on Control, Automation and Information Sciences (ICCAIS), Ansan, Republic of Korea, 27–29 October 2016; pp. 132–136.
8. Stuhne, D.; Vasiljević, G.; Bogdan, S.; Kovačić, Z. Design and Validation of a Wireless Drone Docking Station. In Proceedings of the 2023 International Conference on Unmanned Aircraft Systems (ICUAS), Warsaw, Poland, 6–9 June 2023; pp. 652–657.
9. Autonomous Drones In a Box | Dronehub. Available online: https://dronehub.ai/ (accessed on 31 July 2024).
10. D135 Drone Dock—HEISHA Robot Charging Station. Available online: https://www.heishatech.com/autonomous-drone/d135-drone-dock/ (accessed on 30 July 2024).
11. Bláha, L.; Severa, O.; Goubej, M.; Myslivec, T.; Reitinger, J. Automated Drone Battery Management System—Droneport: Technical Overview. *Drones* **2023**, *7*, 234. [CrossRef]
12. Vilkki, A. Functionalities in Micro-Airports. Master's Thesis, University of Oulu, Oulu, Finland, 13 December 2023.
13. OpenCV: Detection of ArUcO Markers. Available online: https://docs.opencv.org/4.x/d5/dae/tutorial_aruco_detection.html (accessed on 30 July 2024).

machines

MDPI

Article

A New Reconfigurable Agricultural Vehicle Controlled by a User Graphical Interface: Mechanical and Electronic Aspects

Supaphon Kamon [1], Enrico di Maria [1], Nicola Ivan Giannoccaro [2,*] and Kazuo Ishii [1]

1 Graduate School of Life Science and Systems Engineering, Kyushu Institute of Technology, Kitakyushu 808-196, Japan
2 Department of Innovation Engineering, University of Salento, 73100 Lecce, Italy
* Correspondence: ivan.giannoccaro@unisalento.it; Tel.: +39-0832297813

Abstract: The use of innovative mobile vehicles with increasingly advanced mechatronic aspects in the agricultural sector is becoming, in recent years, a stimulating field of research and comparison. In particular, the problem addressed in the present work refers to improving the locomotion of mobile vehicles on agricultural terrain by reducing the soil damage and improve the overall performance. Agricultural vehicles generally use tracks and wheels for locomotion; the main difference between the two systems is the contact area with the ground and, consequently, the pressure distribution. The present work presents a new reconfigurable agricultural vehicle that can switch from one locomotion system to another, choosing the suitable configuration according to the terrain conditions. All the mechanical and electronic aspects of the prototype developed are analyzed together with an in-depth analysis of the management of the innovative functions through a user-friendly graphical interface able to control the vehicle.

Keywords: reconfigurable locomotion system; robots for agriculture; mechatronics applications; control of mechanical systems

1. Introduction

In the last decade, the agricultural field has seen a significant increase in mechanization and technological solutions, ranging from monitoring agronomic parameters using field sensors to robots and machinery for field operations, as detailed in [1]. For some types of crops, it is possible to create ad hoc structures in highly mechanized greenhouses, as in [2–4], where robots are used for harvesting crops like tomatoes, apples, or strawberries. In other cases, it is necessary to intervene on existing plantations, as in [5], where a smart solar fertigation system is installed in a vineyard to optimize the use of fertilizers. One of the drawbacks of this mechanization is the increase in the size of the machinery [6], which has led to detrimental effects on the environment [7,8].

A factor of interest for the farmers is the soil quality, which directly impacts productivity and soil maintenance costs. It is well known that the use of heavy machinery has caused progressive land degradation, and several studies have been carried out to quantify this damage [9–11].

As a result of this altered soil state, farmers experience a loss of production and higher soil maintenance costs [12]. Moreover, there are correlated environmental issues, such as the loss of water and soil erosion [13].

Compaction and distortion are the mechanisms by which vehicle traffic in the field damages the soil. The first is the result of the normal pressure exerted at the contact patch, which causes a compression of the soil pores and reduces their volume. The shear forces cause the distortion of the soil during the vehicle motion, which destroys the pores by a constant volume displacement [14]. Without a proper amount of pore space, the soil health is compromised.

The most common running gears with which agricultural vehicles are equipped are tracks and wheels. Given a tracked and a wheeled vehicle comparable in size, their main difference will be in the contact area at the ground. Tracks provide a lower soil compaction because of the lower pressure distribution over a larger area but higher soil distortion due to the higher traction force and the peaks of pressure under the rollers [15]. Instead, wheels have a smaller contact patch and cause higher compaction but are also more agile and cause less distortion, especially during the turning maneuvers. Reducing the contact area sliding during the skid-steering maneuvers requires lower torque and energy consumption. In the literature, there are examples of systems able to change the contact area for improving the steering maneuvers, as in [16], where an eight-wheeled vehicle has a mechanism for lifting one or two pairs of wheels. Vehicle sinkage should also be considered if the soil is highly deformable. The bulldozing effect during the turning maneuver might make a tracked system preferable, despite the longer contact patch.

The different contact area also impacts tracked and wheeled vehicles' traction performance and energy consumption. If the soil has cohesive properties (i.e., contains clay), a larger contact patch ensures a higher traction force [17], while on frictional soils, a higher normal stress can increase the soil strength and might lead to a higher draught. Moreover, the volume of soil deformed is the main contribution to the running gear rolling resistance. If the contact area is small, we can expect a high sinkage and rolling resistance on highly deformable soil. Conversely, a large area would reduce the sinkage.

To combine the benefits of wheels and tracks and adjust the contact area based on the soil conditions, we proposed a locomotion system to change the contact patch [18], and we realized "Hadrian" a new wheel/track reconfigurable vehicle. The vehicle is equipped with a locomotion system that can switch between a wheeled and a half-tracked mode by partially lifting the tracks' idlers and leaving only the sprockets in contact with the soil to have a wheel shape interacting with the ground. This concept is depicted in Figure 1.

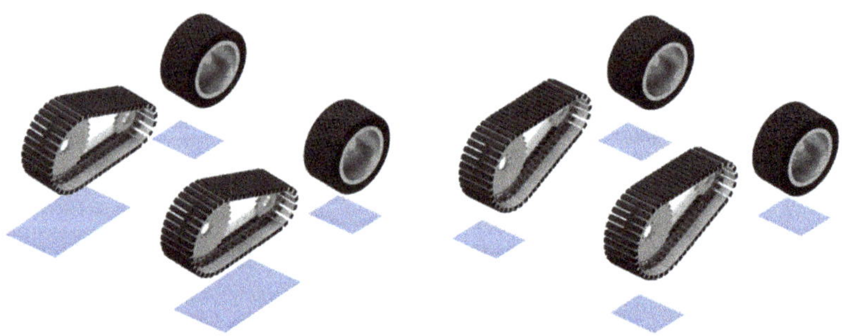

Figure 1. Concept of the reconfigurable locomotion system. The contact area is adjusted based on the track's configuration.

Such a system aims to minimize soil damage during farming operations and adapt to vehicle performance (traction, rolling resistance, and steering torque).

As a testbed for our system, we selected the vineyard, a good example of a field experiencing soil damage [13], especially during the harvesting period, when the farmers enter the field with heavy tractors, regardless of the soil conditions. Our test fields are the training systems used in Italy, one of the countries with the highest production of wine in the world [19], where innovative systems for the soil damage issue are required. Hadrian aims to transport the grape in the vineyard during harvesting, by avoiding using tractors. The vehicle will have to decide the locomotion mode autonomously based on proprioceptive sensors' data (e.g., motors' velocity and current) and exteroceptive sensors' data (e.g., soil moisture sensors in wireless communication with the vehicle, Figure 2). Similar strategies are adopted in other works, as in [20], where exteroceptive and proprioceptive sensor data are used for terrain estimation, improving the mobility of a mobile base.

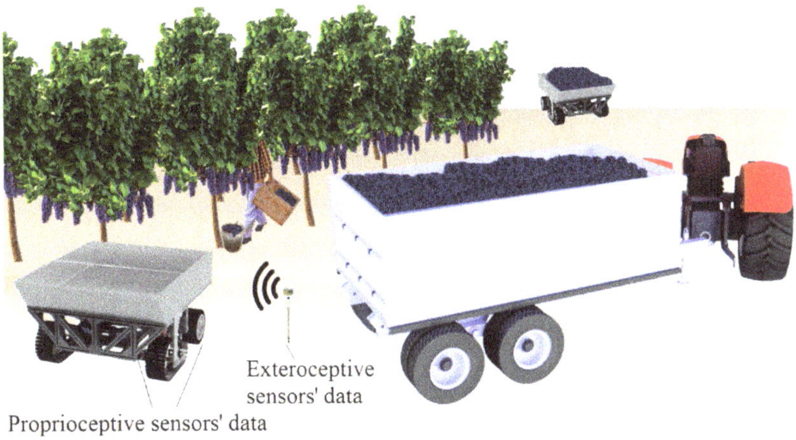

Figure 2. Vehicles operating in the field by using proprioceptive and exteroceptive sensors' data.

Implementing robotics in agriculture and beyond has to consider ease of use. The final user will probably not be a technician, and an easy interface becomes an essential tool for the diffusion and acceptance of new machinery. For this reason, we developed a straightforward user interface to control Hadrian remotely. Although a final version of the vehicle will be able to decide autonomously the locomotion mode based on the soil conditions, we can expect the navigation to be still performed or supervised by a human operator because of the complexity of the environment and the presence of humans in the field.

In this paper, we provide a detailed description of the vehicle and the solution adopted for its realization, specifically focusing on the system architecture and user interface developed for its control. First, we will briefly describe the environment in which the mobile base will operate and how this affects the overall size. Then, we will describe the mechanical and electrical systems, and finally, we will show in detail the user interface we developed for the easy remote control of the vehicle.

2. Materials and Methods

2.1. Reconfigurable Grape-Transporting Vehicle

2.1.1. Environment and Relative Constraints

The vehicle's overall size is decided based on the environment in which it will operate. Since Hadrian is meant to transport grapes during the harvesting period, we have considered the sizes and characteristics of the vineyards. Several types of training systems are used worldwide: head/can, cordon/cane, head/spur, and cordon/spur [21]. The vines are usually placed in rows, leaning on steel wire structures that guide their branches.

The distance among the trees is decided based on the agricultural machinery used in the field. This distance is about 2 m, allowing the pass of tractors.

Because the vehicle must operate in the field with the people collecting the grapes, we need to keep a marginal distance between the vehicle and the people. We assume the vehicle's size is 1.3 m $\times$ 1.3 m square. The maximum weight (including the payload) is 400–450 kg, assuming a grape payload of a maximum of 200 kg.

2.1.2. Running Gear's Size and Traction Motors

The running gear's size, summarized in Table 1, is a trade-off among soil damage, performance, and weight.

We have conducted field experiments in a vineyard and FEM analyses of a single wheel as the size changes [22] and of a track [18] to evaluate the effect of different contact areas in terms of sinkage, rolling resistance, and pressure exerted onto the soil. Unlike the wheel, a track is a much more complex system to analyze and simulate because of the flexible chain and peak of pressures caused by the rollers, as we further discussed in [23].

Table 1. Running gear's size.

Component of Gear	Size (mm)
Sprocket diameter	330
Idler diameter	175
Track width	160
Wheel diameter	300
Wheel width	150

We tested the rolling resistance and sinkage of our designed track module in a soil bin facility built at the Kyushu Institute of Technology [18]. We have mixed clay, silt, and sand in the percentages of 20%, 40%, and 40%, respectively, to obtain loam soil [24] with the frictional and cohesive properties typical of agricultural soil. Then, we carried out several pull-tests in which the track was mounted to a frame, loaded to 100 kg, and pulled for 2.5 m at a constant velocity by a walking tractor through a steel wire. The steel wire was attached to a force gauge sensor to measure the rolling resistance. We tested on different soil conditions: firm, soft, and wet saturated. In Figure 3, some of the test results are shown. As expected, the highest rolling resistance is obtained for the case of deformable soil (soft and wet), although the mechanism of soil deformation in the two cases is different. In soft conditions, the soil contains a higher fraction of pore space, which undergoes a compaction and volume reduction during the track pass. In saturated wet conditions, the pores are filled with water, making the soil incompressible and leading to deformation by distortion. We have a significant volume of deformed soil and a corresponding high rolling resistance in both cases.

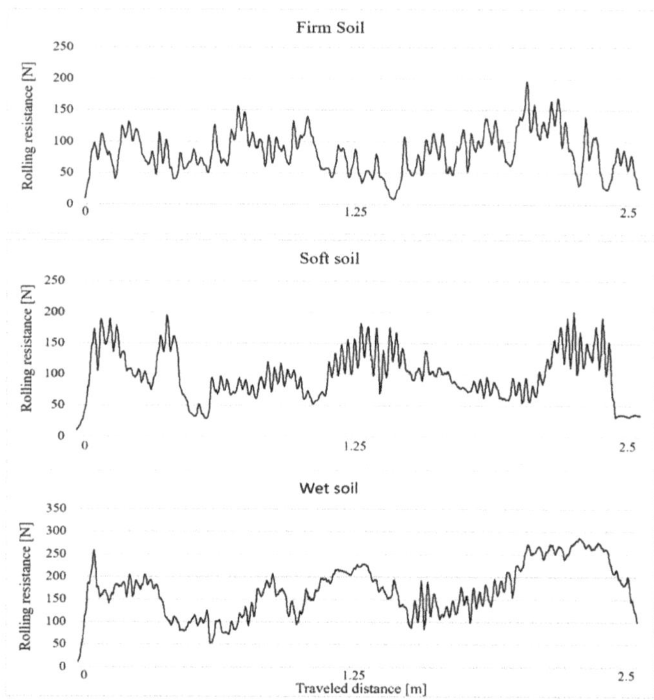

Figure 3. Some of the pull test results carried out with the single-track module.

The results of the pull tests have been used to select the four traction motors after assuming a maximum velocity of 0.5 m/s, comparable with that of a human being walking in the field.

The DC motors selected are BLV620K100S for the front tracks and BLV510K100S for the rear wheels, developed by Oriental Motor. The BLV620K100S has a maximum torque of 52.7 Nm (1–30 rpm) for a rated power of 200 W. The BLV510K100S has a maximum torque of 27.4 Nm (1–30 rpm) for a rated power of 100 W. We expect the tracks to "support" the wheels in more difficult soil conditions (i.e., wet saturated), in which a high sinkage and high rolling resistance may cause the wheels to be stuck. In such a situation, the vehicle is supposed to switch to track mode, reducing the weight sustained by the wheels and providing the traction force to move the vehicle.

2.1.3. Mechanical Design

To describe Hadrian's mechanical solution, we can divide the vehicle into three main groups, all mounted to the chassis: rear axle, front axle, and switching mechanism. Figure 4 shows the full vehicle and the three main groups.

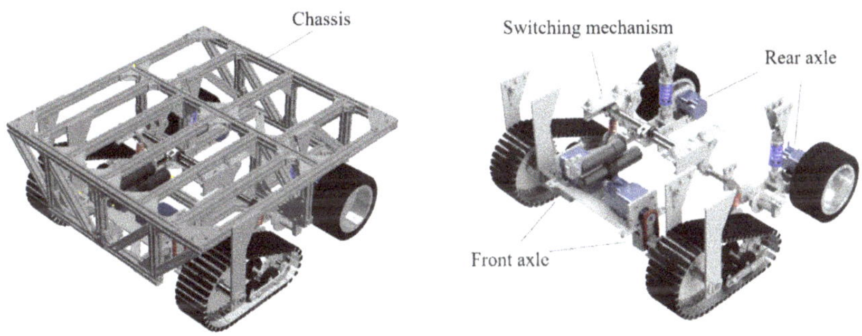

Figure 4. Hadrian's overview and main subsystems.

The rear axle consists of two wheels connected to the chassis through a longitudinal trailing arm suspension system. This type of suspension system has the advantage of being simple and compact. The trailing arm consists of welded steel parts, as described in Figure 5. Each wheel has a traction motor directly connected to the wheel hub. This configuration saves space onboard, leaving room for the batteries and the electrical system components at the vehicle's rear inside the chassis. In Figure 6, the wheel system is shown in detail.

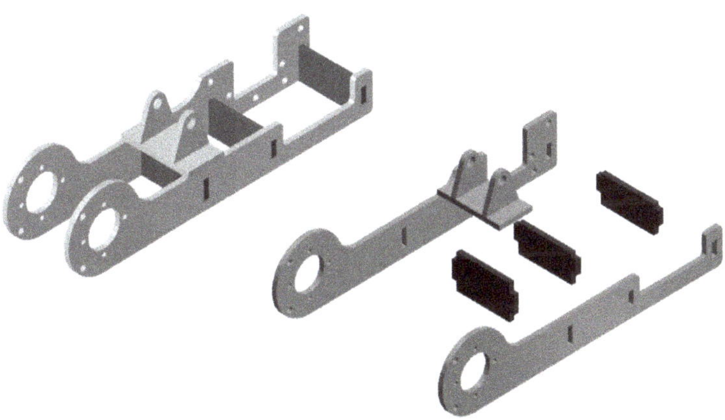

Figure 5. Welded parts in the trailing arm of the wheel suspension system.

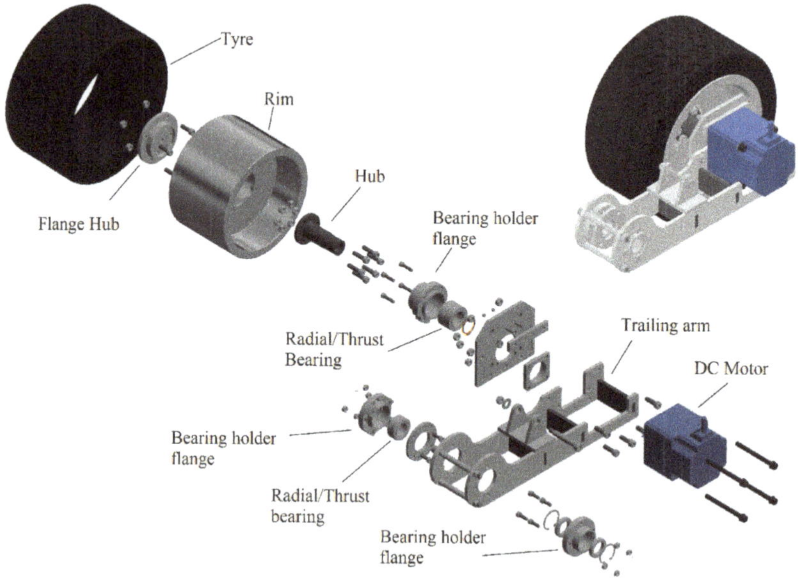

Figure 6. Wheel group's components.

The reason for having the wheels at the rear is that, while moving, the front running gears compact the soil, making the conditions more favorable for the rear running gears. Since the wheels cannot adjust their contact patch, having them on the rear side is more convenient. In addition, by having the sprockets facing the moving direction, we ensure that the contact shape still resembles a wheel when the vehicle is sinking, avoiding contact between the lifted portion of the track and the soil. The two reasons explained here are depicted in Figure 7.

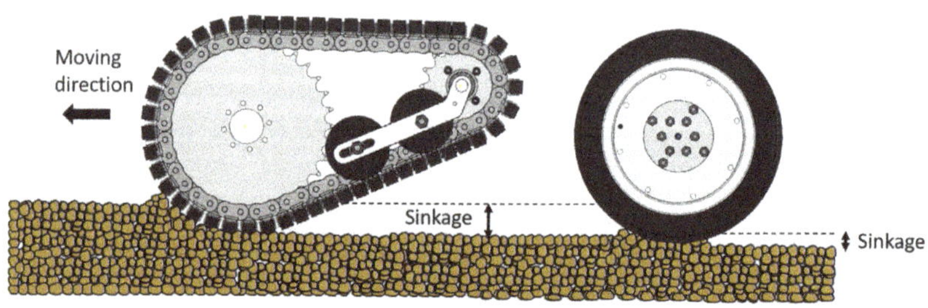

Figure 7. Tracks at the front axle improve the soil conditions for the wheels.

The track system module was designed and realized from scratch using a steel chain with mounted attachments. A U-shaped aluminum channel and a rubber lug are fixed on each attachment by bolts and glue. The chain is connected to two gears, a sprocket, and an idler. The sprocket has a larger diameter than the idler to facilitate the wheel mode configuration when the idler is lifted. A more detailed description of the track module design can be found in [18,23].

Unlike the rear axle, the traction motors of the front axle are positioned within the chassis. The traction motor drives the sprocket gear through a chain coupled with two sprocket gears, and the transmission ratio is 1. Each sprocket gear is rigidly connected to the chassis by a pair of links. The idlers are connected to the switching mechanism by a shock absorber. Two thrust bearings mounted in O-configuration support one end of the driven shaft. Figure 8 shows the design solution adopted for the front axle. The triangular elements connected to the tracks increase the system's overall stiffness during the turning maneuvers, when the tracks experience lateral forces. In Figure 9, the assembled wheel group and track group are shown.

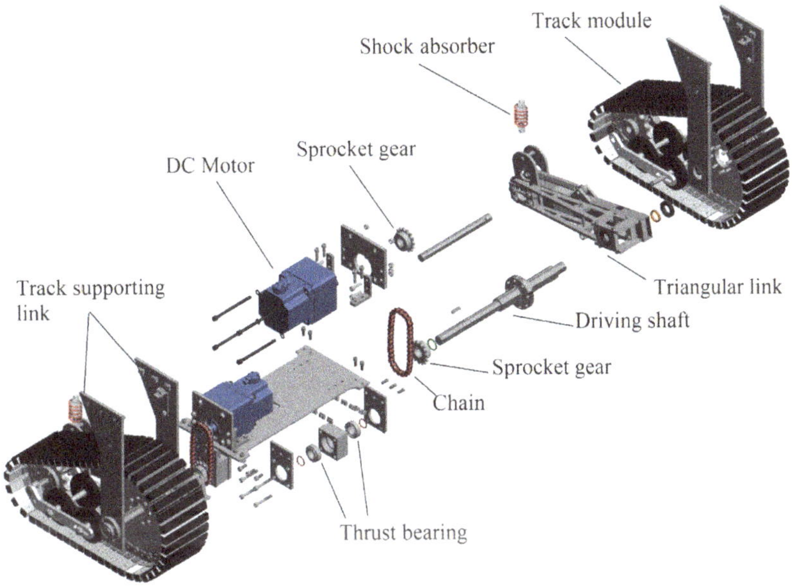

Figure 8. Front axle's components.

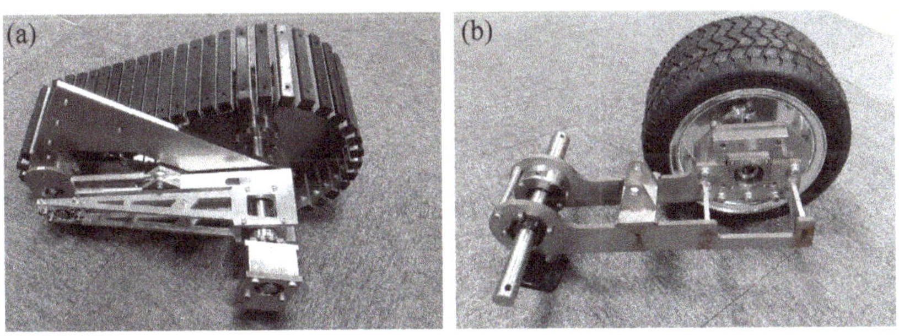

Figure 9. Track group (**a**) and wheel group (**b**).

The last group to be described is the switching mechanism. This is a Scott Russel linkage, which converts the horizontal displacement of an electro-hydraulic actuator (EHA) into a vertical linear movement of the idlers. The mechanism consists of an EHA connected to a shaft (S1) with two bearings mounted at its ends. The bearings are inserted into two parallel linear guides. The shaft is coupled with the longest link of a scissor mechanism where the long link length is double that of the short one. A revolute joint connects the middle of the long link to one end of the short link. The other end of the long link is coupled with a second shaft (S2) connected to the idlers' shock absorbers, as shown in Figure 10. The EHA displacements cause S1 to move along the linear guide and close or open the scissor mechanism to lift or lower S2 and the shock absorbers connected to the track modules.

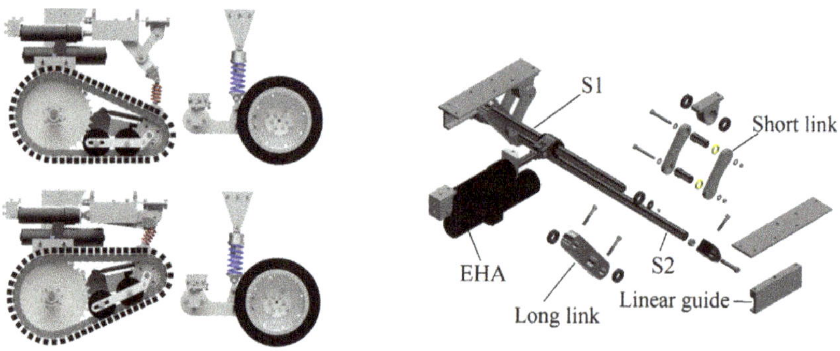

Figure 10. Switching mechanism configurations and components.

The relation between the horizontal displacement of the EHA and the vertical displacement of the idlers is evaluated as $h = b(x) \times tang(\alpha)$, where $b = b_{max} - x$, and b_{max} is the maximum distance between the roller shaft and the revolute joint of the short scissor link, and α is the angle between scissor's links and the horizontal, as in Figure 11a. We selected the desired vertical displacement of the idler to be in the range of 100–150 mm. With this displacement, we can ensure that, while in wheel mode, the shape in contact with the ground is circular, as if it were a wheel. Figure 11b provides the static analysis of the mechanism. The number 0 indicates the chassis, 1 is the actuator, 2 and 3 are the long and short scissor links, respectively, and 4 is the linear guide. *Fa* is the actuator force, *R* the reaction force on the link and *Fs* the force on the idler. As can be seen from the static analysis, small values of the angle α in track mode would require a high actuator force and a higher reaction force on the links. In wheel mode instead, the actuator only lifts the idler, and relatively small α angles can be accepted. The mechanism synthesis is a trade-off between the space available inside the chassis, the desired vertical displacement, and the necessity to avoid small angles. In the final design, the α angle in track mode is 53 degrees.

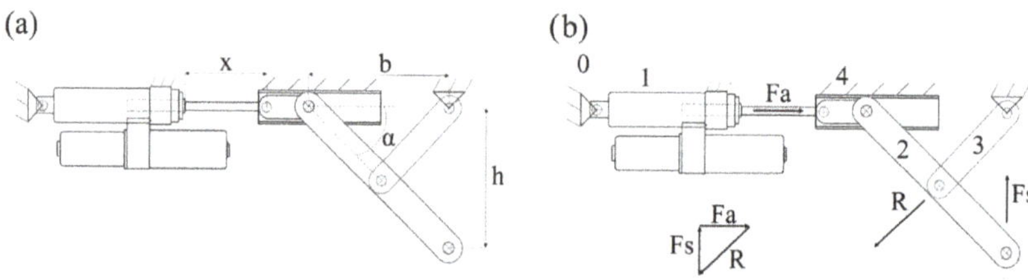

Figure 11. (**a**) Relation between EHA stroke and vertical displacement. (**b**) Static analysis.

The EHA selected is the mini motion package MMP3-B1A100BA-GH, developed by KYB corporation, whose stroke is 0–100 mm, with a total minimum length of 280 mm and a maximum length of 380 mm. The maximum force is 5800 N, with a piston pulling velocity of 33 mm/s and a pushing velocity of 19 mm/s. Although this application's maximum force is oversized, we preferred the Mini Motion Package because of its robustness, reliability in dirty outdoor environments, and capacity to act as a rigid link once the desired position is reached. These kinds of actuators are often used on agricultural machinery, for example, for adjusting the position of a plowing machine mounted behind a tractor.

An alternative solution we considered is using a lead screw mechanism driven by an electrical motor. Eventually, we preferred to adopt the EHA because of its higher robustness, especially under dynamic loads.

3. Results

3.1. Mechanical Final Design

The three main groups previously discussed are connected to the chassis, which consists of a ladder frame made of squared aluminum bars connected through triangular brackets. Wherever possible, the sections of the chassis have been triangulated by using diagonal elements to improve the stiffness, avoiding rectangular arrangements of bars, which would have low stiffness, especially when subjected to torsional loads [25]. The final assembly of the vehicle is shown in Figure 12, and Figure 13 shows the vehicle in the field.

Figure 12. Final assembly of the vehicle.

Figure 13. Final vehicle while operating in the field.

The total weight of the vehicle without any payload is 237.5 kg. The weight distribution has been measured using a weight scale, as in Figure 14, and those measurements have been used to locate the center of gravity. In wheel mode, the weight measured is 65.5 kg and 69 kg for the left and the right tracks, respectively, and 52.5 kg and 50.5 kg for the left and right wheels, respectively. Although the track mode's weight distribution is measured similarly, the result depends on the EHA final extension and the characteristics of the shock absorber, as detailed in our previous work [23]. With a maximum extension of the EHA (100 mm), the vertical load on the wheels becomes 31.5 kg and 33.5 kg for the left and right wheels, respectively. Future works will investigate the effects of different EHA extensions and weight distributions in a real environment. The main characteristics of the vehicle are summarized in Table 2.

Figure 14. Weight distribution measurement by weight scale.

Table 2. Vehicle's main specifications.

Vehicle Characteristics	Specification
Mass	237.5 kg
Size	1230 mm × 1260 mm × 540 mm
Max velocity	0.5 m/s
Battery	Li-ion 25.2 V 27 Ah ×2, mobile battery 5 V
EHA	MMP3 12 V, 5800 N
Motors	DCM 200 W ×2, DCM 100 W ×2

3.2. Electronic System Final Design

In Figure 15, the flowchart of Hadrian's power system is provided. This includes two Li-ion batteries of 25.2 V for the four brushless DC motor driving tracks and wheels and a DC–DC converter from 25.2 V to 12 V for supplying energy to the electro-hydraulic actuator (EHA). The vehicle is equipped with a microcontroller ESP32, which has Wi-Fi integrated, used for receiving the soil moisture information from the field. The ESP32 microcontroller (receiver) is connected to a PC by USB. The relay and UART/RS485 transceiver is powered through an ESP32 (receiver) with 5 V. Each soil moisture sensor in the field (three in the example of Figure 13) is connected to a separate ESP32 (transmitter) in wireless communication with the ESP32 onboard the vehicle. Mobile batteries of 5 V power the moisture sensors. The final number of soil moisture sensors will depend on the size of the field and the expected soil moisture variability.

In our preliminary tests, we had the vehicle moving on straight lanes, and the slip of wheels and tracks is evaluated by using a laser distance sensor located at the end of the lane and pointing to a flat black panel mounted on the front of the vehicle. With the laser distance sensor (LDS), we measure the vehicle position (the distance from the sensor at the end of the lane), and we can estimate the moving velocity. By monitoring the motors' rotational velocity provided by the encoders, we evaluate the slip i as $i = (wr - V)/wr$, where w is the angular speed measured by the encoders, r is the wheel's radius, and V is the estimated vehicle moving velocity.

The laser distance sensor is connected to an ESP32 (transmitter), and a battery of 3.3 V powers it. The LDS is only a temporary solution for the preliminary tests because it's simple and cost-effective. In the future, we will adopt a GPS or DGPS.

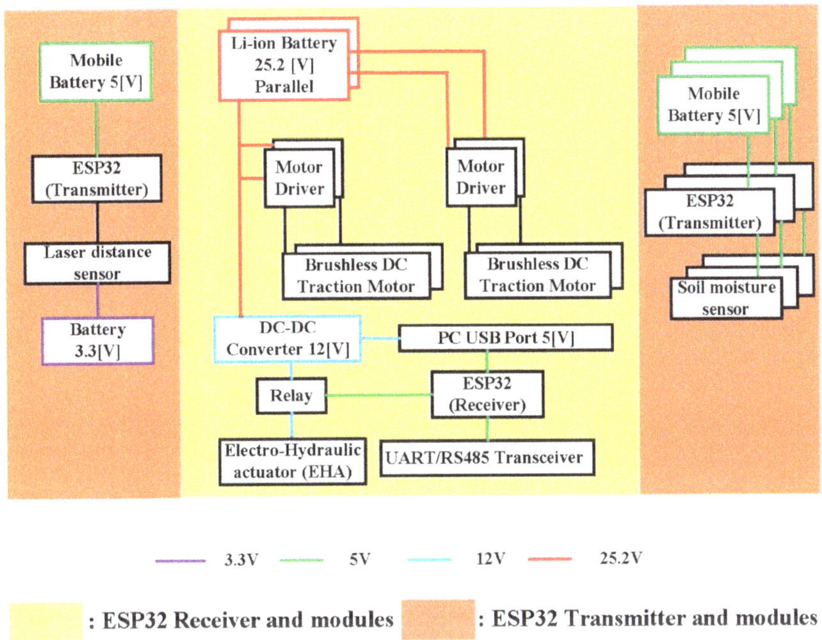

Figure 15. Vehicle power system.

3.3. System Architecture and User Interface

A user interface has been developed for easily operating the vehicle. We first describe the system's architecture, Figure 16, to understand how this interface works.

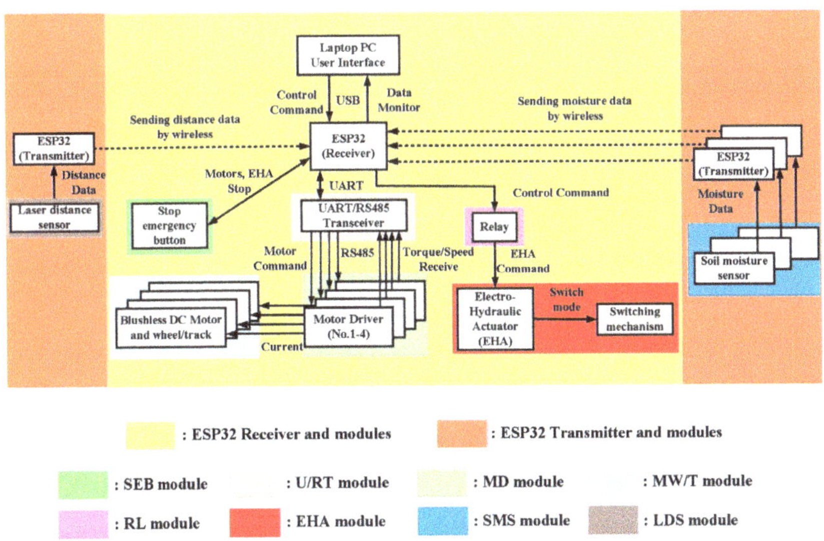

Figure 16. Hadrian's system architecture.

Hadrian is equipped with a laptop PC from which it can be operated using the user interface. As explained in Section 4, a microcontroller ESP32 receives the soil moisture information from the ESP32 transmitters connected to the sensors in the field. The ESP32 is also interfaced with the four motor drivers, the EHA, and the emergency button on the vehicle to stop all the motors in an emergency. The information about the moisture will be used, together with proprioceptive sensors' data, to autonomously assess whether the locomotion mode should be wheels or tracks. At first, we will conduct a series of tests on different soil conditions (firm, soft, wet saturated) and with different payloads, and the soil damage will be assessed by our group. The compaction will be evaluated by manually measuring the Cone Index [26] and the ruts' depth, and the distortion by using image processing and comparing the soil displacement before and after the vehicle passes. We will use these data to compare the two locomotion modes and assess the best one in the various cases. The measurements and our assessment will represent the teaching data for the design of the controller. At the same time, we will measure the motor torque, the current, the vehicle's slip, and the soil moisture during each test. For DC motors, when the vehicle is moving at a constant velocity, the motors' torque is proportional to the rolling resistance, which is directly related to the soil deformation [26] (in particular to the sinkage). Soil distortion depends on the shear forces exerted onto the ground, and these depend on the load distribution at the contact patch, the torque, the slip, and the soil state. A neural network-based controller will be trained with the experimental data to autonomously assess the best contact area, depending on the operating conditions. For example, during a supervised learning phase, the neural network will learn to assign a locomotion mode to a certain combination of proprioceptive and exteroceptive sensors' data. Once the network is properly trained, we will test its performance to choose the locomotion mode on soil conditions different from the one used for the teaching phase. As the field tests are still ongoing, the design of such a controller represents the next step of this research.

The locomotion mode can also be decided manually by the user, and a switch relay controls the EHA. The traction motors use the RS485 protocol, and the UART/RS485 converter communicates with the ESP32.

In Figure 16, the system components are grouped into modules depending on the function. The SEB is the stop emergency button, U/RT is the UART/RS485 transceiver, MD is the module consisting of the motor drivers, MW/T module is the mechanical connection among motors and running gears (wheels and tracks), RL is the relay, EHA module is the electro-hydraulic actuator connected to the switching mechanism, and the SMS module consists of the soil moisture sensors. Finally, the LDS is the laser distance sensor module.

The user interface has been developed using Visual Basic 2022 [27], whose toolboxes allow for the easy implementation of the required functions. Through the interface, we want to send commands to the ESP32 to control the velocity and direction of the four traction motors and extend or shorten the EHA while monitoring the motors' torque and running gears' slip. The main blocks are "Button", "TexBox", "Chart", and "SerialPort". These are responsible, respectively, for operating the scan port and sending commands from visual basic.NET to the controller, displaying the sensors' values, displaying the graphs of sensors' values in a chart, and creating a serial port component for the synchronous and event-driven I/O to access serial driver properties. Additional blocks are used for adding labels and pictures to the graphic interface. The designed GUI is shown in Figure 17.

The buttons "WHEEL" and "TRACK" allows the user to decide on the locomotion system. The relay controls the direction to extend or shorten the piston's position. When the "WHEEL" button is pressed, the actuator shortens until it reaches the shortest length (zero position). When the "TRACK" button is pressed, the actuator extends until it reaches the maximum extension. The image of the vehicle is used to confirm the locomotion system in use, and the switch must take place with the vehicle at a standstill. The operator must press the "Stop" button before switching the mode to ensure the vehicle is not moving. To tune the maximum possible extension of the EHA, we placed on the chassis two position switches. In particular, the shaft driven by the EHA can activate the position switches and stop the piston's extension when the extension of the EHA is the one decided by the position of the switches (in our preliminary tests, we used 80% of the stroke). The user manually adjusts the position of the switches before using the vehicle.

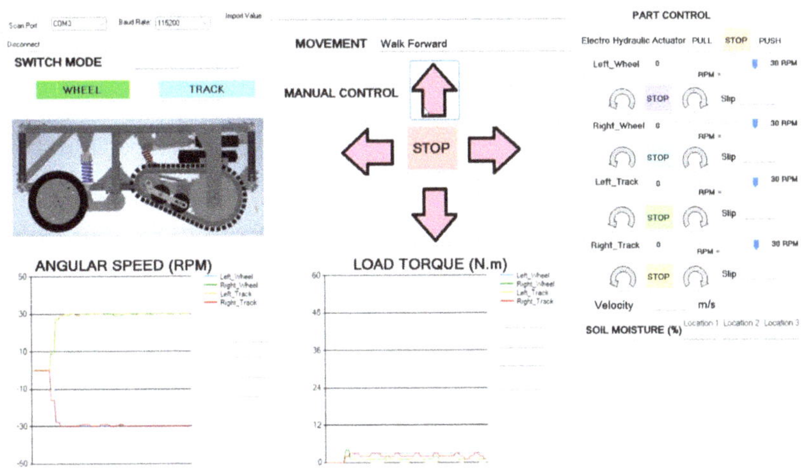

Figure 17. The graphical user interface for operating the robot.

Since the sprockets are connected to the motor, while the idlers can rotate freely, only the idlers rotate during the switch. This prevents the track belt from sliding onto the soil during the switch, which would cause additional forces to be sustained by the EHA and soil damage. The four arrows allow the user to decide the direction of movement. The traction motors receive a velocity input decided by the user. A velocity input is sent to wheels and tracks when the forward or back arrow is pressed. The input velocity for the front and back axles is different since the rolling radii of wheels and tracks are not the same. When an input velocity is set, we can monitor the slip of each running gear and adjust the velocity accordingly. This is possible by using the "Part Control". The "Part Control" controls each actuator individually by selecting the rotation direction and velocity. The arrow buttons indicate the rotations' directions, while the stop buttons can stop the corresponding running gear. The interface also allows us to display in real-time the angular velocity of each running gear and the torque exerted by every single motor. In the case of EHA, the "PULL" and "PUSH" allow the user to decide the actuator's extension.

The vehicle uses the skid-steering to turn. The turning maneuver is carried out with the vehicle at a standstill. When the left or right arrow is pressed, the left and right side running gears spin in opposite directions, causing the vehicle to rotate. Because the skid-steering might require a high torque, especially on soft soils, we set the velocity to a maximum of 15 rpm to not overstress the motors and tackle the torque peaks due to small obstacles or the bulldozing effect.

In the "Soil moisture" section, the user can monitor the soil moisture level in the field. In our preliminary tests, we are using only three soil moisture sensors. We can expect soil moisture sensors to cover the field in a real application. In that case, the field should be divided into sub-areas, and each sub-area's average soil moisture value should be monitored in the interface.

Although for the current prototype, the human operator can control the robot through the interface and decide the locomotion system, the final goal is to have the vehicle decide the locomotion system based on the sensors' data collected during its operation. To achieve this, the next step of this research is to test the robot in the field under several conditions (different soils and payloads) and use the measurements of internal and external sensors as teaching data to train a neural network for the control strategy.

The realized vehicle is presented in Figure 18 during the preliminary tests, both in the track and wheel modes.

Figure 18. Vehicle in track mode (**left**) and wheel mode (**right**).

4. Discussion

In this paper, we have presented our reconfigurable vehicle, Hadrian. This system can change its contact area by partially lifting the front tracks through a Scott Russel mechanism driven by an electro-hydraulic actuator. We described the mechanical solution adopted and the system architecture, and we finally described in detail the graphical user interface developed for easy remote control of the vehicle. Such a vehicle aims to transport the grape in the vineyard during harvesting while minimizing soil damage and adapting the performance based on the need.

The control interface allows the control of every actuator while monitoring the slip, the rotational velocity, and the torque. The graphical interface represents an important tool for user-friendly vehicle control. Moreover, in the context of the growing development of IoT technologies for agriculture, an interface of this type can be implemented via a smartphone app, not only to facilitate the control of the vehicle during the harvesting period, but also to use the same for field monitoring, collecting data on the state of the soil.

The next step of this research is to test the vehicle in the field and measure the soil damage and performance in several soil conditions. The experimental data will serve as teaching data of a controller, such as a neural network, for making autonomous choices on the best locomotion mode from the vehicle, based on the onboard sensors data and field sensors data. It can be foreseen that the forthcoming improvements to the control system will be able to further improve the behavior of the vehicle developed, making it an important and helpful tool in the field of grape harvesting.

Author Contributions: Conceptualization, E.d.M. and S.K.; methodology, S.K. and E.d.M.; software, S.K.; validation, S.K., E.d.M., N.I.G. and K.I.; formal analysis, S.K. and E.d.M.; investigation, S.K. and E.d.M.; resources, K.I.; data curation, S.K., E.d.M. and N.I.G.; writing—original draft preparation, E.d.M., S.K. and N.I.G.; writing—review and editing, E.d.M. and N.I.G.; visualization, S.K. and E.d.M.; supervision, N.I.G. and K.I.; project administration, E.d.M.; funding acquisition, N.I.G. and K.I. All authors have read and agreed to the published version of the manuscript.

Funding: This research received no external funding.

Data Availability Statement: The data presented in this study are available on request from the corresponding author.

Acknowledgments: We thank our friends Raji Alahmad and Yoshiki Tanaka for their precious support during the realization of the vehicle.

Conflicts of Interest: The authors declare no conflict of interest.

References

1. Bechar, A.; Vigneault, C. Agricultural robots for field operations: Concepts and components. *Biosyst. Eng.* **2016**, *149*, 94–111. [CrossRef]
2. Su, L.; Liu, R.; Liu, K.; Li, K.; Liu, L.; Shi, Y. Greenhouse Tomato Picking Robot Chassis. *Agriculture* **2023**, *13*, 532. [CrossRef]

3. Yu, X.; Fan, Z.; Wang, X.; Wan, H.; Wang, P.; Zeng, X. A lab-customized autonomous humanoid apple harvesting robot. *Comput. Electr. Eng.* **2021**, *96*, 107459. [CrossRef]
4. Zhou, C.; Jun, H.; Xu, Z.; Yue, J.; Ye, H.; Yamg, G. A Novel Greenhouse-Based System for the Detection and Plumpness Assessment of Strawberry Using an Improved Deep Learning Technique. *Front. Plant Sci.* **2020**, *11*, 559. [CrossRef] [PubMed]
5. Visconti, P.; de Fazio, R.; Primiceri, P.; Cafagna, D.; Strazzella, S.; Giannoccaro, N.I. A solar-powered fertigation system based on low-cost wireless sensor network remotely controlled by farmer for irrigation cycles and crops growth optimization. *Int. J. Electr. Telecom.* **2020**, *66*, 59–68.
6. Keller, T.; Sandin, M.; Colombi, T.; Horn, R.; Or, D. Historical increase in agricultural machinery weights enhanced soil stress levels and adversely affected soil functioning. *Soil Tillage Res.* **2019**, *194*, 104293. [CrossRef]
7. Eriksson, J.; Hakansson, I.; Danfors, B. *The Effect of Soil Compaction on Soil Structure and Crop Yields*; Swedish Institute of Agricultural Engineering: Uppsala, Sweden, 1974.
8. Lal, R. Soil Structure and Sustainability. *J. Sustain. Agric.* **1991**, *1*, 67–92. [CrossRef]
9. Keller, T. Soil Compaction and Soil Tillage-Studies in Agricultural Soil Mechanics. Ph.D. Thesis, Department of Soil Science, Swedish University of Agricultural Science, Uppsala, Sweden, 2004.
10. Keller, T.; Berli, M.; Ruiz, S.; Lamandè, M.; Arvidsson, J.; Schionning, P.; Selvadurai, A.P.S. Transmission of vertical soil stress under agricultural tyres: Comparing measurements with simulations. *Soil Tillage Res.* **2019**, *140*, 106–117. [CrossRef]
11. ten Damme, L.; Stettler, M.; Pinet, F.; Vervaet, P.; Keller, T.; Munkholm, L.J.; Lamandè, M. The contribution of tyre evolution to the reduction of soil compaction risks. *Soil Tillage Res.* **2019**, *194*, 104283. [CrossRef]
12. Chamen, W.C.; Moxey, A.P.; Towers, W.; Balana, B.; Hallet, P.D. Mitigating arable soil compaction: A review and analysis of available cost and benefit data. *Soil Tillage Res.* **2015**, *146*, 10–25. [CrossRef]
13. White, R.E. *Understanding Vineyard Soils*; Oxford University Press: Oxford, UK, 2015.
14. Horn, R. Stress-strain effects in structured unsaturated soils on coupled mechanical and hydraulic processes. *Geoderma* **2003**, *116*, 77–88. [CrossRef]
15. Lamandé, M.; Greve, M.H.; Schjonning, P. Risk assessment of soil compaction in Europe—Rubber tracks or wheels on machinery. *Catena* **2018**, *167*, 353–362. [CrossRef]
16. Visconte, C.; Cavallone, P.; Carbonari, P. Mechanism for the Locomotion Layout Reconfiguration of the Agri_q Mobile Robot. In *Advances in Service and Industrial Robotics: Results of RAAD*; Springer: Berlin/Heidelberg, Germany, 2020; pp. 390–399.
17. Gerhart, G.R. *The Bekker Model Analysis for Small Robot Vehicles*; US Army TACOM: Warren, MI, USA, 2004.
18. di Maria, E. Running Gear Size Selection of a Wheel/Track Reconfigurable Grape Transporting Vehicle by FEM Analysis. Ph.D Thesis, Graduate School of Life Science and Systems Engineering, Kyushu Institute of Technology, Kitakyushu, Japan, 2021.
19. Statista. Available online: www.statista.com (accessed on 20 June 2023).
20. Reina, G.; Milella, A.; Galati, R. Terrain assessment for precision agriculture using vehicle dynamic modelling. *Biosyst. Eng.* **2017**, *162*, 124–139. [CrossRef]
21. Reynolds, A.G.; Vanden Heuvel, J.E. Influence of Grapevine Training Systems on Vine Growth and Fruit Composition: A Review. *Am. J. Enol. Vitic.* **2009**, *60*, 251–268. [CrossRef]
22. di Maria, E.; Reina, G.; Ishii, K.; Giannoccaro, N.I. Rolling resistance and sinkage analysis by comparing FEM and experimental data for a grape transporting vehicle. *J. Terramechanics* **2021**, *97*, 59–70. [CrossRef]
23. Grazioso, A.; di Maria, E.; Giannoccaro, N.I.; Ishii, K. Multibody Modeling of a New Wheel/Track Reconfigurable Locomotion System for a Small Farming Vehicle. *Machines* **2022**, *10*, 1117. [CrossRef]
24. Kaufmann, R. *Environmental Science*; McGraw-Hill: New York, NY, USA, 2008.
25. Adams, H. *Chassis Engineering. Chassis Design, Building & Tuning for High-Performance Cars*; HP Books: New York, NY, USA, 1992.
26. Wong, J.Y. *Terramechanics and Off-Road Vehicle Engineering: Terrain Behaviour, Off-Road Vehicle Performance and Design*, 2nd ed.; Butterworth-Heinemann: Oxford, UK, 2010.
27. Visual Basic Documentation. Available online: https://learn.microsoft.com/en-us/dotnet/visual-basic/ (accessed on 20 June 2023).

machines

Article

Performance Optimization of Ultralow-Frequency Electromagnetic Energy Harvester Driven by Eccentric mass

Jintao Liang *, Chao Zhang and Kangqi Fan

School of Mechano-Electronic Engineering, Xidian University, Xi'an 710071, China; zhangchao@xidian.edu.cn (C.Z.); kangqifan@gmail.com (K.F.)
* Correspondence: jtliang@xidian.edu.cn; Tel.: +86-137-7210-4018

Abstract: Driven by an eccentric mass through a two-layered cantilevered plectrum, the electromagnetic energy harvester (EEH) can convert low-frequency mechanical vibrations into continuous uni-directional rotation. To optimize the performance of the EEH, electromagnetic analysis of the EEH was conducted. Three-phase winding permanent magnet (PM) topology was employed, and combinations of different coils and magnet pole numbers were designed. Then, the finite element method (FEM) was applied to analyze the influence of the combinations of the coils and pole numbers as well as the PM dimensions on the three-phase induced voltage. Prototypes with different configurations were fabricated and the analysis effectiveness was confirmed. Furthermore, different types of stator yokes were designed to enhance the magnetic field. Compared to the original prototype, the output voltage of the optimal prototype increased by 0.5 V with the same rotation speed, and the harmonic components were sufficiently low. Then, experiments with excitation by linear reciprocating motions and swing motions were conducted. Under different exciting conditions, the optimal prototype can also induce the highest voltage amplitude. With an increase in the weight of the eccentric mass, a long duration can be reached that lasts up to 12 s. In summary, the proposed optimization can achieve a high-efficiency and high-power density EEH.

Keywords: electromagnetic energy harvester (EEH); three-phase machine; finite element method (FEM); harmonic analysis; ultralow-frequency vibration

1. Introduction

Harvesting energy from the surrounding environment and transforming it into electric energy is a feasible process that can supply power to wearable, portable and implantable devices, including wireless miniature sensors, etc. [1–4]. Low- and ultralow frequency vibrations widely exist within the natural environment and are considered to be an attractive sustainable resource for replacing traditional chemical batteries. Mechanical vibration contains various kinds of linear and swing vibrations. However, how to capture these mechanical energies efficiently is a great challenge, especially regarding ultralow frequency vibrations [5,6]. The main principles of mechanical energy harvesting include the piezoelectric effect [7,8], electromagnetic effect [9,10], electrostatic effect [11,12], triboelectric effect [13,14], etc. By contrast, electromagnetic harvesters exhibit higher power and longer service time.

Many research studies have been conducted to design various electromagnetic energy harvesters (EEH) for low- and ultralow-frequency vibrations. For instance, Halim et al. [15] proposed a non-resonant EEH, where a free non-magnetic ball in a cylinder periodically hits two magnets suspended on two springs at either ends of the cylinder, and the voltage is induced in the coil wrapped around the cylinder. The prototype generated 110 μW of average power with a 15.4 μWcm^{-3} average power density. Gutierrez et al. [16] designed a 2D magnetic levitation kinetic energy harvester, where a disk magnet lies on a 2D plane, the boundary of which is constrained by a circular sidewall, and cuboidal magnets are distributed around the sidewall to provide a spring force. The prototype obtained a resonant frequency of 8.2 Hz and an output power of 41.0 μW and 101 μW at 0.1 g and 0.2 g, respectively. Zhao et al. [17] explored an EEH with a magnetic orbit (EMH-MO) that has a circular uniform and low potential energy orbit, and where the moving magnet is magnetically modulated to move regularly along the magnetic orbit under arbitrary excitation. Under a reciprocating motion of 10 Hz and 20 mm, the open-circuit peak-peak voltage was 4.3 V, and the average power was 0.33 mW. Zhang et al. [18] presented a rotational EEH that comprises a twist driving rod and a ratchet-clutch. When the twist rod is compressed, the ratchet can keep rotating inertially for about 20 s. An overall energy of 85.2 mJ can be captured, and a peak power output of 32.2 mW was achieved. Although the latter two rotational EEHs can increase the level of electric power prominently, a relatively large vibration amplitude is generally required to make these harvesters operate efficiently.

For swing motion harvesting, an eccentric mass or pendulum structure is needed to scavenge for the swing vibrational energies [19]. For example, Smilek et al. [20] presented a Tusi couple energy harvester based on a proof mass with permanent magnets rolling in a circular cavity, where an average power of 5.1 mW from a 1.3 g periodic acceleration waveform at 2.78 Hz can be obtained. Hou et al. [21] introduced a rotational pendulum triboelectric-electromagnetic hybrid generator. With the rotation of the magnet pendulum, the four coils induce the electromagnetic power, and the Cu ring around magnets periodically contacts with the FEP blades to generate the triboelectric induced power. Through the two modes, the maximum power density of 3.25 W/m^2 and 79.9 W/m^2 was obtained at a vibration amplitude of 14 cm and frequency of 2 Hz. Li et al. [22] designed an eccentric pendulum-based EEH, where the eccentric rotor is arranged with permanent magnets and two sets of coils are symmetrically distributed on both axial sides of the eccentric rotor. The center working frequency is 2.0 Hz with an average output power of 8.37 mW. Halim et al. [23] employed a torsional spring to enhance the harvesting capacity of the eccentric magnetic rotor. The optimal prototype obtained a maximum 61.3 μW average power under a pseudo-walking motion of 1 Hz and $\pm 25^\circ$ amplitude. Liu et al. [24] developed a non-resonant rotational EEH with a cylindrical stator and a disk-shaped rotor. The rotor magnet can easily rotate around the stator due to external vibrations, and four wound coils arranged around the stator magnet can induce the electrical voltage. Under an 8 Hz excitation, the EEH can supply a maximum power of 10.4 mW at a load resistance of 100 Ω. Since the above eccentric masses or pendulums are excited to swing at small angle and low speed, these harvesters generally produce low electric power and voltage.

In order to capture low- and ultralow- frequency vibrations efficiently, an innovative eccentric mass combined with a two-layered cantilevered plectrum was proposed in our previous study [25]. However, the waveform and the efficiency of the electromagnetic induction effect was not considered and optimized. Therefore, in this study analysis, an optimization design of the electromagnetic performance is carried out to obtain a high efficiency, high power density, and low harmonic components of the inducted voltage. Firstly, a three-phase PM topology with fractional-slot concentrated-winding stator is employed, which can generate three-phase symmetric sinusoidal voltage. Next, how to design the combinations of coils and PM pole numbers is discussed. Then, to enhance the voltage amplitude and reduce the harmonic components, finite element method (FEM) is conducted, and the influence of the coil-pole combinations, the dimensions of PMs, and the types of stator yoke are analyzed. The corresponding prototypes are manufactured and tested, resulting in numerical FEM results that show good agreement with the experimental results. Moreover, the prototypes are installed to a crank-link sliding mechanism for linear reciprocating excitation and a crank-link swaging mechanism for swing motion. Different frequencies and amplitudes of vibrations are loaded, and the induced voltage waveforms are compared.

2. Design of EEH

2.1. Vibrational Exciting Principle

The structure of the proposed EEH is shown in Figure 1. The eccentric mass can be excited by external mechanical vibrations and transfer the kinetic energy to the rotor via a two-layered cantilevered plectrum that pushes the ratchet set at the inner surface of the rotor. The two-layered cantilevered plectrum is inserted into the central arc of the eccentric mass, which includes a left elastic layer and a right rigid layer. The effective length of the elastic layer (from its clamped end to the tip) is long enough to touch the tooth of the ratchet, whereas the rigid layer is a little shorter, avoiding connection with the tooth. As shown in Figure 2, when the eccentric mass is excited and spins clockwise, the cantilevered plectrum exhibits low stiffness and low frictional resistance against the rotor. On the contrary, when the eccentric mass is excited and spins anticlockwise, the cantilevered plectrum exhibits sufficient stiffness to drive the rotor homodromy. Thus, the rotor can be driven and accelerated to a high unidirectional speed by the eccentric mass, even though the excitation is ultralow-frequency, irregular or intermittent.

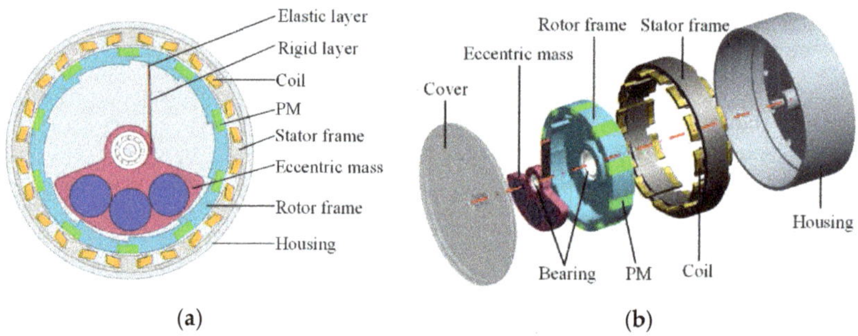

Figure 1. Structure of the proposed EEH: (**a**) cross-section view; (**b**) Exploded view.

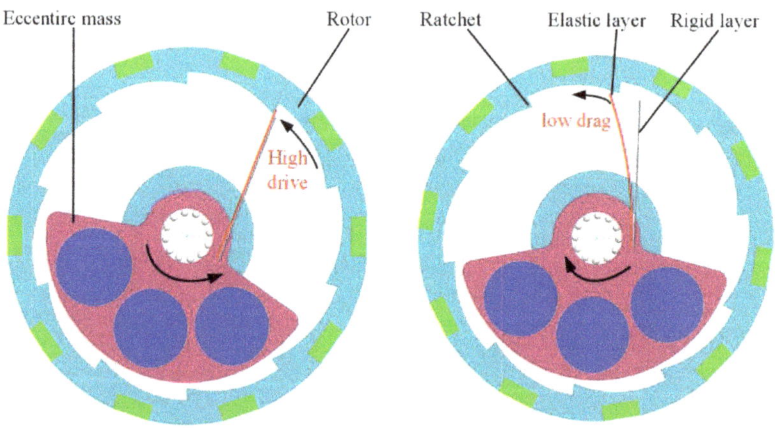

Figure 2. Principle of the proposed EEH.

2.2. Stator and Rotor Topology

As shown in Figure 1, the outer surface of the rotor frame is uniformly embedded with a sequence of rectangular PMs that are radially magnetized. The polarity of the adjacent two PMs is the opposite. The stator frame has a circumferential teeth-slot structure, in which three-phase individual concentrated coils are wound around a tooth in sequence (A, B, C), and each phase winding consists of several concentrated coils in a series (A1-A2-A3-A4, B1-B2-B3-B4, and C1-C2-C3-C4). When the rotor spins, the rotational magnetic field is established, and then the three-phase sinusoidal voltage can be induced in the three-phase winding and conveniently transferred to the DC voltage through a three-phase rectifier bridge. For this three-phase PM topology, the number of coils Q must be a multiple of 3 (three phase); an even number can avoid the magnetic unbalance. Thus, Q can be set to 6, 12, 18, etc. When Q is less than 12, the leakage flux would become more, and when Q more than 18, the structure would become complicated. As such, the preferable number of Q is 12 or 18. Correspondingly, the number of PM poles p should be an even number for the alternant polarity. When the rotor spins across two adjacent PMs, it represents one electrical period (360° electrical angle). Therefore, the electrical angle between adjacent coils can be expressed as:

$$\alpha_c = \frac{180^\circ p}{Q} \tag{1}$$

The number of coils per pole and per phase can be expressed as:

$$q = \frac{Q}{pm} = \frac{z}{d} \tag{2}$$

where m is the number of phases. When q is a fraction, it is called fractional-slot winding. $q = z/d$ represents an irreducible fraction, which means one phase has z coils in d poles.

As such, each coil can use a 2D vector to indicate its position and magnitude, and the coil vectors in the same phase can be combined as the vector of one-phase winding. Then, three-phase vectors are mutually separated at a 120° electrical angle, which can be satisfied by configuring the number of Q and p properly. For instance, if p is set to 10 and Q is set to 12, then the adjacent coil vectors are staggered where α_c = 150° electrical angle. The coil and phase vectors are shown in Figure 3a. The No. 1 and 2 coils are combined as vector A+, while No. 7 and 8 coils are combined as vector A−. If the No. 7 and 8 coils are wound reversely, then the direction of vector A− becomes the same as vector A+, combining together as the phase A vector. This is the same for the phase B and Phase C vectors, which are staggered at a 120° electrical angle respectively.

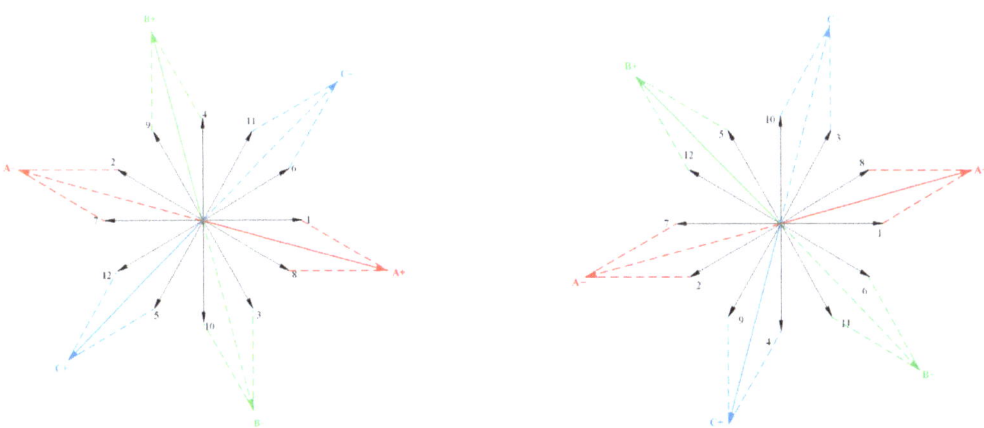

(**a**) (**b**)

Figure 3. *Cont.*

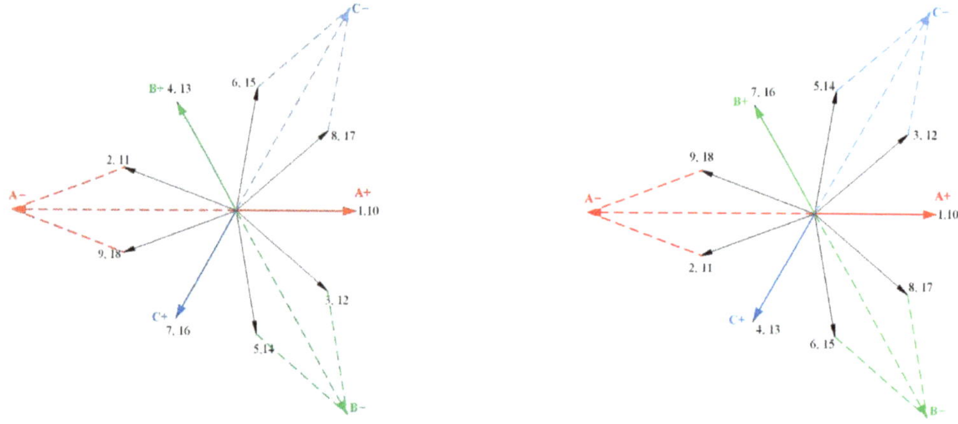

(c) (d)

Figure 3. Coil and phase vectors with different configurations: (**a**) 12C/10P; (**b**) 12C/14P; (**c**) 18C/16P; (**d**) 18C/20P.

Because the coil vectors of the same phase are not in the same direction, the winding factor should be considered to enhance the output voltage. The winding factor k_w is defined as the product of the winding distribution factor (k_d) and the short pitch factor (k_c). The distribution factor measured the resultant voltage of the distributed winding regards concentrate winding. k_d can be expressed as:

$$k_d = \frac{\sin(\frac{\pi}{2m})}{z\sin(\frac{\pi}{2mz})} \tag{3}$$

The short pitch factor k_c is the measure of the number of armature slots between the two sides of a coil. k_c can be expressed as:

$$k_c = \sin(\frac{\alpha_c}{\pi}\frac{\pi}{2}) = \sin(\frac{\alpha_c}{2}) \tag{4}$$

Thus, the winding factor k_w is:

$$k_w = k_d k_c \tag{5}$$

A higher winding factor can induce larger voltage amplitude. Initial 12C/10P (k_w = 0.933), 12C/14P (k_w = 0.933), 18C/16P (k_w = 0.945) and 18C/20P (k_w = 0.945) are also suitable for configuration, following the phase vector arrangement as shown in Figure 3b–d.

2.3. Dimensions and Materials

In order to reduce the weight of the EEH for a wearable device, the structure components, including the eccentric mass frame, rotor frame and stator frame, were fabricated using a resin material via a 3D printer. To enhance the magnetic flux and generate a larger voltage, the yoke section of the stator frame can be produced using a permeability magnetic material, such as a laminated silicon steel sheet. The following electromagnetic analysis was used to compare this difference. For manufacturing the experimental prototype, the outer and inner diameters of the stator were set to 60 mm and 50 mm, respectively, and the outer diameter of the rotor was set to 49 mm. Initially a 12/10 coil-pole configuration was selected. The dimensions of each PM were 10 mm × 5 mm × 2 mm, and the number of turns per coil was 60.

The fan-shaped resin frame of the eccentric mass with a central angle of 130° and radius of 20 mm was embedded with three identical cylindrical copper blocks (thickness:

4 mm, diameter: 10 mm). For comparison, the eccentric mass can be fabricated as a whole block of steel to increase its rotational inertia. The rigid layer of the cantilevered plectrum with a size of $18.5 \times 4 \times 0.2$ mm^3 was made by a copper sheet, while the elastic layer was cut from polyethylene terephthalate (PET) film that has a low friction coefficient of 0.04 to form a size of $19.5 \times 4 \times 0.2$ mm^3.

The exploded view of the whole EEH is shown in Figure 1b. The stator was embedded in the housing, which includes a step shaft at the rear central. The eccentric mass and the rotor were connected to the shaft through two different ceramic bearings. A cover was attached to the front of the housing to prevent air resistance. Then, the initial prototype was implemented, exhibiting a 63 mm outer diameter, 22.5 mm axial length and 67.6 g weight, as shown in Figure 4.

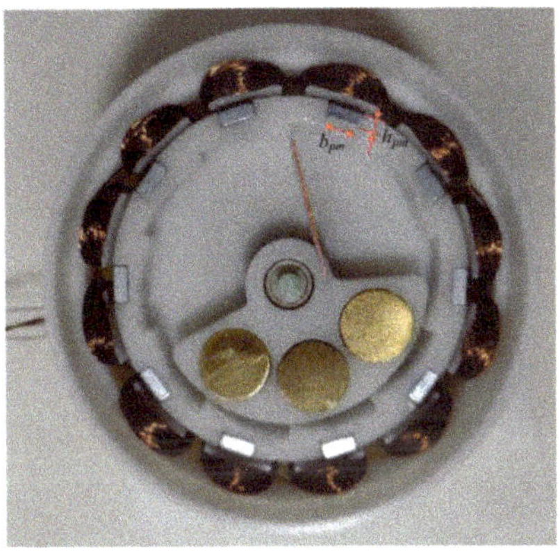

Figure 4. Initial prototype of the proposed EEH.

3. Electromagnetic Analysis

3.1. Comparison of FEM and Experimental Results

The electromagnetic performance of the proposed EEH can be simulated by FEM through the opensource software FEMM, MaxFem, and the commercial software Anasys, Comsol, etc. The general processes are approximately the same for each of these software packages. The initial prototype, including stator and rotor topology, can be established as the 2D electromagnetic geometric model for its cross-section, as shown in Figure 5a. The material property of each part is defined, and excitation and boundary conditions are set. Then, the finite element mesh is generated, as shown in Figure 5b. According to the following partial differential equation of magnetic potential vector A_z, numerical computation can be conducted by the FEM solver.

$$\nabla \times \frac{1}{\mu} \nabla \times A_z = -\sigma \dot{A} + J_z \tag{6}$$

where μ is the permeability, σ is the conductivity, J_c is the current density and ∇ represents the Hamiliton operator. On this basis, the other parameters, including magnetic density B, magnetic flux Φ, induced voltage V, etc., can be obtained through postprocessing.

For instance, the rotor can be simulated to spin for an electrical period with different speeds, which enables the induced three-phase voltage waveforms to be obtained. As shown in Figure 6a, the three-phase voltage waveforms are obtained when the rotor speed is set to 600 rpm. To verify the accuracy of the FEM results, the initial prototype was connected to a brushless DC motor, which can be regulated to drive the prototype at different speeds, as shown in Figure 7. The experimental results of the induced voltage waveforms at 600 rpm were obtained, as shown in Figure 6b. By comparison, in one period of Phase A

(Figure 6c), the experimental results are is in a good agreement with the predicted FEM results in regard to both the waveform and peak values.

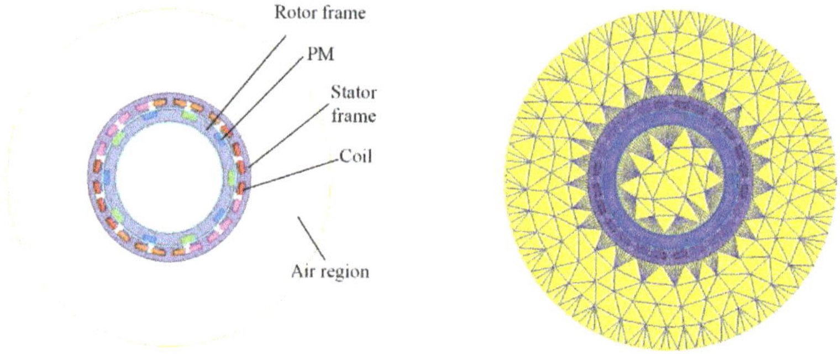

(**a**) (**b**)

Figure 5. Electromagnetic FEM model of the EEH: (**a**) geometric model; (**b**) mesh generation.

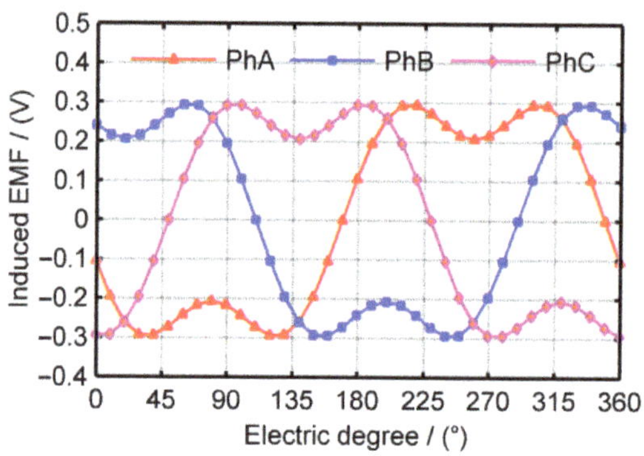

(**a**)

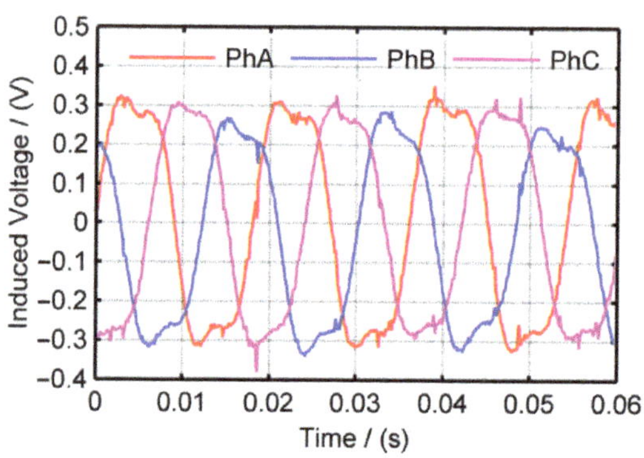

(**b**)

Figure 6. *Cont.*

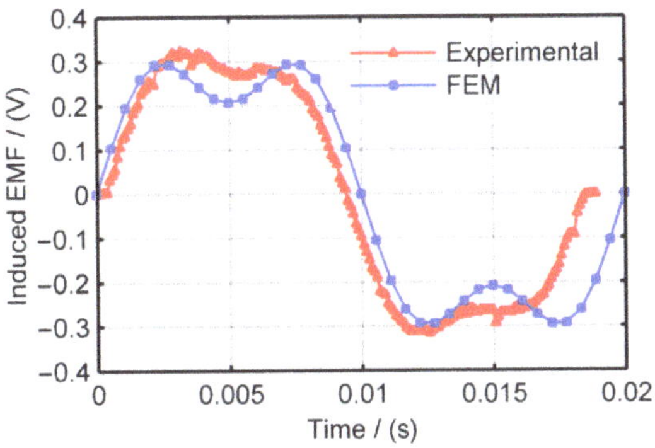

(c)

Figure 6. Three-phase voltage waveforms with initial model at 600 rpm: (**a**) FEM result; (**b**) Experimental result: (**c**) Result comparison in one period of Phase A.

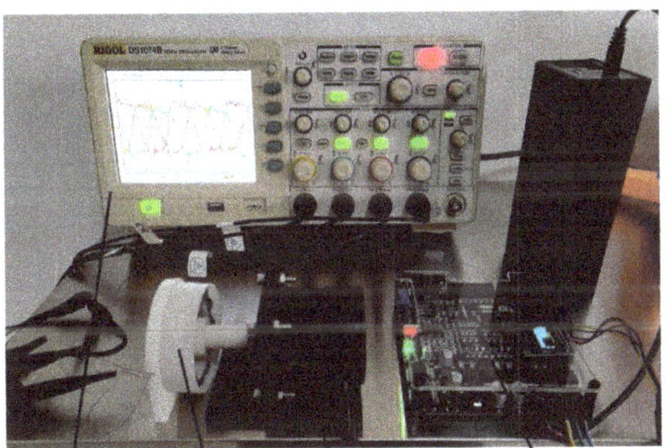

Figure 7. Experiment device for voltage measurement.

Moreover, the induced voltage waveforms have excessive high harmonic components that should be reduced. Therefore, the influences of the design parameters were analyzed through FEM.

3.2. Magnetic Field Optimization

The dimensions of the PM affect the intensity of the magnetic field directly. The axial length is equal to a rotor length of 10 mm. To reduce the magnetic resistance, the thickness of the PM h_{pm} should be set to a small value; this value was determined to be 2 mm, as shown in Figure 4. Thus, the width of the PM b_{pm} is regulated at different values of 5 mm (initial), 6 mm, 7 mm, 8 mm and 9 mm. The corresponding results of the induced voltage waveforms are shown in Figure 8a. To present a distinct display, only the phase-A waveforms are compared.

With the increase in b_{pm}, the voltage amplitude increases, and exhibits a more fundamental sine. Harmonic analysis is shown in Figure 8b. The main harmonic component is the three-time harmonic, when b_{pm} = 5 mm, and the total harmonic distortion (THD) is about 34%; when b_{pm} = 9 mm, the THD reduces to only 2.0%.

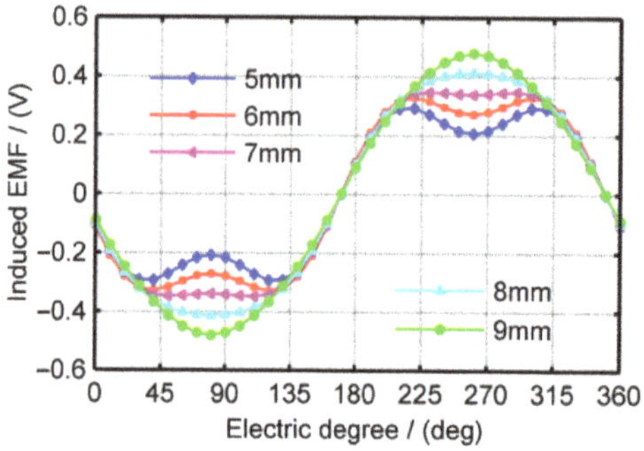

(a)

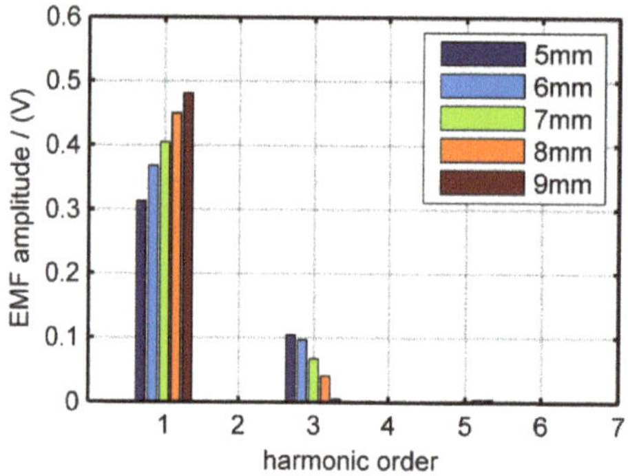

(b)

Figure 8. FEM results of initial 12C-10P model with different b_{pm}: (**a**) Voltage waveforms; (**b**) Harmonic magnitude of voltage.

Employing a permeability magnetic material to produce the stator core can enhance magnetic density and reduce the flux leakage. The stator core is composed of the teeth and the yoke. The yoke is a magnetic frame between teeth that can pass the magnetic flux, and also houses the coil windings. However, if the whole stator is laminated by silicon steel sheet, cogging torque generated between the stator teeth and PM develops tremendous spinning resistance. Thus, a non-teeth stator core is designed, including a resin skeleton and a yoke laminated by silicon steel sheets, as shown in Figure 9. Compared with the initial FEM model, the voltage amplitude increases by about 65%, as shown in Figure 10.

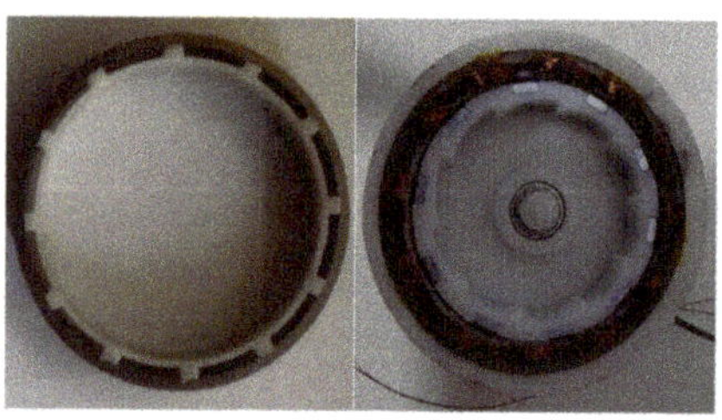

Figure 9. Non-teeth stator core structure.

3.3. Configuration of Coil and Pole Number

Based on the advanced stator core, the influence of different coil-pole configurations was analyzed. Figure 11 shows the voltage performance of the 12C-14P FEM model. Compared with the 12C-10P model, the voltage waveforms with different b_{pm} have favorable fundamental sinusoidal voltages, and the harmonic components are sufficiently low. Moreover, the values of voltage amplitude are larger than the 12C-10P. When b_{pm} = 6 mm with 14 poles (10 mm × 6 mm × 2 mm × 14 × 7500 kg/m^3 = 0.0126 Kg), the amplitude is about 0.95 V. By contrast, when b_{pm} = 9 mm with 10 poles (10 mm × 9 mm × 2 mm × 10 × 7500 kg/m^3 = 0.0135 Kg), the amount of PMs is larger, but the amplitude is only 0.80 V.

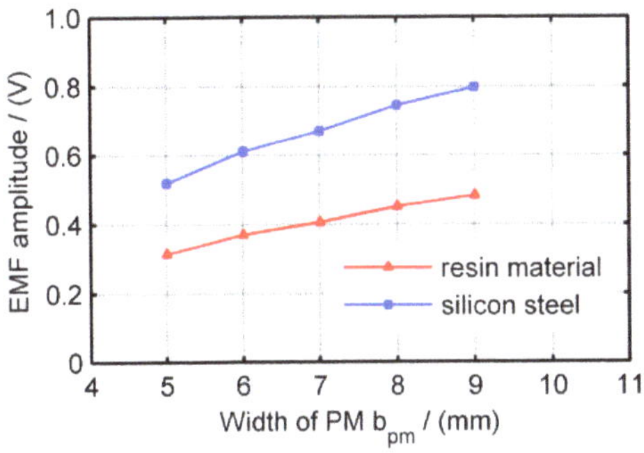

Figure 10. Voltage amplitude of 12C-10P FEM model with different material.

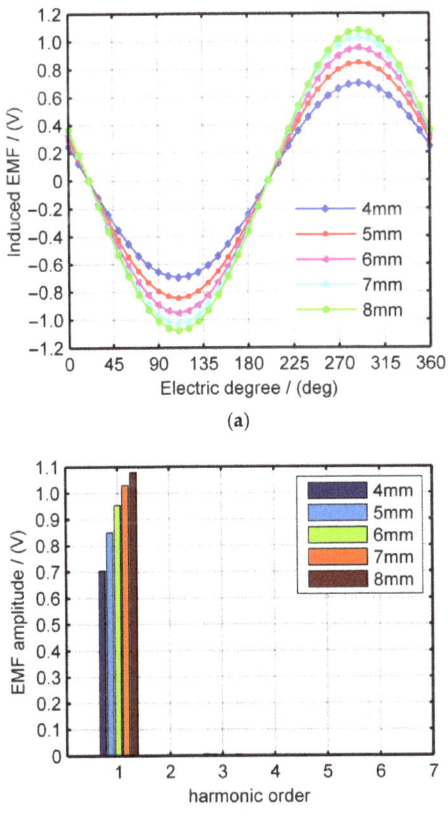

Figure 11. FEM results of steel-yoke 12C-14P model with different b_{pm}: (**a**) Voltage waveforms; (**b**) Harmonic magnitude of voltage.

The 18C-16P and 18C-20P configurations were also analyzedand the voltage amplitude results of different configurations were compared, as shown in Figure 12a. In addition, to indicate the magnetic efficiency of the PM, the ratio of the voltage amplitude versus the PM mass was also compared, as shown in Figure 12b. Moreover, the values of voltage amplitude, harmonic THD and voltage per PM mass with different configuration are listed for comparison in Table 1.

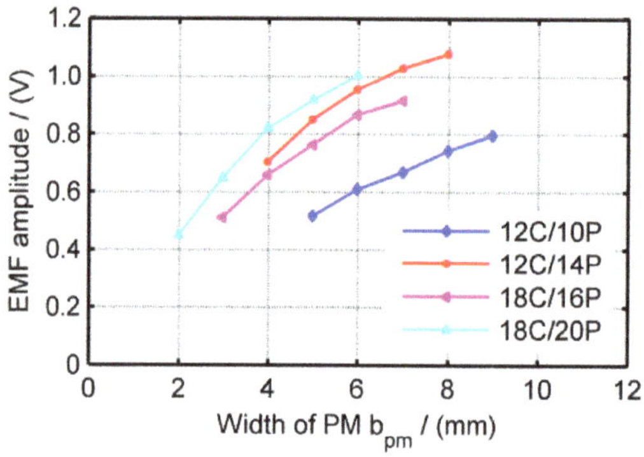

(**a**)

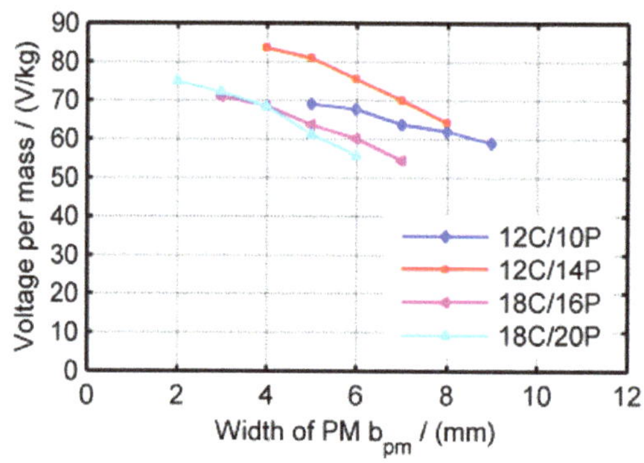

(**b**)

Figure 12. Comparison of voltage performance with different configurations: (**a**) Voltage amplitude; (**b**) Voltage amplitude versus PM mass.

In general, the 12C-10P and 18C-16P configurations exhibit a high voltage hamonic component. By contrast, the harmonic THD of 12C-14P and 18C-20P are sufficiently low, all less than 1%. With the increase of b_{pm}, the voltage amplitude values of every configuration would increase uniformly, but the magnetic efficiency of every configuration would decrease. This is because although the 12C-14P and 18C-20P configurations exhibit a larger value (over 1.0 V), 18C-20P has a low ratio of voltage amplitude to PM mass, while the ratio of 12C-14P is the highest. This is due to the magnetic resistance of the non-magnetizing resin material being large, and the flux leakage increasing when the PM intensity increases. To reduce the cost and weight, 12C-14P with b_{pm} = 6 mm is determined as the optimal configuration, as shown in Figure 13. Meanwhile, 18C-20P with b_{pm} = 4 mm was also manufactured to verify the effectiveness of the FEM results.

Table 1. Comparison of voltage values with the different configurations and b_{pm}.

Configurations	Voltage Amplitude	Harmonic THD (%)	Voltage per Mass (V/Kg)
12C10P b_{pm}5 mm	0.517	26.86	68.997
12C10P b_{pm}6 mm	0.609	21.09	67.684
12C10P b_{pm}7 mm	0.669	13.74	63.739
12C10P b_{pm}8 mm	0.744	7.81	61.983
12C10P b_{pm}9 mm	0.796	1.44	58.971
12C14P b_{pm}4 mm	0.703	0.74	83.682
12C14P b_{pm}5 mm	0.850	0.56	80.923
12C14P b_{pm}6 mm	0.953	0.29	75.611
12C14P b_{pm}7 mm	1.029	0.21	70.002
12C14P b_{pm}8 mm	1.078	0.36	64.164
18C16P b_{pm}3 mm	0.511	9.72	70.926
18C16P b_{pm}4 mm	0.658	6.46	68.554
18C16P b_{pm}5 mm	0.763	3.04	63.569
18C16P b_{pm}6 mm	0.866	0.66	60.149
18C16P b_{pm}7 mm	0.915	3.12	54.457
18C20P b_{pm}2 mm	0.450	0.17	75.034
18C20P b_{pm}3 mm	0.649	0.11	72.138
18C20P b_{pm}4 mm	0.820	0.12	68.368
18C20P b_{pm}5 mm	0.917	0.06	61.153
18C20P b_{pm}6 mm	1.004	0.08	55.767

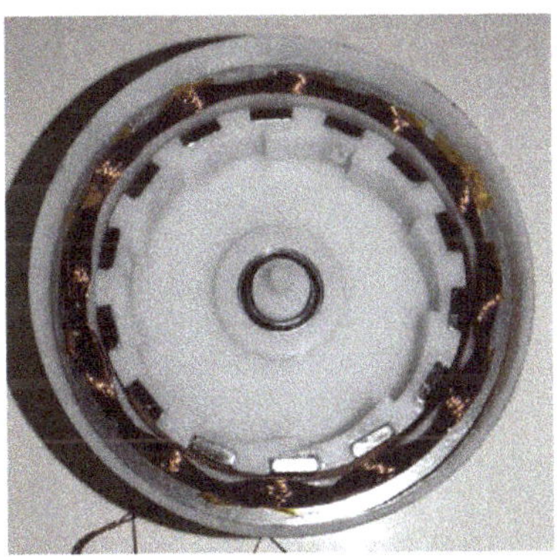

(**a**)

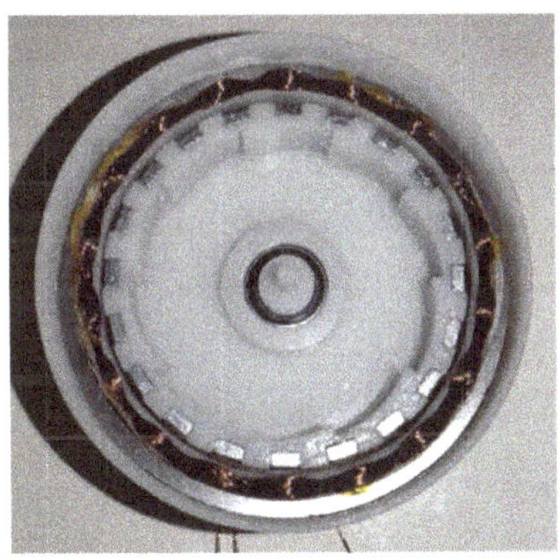

(**b**)

Figure 13. Prototypes with different configurations: (**a**) 12C-14P with b_{pm} = 6 mm; (**b**) 18C-20P with b_{pm} = 4 mm.

The experimental results of the three-phase voltage waveforms with different configurations are shown in Figure 14. All voltage waveforms and their amplitudes correspond with the FEM results. After optimization, the voltage amplitude increases from the initial 0.5 V to about 1.0 V, and the PM mass slightly increases to 5.1 g. Furthermore, the harmonic components of the voltage waveforms are reduced, and exhibit a favorable fundamental sine.

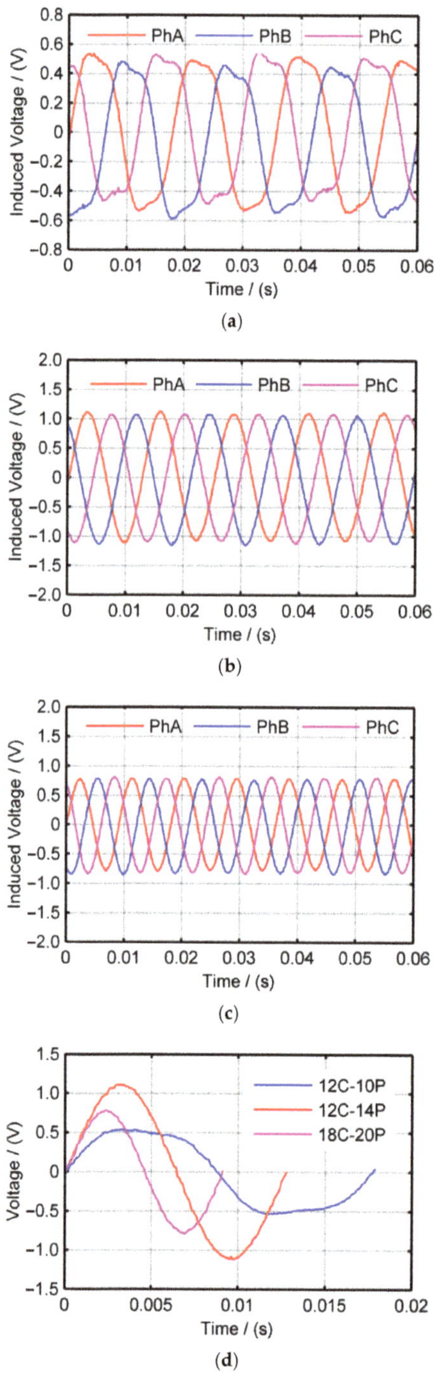

Figure 14. Experiment results of three-phase voltage waveforms at 600 rpm: (**a**) 12C-10P with b_{pm} = 5 mm (initial); (**b**) 12C-14P with b_{pm} = 6 mm (optimal); (**c**) 18C-20P with b_{pm} = 4 mm; (**d**) Result comparison in one period of Phase A.

4. Experiment under Swing Motion

4.1. Test Bench for Swing Motion

With the optimal electromagnetic performance, the EEH prototypes were excited by different kinds of vibrations to validate their vibration energy-harvesting performance. The experimental setup for swing motion exciting is shown in Figure 15, where the swing motion is provided by a crank-rocker mechanism driven by a BLDC motor. Different conditions of swing excitation were supplied. The swing frequency of the crank-rocker mechanism was regulated by controlling the rotation speed of the motor (0~3000 rpm). Additionally, the swing angle can be changed by varying the length of the crank in a range from 20° to 80°. The length of the rocker, which is the distance from the pivot point to the center of the EEH, can be set to 132 mm, 102 mm or 72 mm. The electric outputs of the EEH prototype were displayed and stored through an oscilloscope.

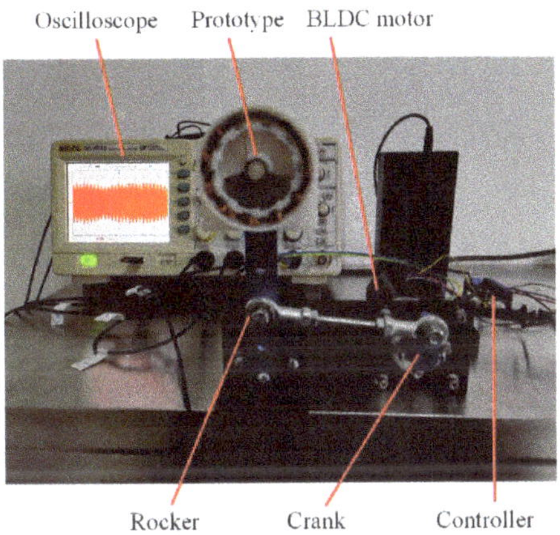

Figure 15. Experiment device for swing excitation.

4.2. Performance under Swing Motion

The length of the rocker was set to 132 mm, with a swing angle of 60° and a swing frequency of 2.5 Hz. Excited by the same swing motion, the open-circuit voltage outputs of the three manufactured prototypes are shown in Figure 16. Continuous voltage waveforms were generated. The optimal 12C-14P prototype produced the maximum amplitude voltage. The swing frequency was varied to explore the harvesting response performance, from 1.6 Hz to 3.5 Hz. Then, the swing angle was changed to 40°, and the frequency variation was conducted again.

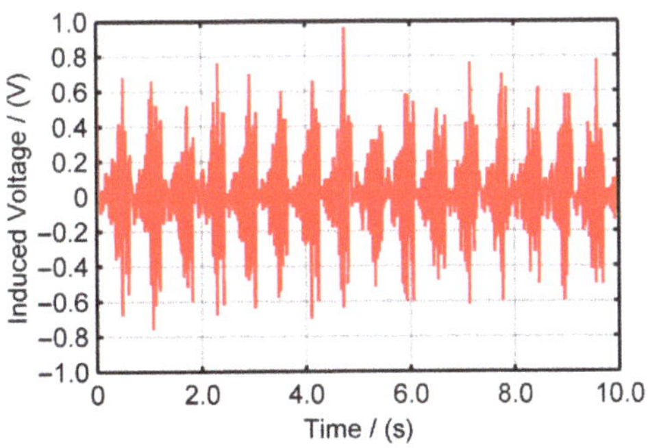

(**a**)

Figure 16. *Cont.*

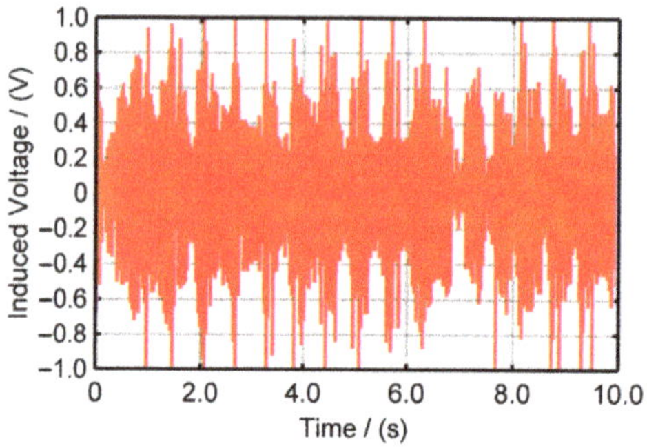

(**b**)

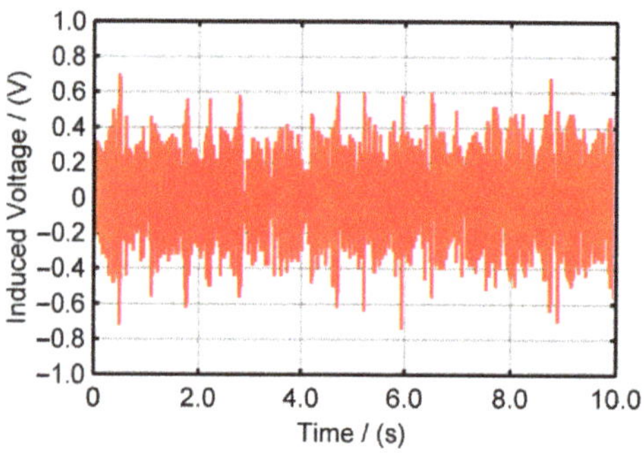

(**c**)

Figure 16. Output voltage of the three prototypes under swings with 2.5 Hz, 60°: (**a**) 12C-10P with b_{pm} = 5 mm (initial); (**b**) 12C-14P with b_{pm} = 6 mm (optimal); (**c**) 18C-20P with b_{pm} = 4 mm.

The approximate values of the voltage amplitude under different conditions were compared, as shown in Figure 17. Generally, the voltage amplitude is enhanced with the swing angle and frequency, particularly when the frequency shifts between 2.5 Hz and 3.2 Hz. Under all swing conditions, the optimal 12C-14P prototype gained the highest amplitude when compared to the other two prototypes. This is because, when under the same rotor speed excited by the swing motion, the optimal 12C-14P prototype induces the maximum voltage amplitude. Moreover, with the increase of the swing frequency, the eccentric mass would exhibit a continuous circular motion that is opposite to the rotor direction, which cannot contribute to increasing the rotor speed. Therefore, the voltage amplitude trend reaches the peak value.

As shown in Figure 15, the swing vibration is provided by a crank-rocker mechanism. When the rocker swings at a low frequency, the harvesting energy is small if the weight of the eccentric mass is low. With the increased weight of the eccentric mass, the inertia and momentum increase, and the harvesting energy is enhanced. The initial eccentric mass with three copper blocks was 20 g. To detect the influence of the weight of the eccentric mass, a whole steel block of eccentric mass was fabricated, the weight of which was 60 g. After equipping the steel eccentric mass to the 12C-14P prototype, the swing motions with a 60° angle and different frequencies were conducted in sequence. It can be reasonably assumed that the whole steel structure can output a higher voltage amplitude. Furthermore, the significant characteristic is that an ultra-low frequency swing can be harvested with the higher inertia eccentric mass, as shown in Figure 18. Under the 1 Hz swing excitation, the initial eccentric mass hardly transfers to the rotor motion, and thus the continuous output voltage cannot be generated. In contrast, the whole steel eccentric mass can capture the ultra-low frequency swing, and then the speed of the rotor can be gradually enhanced to reach stability. Moreover, after the swing is removed, the output voltage can persist over 12 s, which represents a long duration of the swing excitation. Therefore, the higher inertia eccentric mass should be considered under ultra-low, unstable or complicated vibrations.

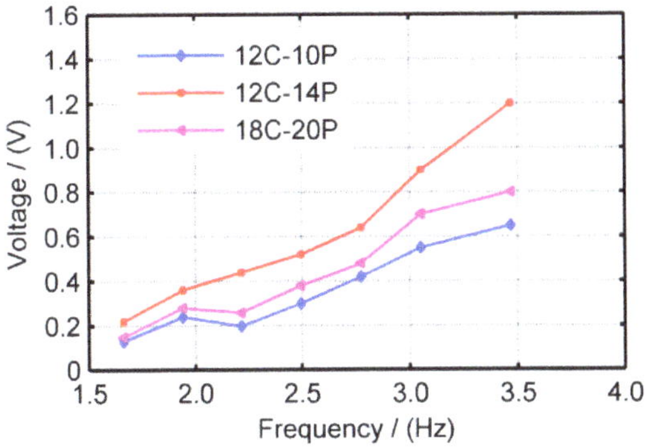

(a)

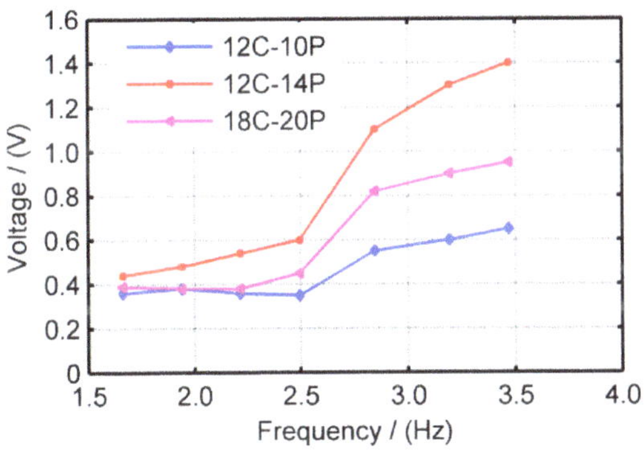

(b)

Figure 17. Frequency response of output voltage under different swing angles: (**a**) Swing 40°; (**b**) Swing 60°.

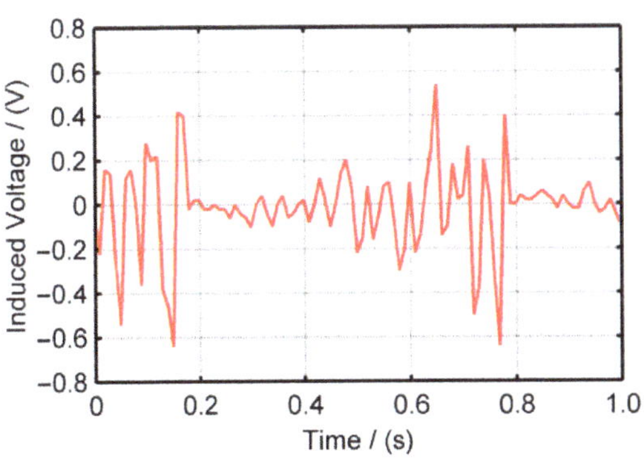

(a)

Figure 18. *Cont.*

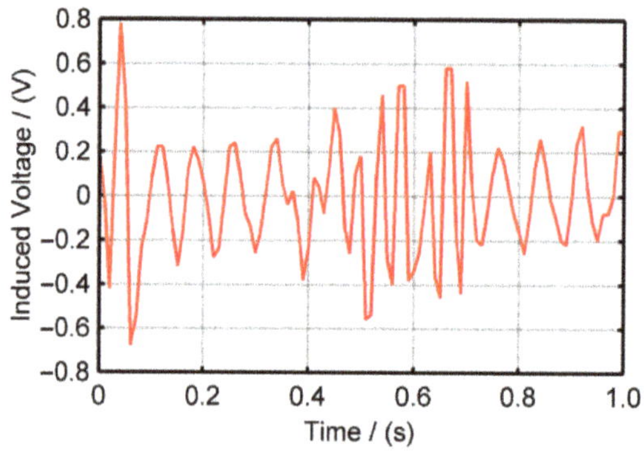

(**b**)

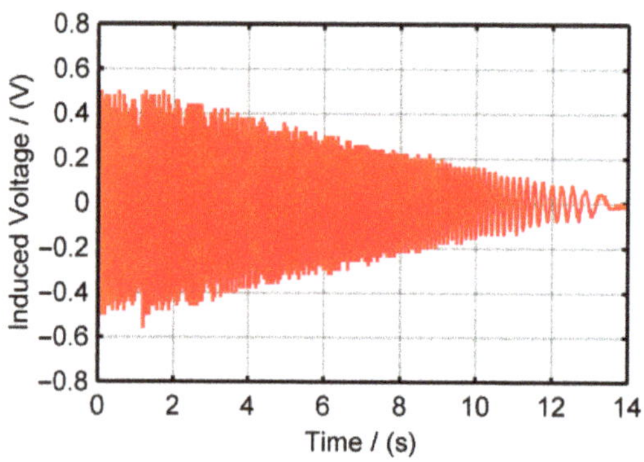

(**c**)

Figure 18. Output voltage of the 12C-14P prototype with two eccentric mass structures in ultra-low swing frequency (1Hz): (**a**) With initial structure under swing; (**b**) With whole steel block structure under swing; (**c**) With whole steel block structure after swing remove.

5. Experiment under Linear Reciprocation

5.1. Test Bench for Linear Reciprocation

The experimental setup for linear reciprocation exciting is shown in Figure 19, where the EEH is excited by a crank-slider mechanism driven by a BLDC motor. Different conditions of linear vibration were supplied. The motion frequency can be regulated by controlling the speed of the motor (0~3000 rpm), and the reciprocating stroke can be tuned by adjusting the effective length of the crank from 20 mm to 60 mm.

5.2. Performance under Linear Reciprocation

In the test, the length of the crank was set to 40 mm and was unchanged, and several frequencies from 2.2 Hz to 3.3 Hz were conducted. The output voltage amplitudes of different prototypes are shown in Figure 20. With the increase of frequency, the amplitudes increased in all three prototypes. The optimal 12C-14P prototype exhibits a larger amplitude than the others, especially over 2.7 Hz. However, when the frequency is higher than 3.0 Hz, the voltage amplitude does not increase. This is mainly because the eccentric mass in the EEH has a complex irregular motion characterized by the random shift between the swing and the multi-circle rotation.

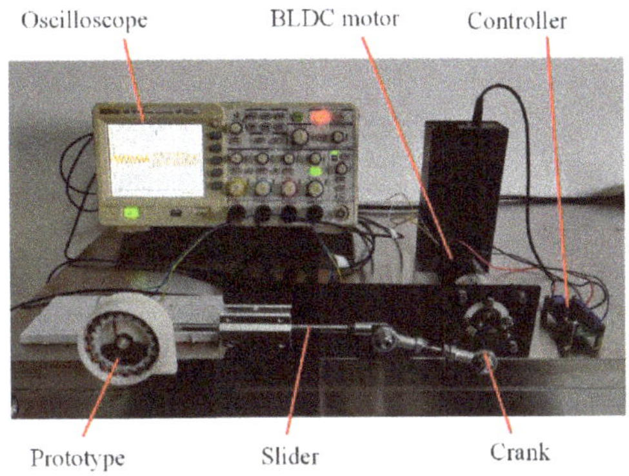

Figure 19. Experiment device for linear reciprocation.

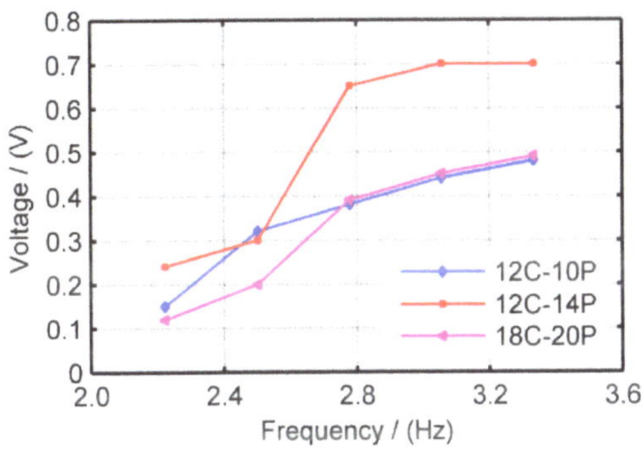

Figure 20. Frequency response of output voltage under 50 mm linear motion.

Furthermore, to indicate the effectiveness of the proposed optimization, the original device, presented in [25], which has been compared with a representative swing-based energy harvester and demonstrated the superiority, was tested under the same linear reciprocation experiment. As shown in Figure 21, the optimal prototype can induce larger volage amplitude and more continuous voltage waveform.

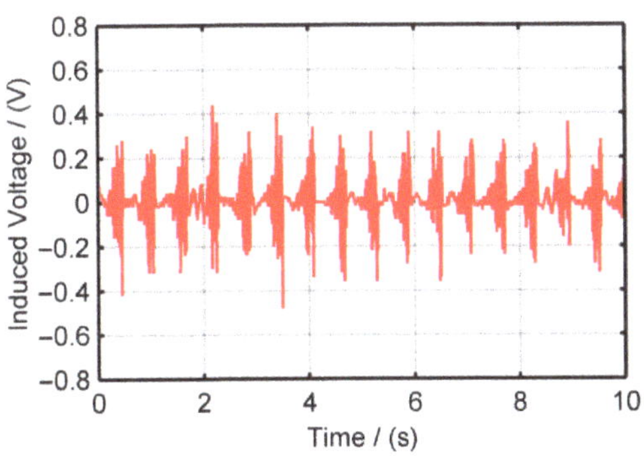

(**a**)

Figure 21. *Cont.*

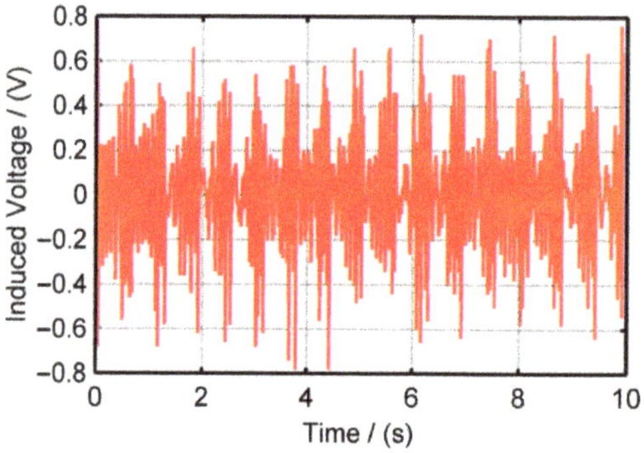

(b)

Figure 21. Output voltage of the original and optimal prototypes under linear reciprocation with 2.5 Hz (**a**) With original prototype; (**b**) With optimal prototype.

6. Conclusions

Electromagnetic characteristic analysis was carried out to optimize the output performance of the proposed EEH. A three-phase fractional coil per magnet pole was the basis structure, and the configurations of coil and pole, including 12C-10P, 12C-14P, 18C-16P and 18C-20P, were the four available candidates. Through FEM, the voltage waveforms and their harmonic amplitude are compared. Meanwhile, the effectiveness of FEM analysis effectiveness was confirmed by the prototype experimental results. Then, the optimal configuration and the PM dimension were determined to be 12C-14P with b_{pm} = 6 mm, which induce over voltage 1.0 V and uses fewer PMs. On this basis, the vibration experiments were conducted, including swing and linear motion. The optimal prototype always generated the highest amplitude of output voltage. The vibration frequency existed within a favorable range that did not exceed 3.0 Hz, otherwise the eccentric mass would undergo multi-circle rotation. Moreover, the influence of the weight of eccentric mass was analyzed. Higher inertia eccentric mass should be considered under ultra-low, unstable or complicated vibrations. This EEH can be applied to exploit the ambient swing and linear vibration energy. For instance, the human arm and leg motion and the generated electric energy can be supplied to the wearable devices. However, low power applications and miniature devices are available. In further research, comparison with different types of EEH would be carried out for quantitative characteristic analysis.

Author Contributions: Conceptualization, J.L. and K.F.; methodology, J.L. and K.F.; software, C.Z.; validation, J.L. and C.Z.; formal analysis, J.L.; investigation, J.L.; resources, K.F.; data curation, J.L. and C.Z.; writing—original draft preparation, J.L.; writing—review and editing, K.F.; visualization, C.Z.; supervision, K.F.; project administration, K.F.; funding acquisition, J.L. All authors have read and agreed to the published version of the manuscript.

Funding: This research work is financially supported by the Project funded by Natural Science Foundation of China (No. 51605363), the Project supported by Natural Science Basic Research Plan in Shaanxi Province of China (No. 2021JM-124), and the Fundamental Research Funds for the Central Universities (No. XJS210403).

Data Availability Statement: The datasets used and analyzed during the current study are available from the corresponding author on reasonable request.

Conflicts of Interest: The authors declared no potential conflict of interest with respect to the research, authorship and publication of this article.

References

1. Ali, F.; Raza, W.; Li, X.; Gul, H.; Kim, K.H. Piezoelectric energy harvesters for biomedical applications. *Nano Energy* **2019**, *57*, 879–902. [CrossRef]
2. Hasan, M.N.; Sahlan, S.; Osman, K.; Mohamed Ali, M.S. Energy harvesters for wearable electronics and biomedical devices. *Adv. Mater. Technol.* **2021**, *6*, 2000771. [CrossRef]
3. Zhou, X.; Liu, G.; Han, B.; Liu, X. Different kinds of energy harvesters from human activities. *Int. J. Energy Res.* **2021**, *45*, 4841–4870. [CrossRef]
4. Sun, F.; Su, X.; Song, H.-C.; Kang, C.-Y. Energy harvesting performance of a novel nonlinear quad-stable piezoelectric energy harvester with only one external magnet. *Machines* **2022**, *10*, 803. [CrossRef]
5. Jackson, N.; Olszewski, O.Z.; O'Murchu, C.; Mathewson, A. Ultralow-frequency Piezo MEMS energy harvester using thin-film silicon and parylene substrates. *J. Micro-Nanolithography MEMS MOEMS* **2018**, *17*, 015005.
6. Wang, K.; Ouyang, H.; Zhou, J.; Chang, Y.; Xu, D.; Zhao, H. A nonlinear hybrid energy harvester with high ultralow-frequency energy harvesting performance. *Meccanica* **2021**, *56*, 461–480. [CrossRef]
7. Zhou, H.; Zhang, Y.; Qiu, Y.; Wu, H.; Qin, W.; Liao, Y.; Yu, Q.; Cheng, H. Stretchable piezoelectric energy harvesters and self-powered sensors for wearable and implantable devices. *Biosens. Bioelectron.* **2020**, *168*, 112569. [CrossRef]
8. Guo, L.; Wang, H.; Braley, J.; Venkiteela, G. Field evaluation of piezoelectric energy harvesters on bridge structure. *Machines* **2023**, *11*, 462. [CrossRef]
9. Muscat, A.; Bhattacharya, S.; Zhu, Y. Electromagnetic vibrational energy harvesters: A review. *Sensors* **2022**, *22*, 5555. [CrossRef]
10. Carneiro, P.M.R.; Ferreira, J.A.F.; Kholkin, A.; Soares dos Santos, M.P. Towards self-adaptability of instrumented electromagnetic energy harvesters. *Machines* **2022**, *10*, 414. [CrossRef]
11. Naito, Y.; Uenishi, K. Electrostatic MEMS vibration energy harvesters inside of tire treads. *Sensors* **2019**, *19*, 890. [CrossRef] [PubMed]
12. Zhang, Y.; Wang, T.; Luo, A.; Hu, Y.; Li, X.; Wang, F. Micro electrostatic energy harvester with both broad bandwidth and high normalized power density. *Appl. Energy* **2018**, *212*, 362–371. [CrossRef]
13. Jayasvasti, S.; Thainiramit, P.; Yingyong, P.; Isarakorn, D. Technique for measuring power across high resistive load of triboelectric energy harvester. *Micromachines* **2021**, *12*, 766. [CrossRef]
14. Liu, L.; Shi, Q.; Ho, J.S.; Lee, C. Study of thin film blue energy harvester based on triboelectric nanogenerator and seashore IoT applications. *Nano Energy* **2020**, *66*, 104167. [CrossRef]
15. Halim, M.A.; Park, J.Y. A non-resonant, frequency up-converted electromagnetic energy harvester from human-body-induced vibration for hand-held smart system applications. *J. Appl. Phys.* **2014**, *115*, 094901. [CrossRef]
16. Gutierrez, M.; Shahidi, A.; Berdy, D.; Peroulis, D. Design and characterization of a low frequency 2-dimensional magnetic levitation kinetic energy harvester. *Sens. Actuators A Phys.* **2015**, *236*, 1–10. [CrossRef]
17. Zhao, L.-C.; Zou, H.-X.; Gao, Q.-H.; Yan, G.; Liu, F.-R.; Tan, T.; Wei, K.-X.; Zhang, W.-M. Magnetically modulated orbit for human motion energy harvesting. *Appl. Phys. Lett.* **2019**, *115*, 263902. [CrossRef]
18. Zhang, Y.; Luo, A.; Wang, Y.; Dai, X.; Lu, Y.; Wang, F. Rotational electromagnetic energy harvester for human motion application at low frequency. *Appl. Phys. Lett.* **2020**, *116*, 053902. [CrossRef]
19. Chen, J.; Bao, B.; Liu, J.; Wu, Y.; Wang, Q. Pendulum energy harvesters: A review. *Energies* **2022**, *15*, 8674. [CrossRef]
20. Smilek, J.; Hadas, Z.; Vetiska, J.; Beeby, S. Rolling mass energy harvester for very low frequency of input vibrations. *Mech. Syst. Signal Process.* **2019**, *125*, 215–228. [CrossRef]
21. Hou, C.; Chen, T.; Li, Y.; Huang, M.; Shi, Q.; Liu, H.; Sun, L.; Lee, C. A rotational pendulum based electromagnetic/triboelectric hybrid-generator for ultra-low-frequency vibrations aiming at human motion and blue energy applications. *Nano Energy* **2019**, *63*, 103871. [CrossRef]
22. Li, M.; Deng, H.; Zhang, Y.; Li, K.; Huang, S.; Liu, X. Ultra-low frequency eccentric pendulum-based electromagnetic vibrational energy harvester. *Micromachines* **2020**, *11*, 1009. [CrossRef] [PubMed]
23. Halim, M.; Rantz, R.; Zhang, Q.; Gu, L.; Yang, K.; Roundy, S. An electromagnetic rotational energy harvester using sprung eccentric rotor, driven by pseudo-walking motion. *Appl. Energy* **2018**, *217*, 66–74. [CrossRef]
24. Liu, H.; Hou, C.; Lin, J.; Li, Y.; Shi, Q.; Chen, T.; Sun, L.; Lee, C. A non-resonant rotational electromagnetic energy harvester for low-frequency and irregular human motion. *Appl. Phys. Lett.* **2018**, *113*, 203901. [CrossRef]
25. Fan, K.; Hao, J.; Wang, C.; Zhang, C.; Wang, W.; Wang, F. An eccentric mass-based rotational energy harvester for capturing ultralow-frequency mechanical energy. *Energy Convers. Manag.* **2021**, *241*, 114301. [CrossRef]

machines

MDPI

Article

An Adaptive Control Method for the Distribution Valve of a Digital Pump

Daling Yue [1], Xiukun Zuo [1], Zengguang Liu [1,*], Yinshui Liu [1,2], Liejiang Wei [1] and Yuyang Zhao [3]

[1] Energy and Power Engineering College, Lanzhou University of Technology, Lanzhou 730050, China
[2] School of Mechanical Science and Engineering, Huazhong University of Science and Technology, Wuhan 430074, China
[3] Taiyuan Research Institute, China Coal Technology and Engineering Group, Taiyuan 030006, China
* Correspondence: liuzg@lut.edu.cn; Tel.: +86-138-9336-4273

Abstract: Perfect flow distribution is extremely important and essential for digital pumps. However, the fluctuation of motor speed and the change of valve dynamic characteristics cause the flow distribution flaw, which generates the backflow of the oil in the piston chamber and the decrease in pump volumetric efficiency. Based on the three-dimensional and mathematical modeling of the digital pump, the perfect distribution state of the digital pump is analyzed. Then, the adverse effects of the variations in motor speed and valve dynamic characteristics on the flow distribution of the digital pump were simulated and investigated by the software AMESim. To overcome the aforementioned problems, we proposed an adaptive control method for the flow distribution valve of a digital pump, which was realized by adopting the axis rotation angle and the pressure difference between the inlet and outlet of the flow distribution valve. The results show that the control signal of the flow distribution valve can be regulated automatically along with the motor speed and the valve dynamic characteristics, achieving the ideal flow distribution of the digital pump designed in this paper. The study can be used as a reference for the optimal design and prototype manufacturing of the digital pump.

Keywords: digital pump; flow distribution; adaptive control; high-speed on/off valve

1. Introduction

Digital hydraulics technology has been rapidly developed and researched for applications in recent years. Its core is the discrete control of flow through the active and intelligent control of digital valves [1]. Digital hydraulic pumps adopt the above solution, and the chamber is independently controlled by using two digital valves. The chamber that is used is activated, and the ones not operated are banned. As a result, more intelligent control and less energy loss are realized [2]. However, the efficiency is affected by the response speed and the timing of the flow distribution valve (FDV). To further improve the operating characteristics and efficiency of the digital pump, improving the response times of the FDV is an effective and convenient manner.

With respect to the dynamic response of FDV, Zhao et al. investigated the effect of control current and voltage on the dynamic characteristics of FDV. Finite element analysis found that there was a balance between energy consumption and improvement of response speed [3,4]. Because the FDV is affected by different supply pressure, carrier frequencies, and coil temperatures, Su et al. proposed a dual voltage adaptive switching method, which can change adaptively according to the oil supply pressure and control port pressure [5]. Gao et al. put forward an adaptive PWM control method consisting of four PWM signals for FDV, the rising time of the pressure was reduced by 84.6%, and the temperature of the coil shell declined by 61.5 % [6]. Zhong et al. proposed a method for FDV which can adaptively select among multiple driving voltage sources when the oil supply pressure changes [7]. Zhong et al. also presented an FDV control method with adaptive switching among five drive voltages, which reduced the open delay time by 66.7%, and lowered the close delay time by 87.5% [8]. Chapple et al. proposed a new type of poppet valve as FDV, and it is applied to high torque and low-speed systems to obtain higher efficiency and displacement control performance, but there are challenges with start-stop, valve timing, overpressure protection, and valve actuators [9]. The solenoid screw-in cartridge valves (SCV) are seen as having the potential to be applied to digital pumps. To improve the dynamic response of SCV and make it meet the flow distribution requirement of digital pumps. Yue et al. adopted positive and negative pulse voltage control for the solenoid coil, and the dynamic response characteristics are significantly enhanced by 56.67% compared to the conventional method [10]. Liu et al. proposed a new hybrid voltage control method consisting of preload voltage, positive pulse voltage, hold voltage, and negative pulse voltage, which reduced the open delay time by 74.24% and reduced the close delay time by 92.06% [11].

In the matter of flow distribution characteristics and control strategies for a digital pump, Pan et al. simulated the flow distribution characteristics of the FDV groups under clockwise and counterclockwise strokes of the motor, the steering variation process of the motor, and the effect of the switching valve frequency on the hydraulic motor [12]. Pederson et al. proposed continuous, discrete, hybrid dynamic approximate methods for digital pumps/motors, but there are still many challenges to be solved [13–16]. Based on the three pistons pump/motor, Breidi proposed a prediction for valve response delay time and mode switching algorithm using pressure fluctuations in high and low pressure lines, realizing an automatic selection of the most efficient control strategy among different displacement control strategies [17,18]. Nordas et al. detailed a simulation of the transient and steady-state response of digital displacement motors with full-stroke, partial-stroke, or sequential partial-stroke [19].

However, due to the changes in oil and coil temperature, the dynamic characteristics of the FDV can not always hold a constant in the operation of digital pumps [20,21]. In addition, the motor speed suffers from variations in the power grid. The above changes result in a flow distribution flaw in the digital pump, failing to achieve the expected flow distribution, which is manifested by fluid backflow and loss of volumetric efficiency during digital pump operation.

Therefore, an adaptive flow distribution control strategy of the digital pump is proposed to deal with the issue. The word "adaptive" refers to the ability of the proposed control method to maintain the perfect flow distribution of the FDV under the above variation. Firstly, we performed three-dimensional and mathematical modeling of the digital pump in Section 2. Secondly, in Sections 3 and 4, we analyzed the working characteristics of the digital pump with a perfect distribution and conducted a model-based simulation to investigate the impact of the changes in the motor speed and the dynamic characteristics of the FDV on a digital pump. Then, in Section 5, the principle of the proposed method is explained and verified. Finally, it was concluded that the adaptive control strategy could eliminate flow distribution error and keep the best efficiency, which not only improves the robustness and stability of the digital pump but also promotes the development of digital hydraulic technology.

2. Working Principle and Mathematical Model of the Digital Pump

2.1. Structure and Working Principles of the Digital Pump

A five-piston constant-speed digital pump with auxiliary check valves is shown in Figure 1. The motor converts electrical energy into mechanical energy, which drives the pump shaft and the eccentric cam to rotate. The piston is forced by the eccentric cam to make a sinusoidal motion and achieve fluid output. When the eccentric cam rotates from 0° to 360°, the piston moves from the top dead center (TDC) to the bottom dead center (BDC) and then returns to the initial TDC position; thus, the volume of the piston chamber completes a continuous variation. The piston and eccentric cam do not detach due to the return spring. When the piston moves from TDC to BDC, the piston chamber volume becomes larger, and the pump enters the suction stage, the FDV, known as the high-speed on/off valve, is actively open. If there is an insufficient oil suction, oil replenishment can be completed through the auxiliary check valve. When the piston moves from BDC to TDC, the volume becomes smaller, and the pump switches to the pumping stage in which the high-pressure valve (HPV) is passively open.

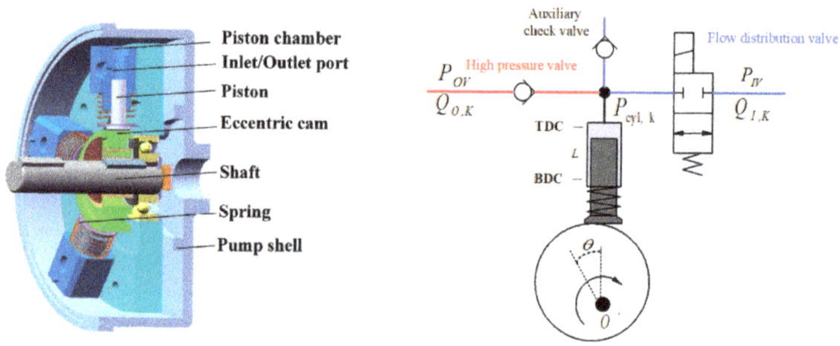

(**a**) (**b**)

Figure 1. Structure and working principle: (**a**) Structural section drawing; (**b**) flow distribution principle.

2.2. *Mathematical Modeling*

The overall structure of the digital pump is divided into two parts, one is the flow distribution valve part, and the other is the piston part. The mathematical modeling of the two parts is as follows.

2.2.1. Force Balance and Response Characteristics of the FDV

The force balance of the spool of the FDV can be expressed as:

$$m\ddot{x}_s = F_e - F_{ss} - F_f - F_{ff} \tag{1}$$

On expression, m represents the total quality of the spool and the armature. X_s is the displacement of the spool, F_e depicts the electromagnetic force of the electromagnetic coil, F_{ss} is spring compression force, F_f demonstrates the total friction force of the spool and the armature, and F_{ff} is the flow force of the spool. Among them, the spring force acting on the spool can be expressed as:

$$F_{ss} = K(x_0 + x_s) \tag{2}$$

In the above expression, K represents the stiffness coefficient of the spring, and x_0 represents the initial compression displacement of the spring. The flow force acting F_{ff} can be divided into steady-state flow force and transient flow force. Steady-state flow force and transient flow force can be expressed as:

$$F_s = 2C_d C_v A_v \cos\alpha\Delta p \tag{3}$$

$$F_t = C_d wl\sqrt{2\rho\Delta p}\dot{x}_s \tag{4}$$

where F_s and F_t are the steady-state and transient hydraulic forces of the spool; C_d and C_v represent the flow coefficient and velocity coefficient of the valve port, respectively; A_v is the flow area of the valve; α and Δp represent the conical spool jet angle and the pressure difference between the inlet and outlet of the valve; w is the area gradient; l is the damping length; and ρ is the density of the oil.

For the convenience of modeling, this paper reduces the transient response of the FDV to a first-order inertia part and a delay part.

The time-domain response expression of the first-order inertial part is:

$$h_{FDV} = 1 - e^{\frac{1}{T} \cdot \tau_{FDV}} \tag{5}$$

where τ_{FDV} is the on/off valve time constant, h_{FDV} is the spool displacement.

2.2.2. Piston Motion Characteristics

Because the five pistons are identical, all five pistons can be described by modeling a single piston. The displacement of the piston is as follows:

$$x = \frac{L}{2}(1 - \cos\theta) \tag{6}$$

where L is the effective length of plunger. The relationship between angular displacement and angular velocity of the shaft is as follows:

$$\theta = \omega t \tag{7}$$

In Equation (7), ω is the angular velocity of pump shaft rotation. By substituting Equation (7) into Equation (6), we can obtain the following:

$$x = \frac{L}{2}(1 - \cos(\omega t)) \tag{8}$$

By differentiating Equation (9), the piston velocity equation is achieved as follows:

$$\dot{x} = \frac{L}{2}\omega \sin(\omega t) \tag{9}$$

The variation in pressure in the piston chamber is as follows:

$$\frac{dP}{dt} = \frac{\beta}{V}\left(q_H - q_L - \frac{dV}{dt}\right) \tag{10}$$

where P is the pressure in the piston chamber, β is the elastic modulus of oil, V is the volume in the piston chamber, q_H is the flow out through HPV, and q_L is the flow through the FDV.

2.2.3. Piston Volume Variation Characteristics

With no consideration of the plunger volume leakage, the change in the volume of the plunger chamber can be expressed as:

$$V = V_0 + \frac{x}{2}(1 - \cos\omega t) \tag{11}$$

where V_0 is the volume of the piston at the initial moment. The derivative of the piston volume is:

$$\frac{dV}{dt} = \frac{x}{2}\omega \sin\omega t \tag{12}$$

Instantaneous angle of piston:

$$\theta_k = \theta_0 + \frac{2 \cdot \pi}{n_{cyl}} \quad k\epsilon\left(1, 2, \cdots, n_{cyl}\right) \tag{13}$$

Flow through the inlet and outlet of the single FDV:

$$Q_{FDV,\,k} = C_d A \frac{\sqrt{2 \cdot \Delta P}}{\rho} \tag{14}$$

The total flow of FDVs:

$$Q_{in} = \sum_{k=1}^{n_{cyl}} Q_{FDV,k} \tag{15}$$

3. The Perfect Flow Distribution on Digital Pumps

The displacement of the digital pump is permitted to be controlled by different control strategies, and the most common and simple operation is the full-stroke control strategy (FSCS) that is adopted in this paper. In the whole pumping stage, the HPV is passively opening, and the FDV is actively closing. The oil in the piston chamber is compressed into the high-pressure circuit. In the entire suction stage, the FDV is actively opening, and the HPV is passively closing so that the digital pump sucks oil from the tank. Another type of control is the partial stroke control strategy (PSCS). Compared to FSCS, the active control of the FDV in the PSCS is strengthened. Whether the PSCS or FSCS of the digital pump, they both achieve the flow distribution depending on the precise opening and closing of the FDV; consequently, the perfect flow distribution of the digital pump has to consider the dynamic response of the FDV. The dynamic response of the opening and closing process of the FDV is shown in Figure 2. When the FDV receives the opening/closing control signal, the FDV goes through the opening/closing delay stage/T_d and the spool opening/closing moving stage/T_m until the response of the FDV is finished.

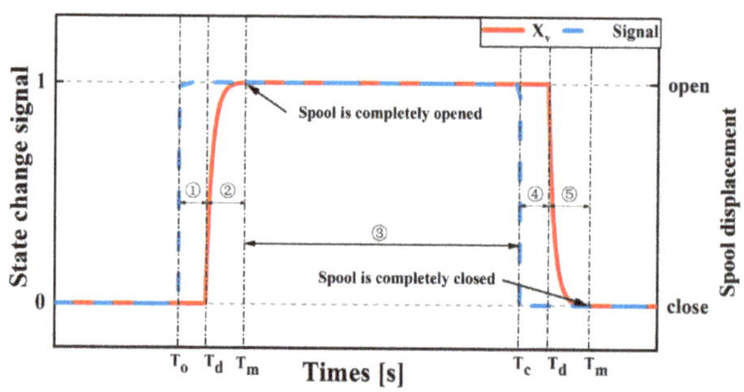

Figure 2. FDV dynamics response characteristic: ①: the spool opening delay stage; ②: the spool opening moving stage; ③: holding opening stage; ④: the spool closing delay stage; ⑤: the spool closing moving stage.

The operating characteristic curve of the digital pump at the perfect flow distribution condition is shown in Figure 3. It can be seen from plot 3 that the signal to open the FDV is made at this point, which is T_{oc} behind the TDC. The decision to close it is made at the moment where T_{cc} is ahead of the BDC. The T_{oc} is added to decompress the piston chamber in advance and make the pressure difference of the FDV equal to zero, which avoids the flow backward of the oil in the piston chamber. The value of T_{oc} is subject to the decompression time of the piston chamber and the spool opening delay stage of the FDV. The T_{cc} is created to ensure that the FDV has been closed completely when the piston reaches the BDC. The settings of T_{oc} must be decided by the sum of the closing delay stage and closing moving stage of the spool. Hence, it is obvious that the T_{oc} and T_{cc} is vital to the perfect flow distribution of the digital pump. From the flow curves of FDV shown in plot 4, it can be seen that the backflow does not occur with the appropriate T_{oc} and T_{cc} that brings the digital pump to the perfect flow distribution.

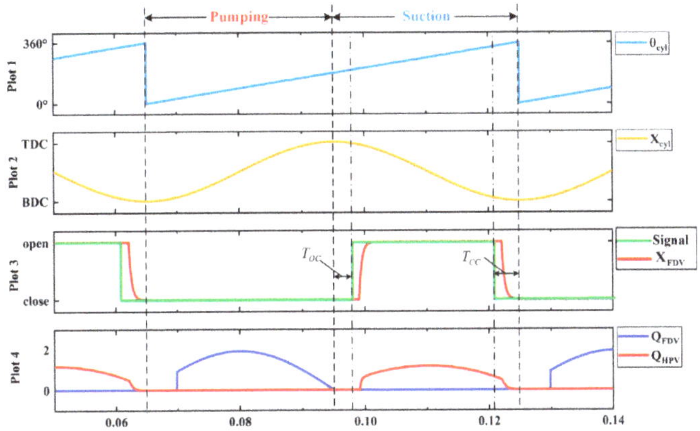

Figure 3. Perfect flow distribution state: plot 1: rotational angle; plot 2: Piston displacement; plot 3: The spool position and control signal; plot 4: FDV and HPV flow rate.

4. Simulation Modeling and Investigation

4.1. Digital Pump Simulation Model

According to the structure and working principle of the digital pump, it can be seen that the modeling of the digital pump involves multidisciplinary, such as mechanics, hydraulic transmission, electrical control, etc.; therefore, the multi-physics coupling modeling and analysis software AMESim is selected as tools to make study convenient.

As shown in Figure 4, the digital pump model with one piston is mainly composed of the hydraulic pump module and the control signal module. The control signal module is composed of the switch control signal module (SC_4) (SC_1) and the FDV adaptive signal module (SC_2). The main parameters of the model have been listed in Table 1. During the investigation of the flow distribution error stage, the adaptive signal control module is not involved in the control work. The correctness of the simulation model is verified by comparing the output flow of FDV and HPV, the backflow in simulation, and experiment results from the literature [14,16,17,22].

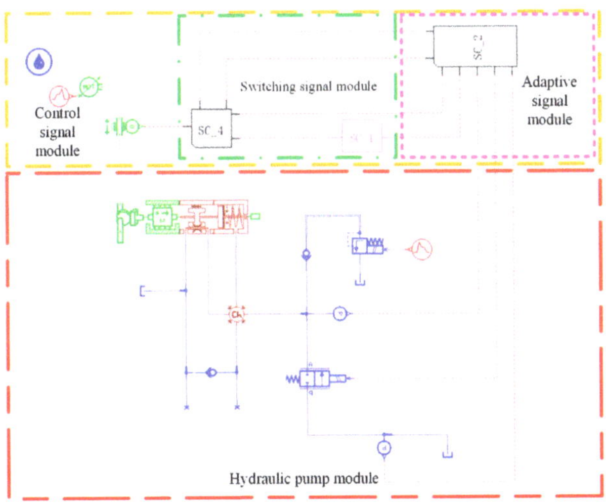

Figure 4. Simulation model of the digital pump.

Table 1. Main performance parameters.

Parameter	Value	Unit
Pump angular speed	1000	rpm
The radius of the eccentric	31	mm
Piston diameter	10	mm
Eccentricity	3.8	mm
Flow coefficient	0.7	
Dead volume of piston	0.08	mm^3
Chamber length	25.1	mm
FDV dynamic response	5	ms
T_{oc}	2	ms
T_{cc}	5	ms

4.2. Effects of Flow Distribution Error on Digital Pumps

The fluctuation of motor speed and the change in the dynamic characteristics of the FDV both make the FDV open and close at an inappropriate time; in other words, the FDV is activated too early or too late. To facilitate the analysis of the effect of incorrect FDV control on the digital pump, the T_{oc} is selected as −4, −2, 0, 2, and 4 ms, and the T_{cc} is −2, −1, 0, 1, and 2 ms. The greater the T_{oc} is, the later the FDV opens. The maximum T_{cc} means that the FDV first begins to close.

4.2.1. Effects of the T_{oc} on the Digital Pump

Figure 5 shows the main performance curves of the digital pump with T_{oc} increasing.

From picture (a), as the T_{oc} increases, the opening time of the FDV spool is transferred from the pumping stage to the suction stage. When T_{oc} is −2 ms, the spool starts moving just at the TDC. It can be seen from picture (b) that the pressure of the piston chamber will fall to zero during at pumping stage with −4 ms, which reduces the output flow of the digital pump. On the contrary, when T_{oc} is equal to 4 ms, the pressure of the piston chamber becomes negative because of the late opening of the FDV. According to the picture (c) and (d), the backflow amplitude gradually becomes smaller with the increase in the T_{oc}. When T_{oc} is greater than 2 ms, no backflow occurs. It can also be apparent that oversized T_{oc} forces the FDV to open too late, which results in the sucking oil having to be executed by the auxiliary check valve.

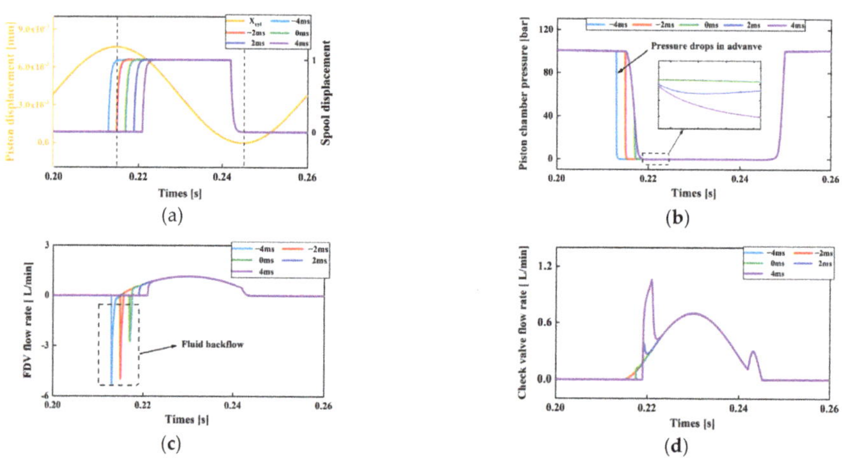

Figure 5. Characteristics with different T_{oc}: (**a**) spool displacement; (**b**) piston chamber pressure; (**c**) FDV flow rate; (**d**) auxiliary check valve flow rate.

4.2.2. Effects of the T_{cc} on the FDV

As shown in Figure 6a, the FDV began to close at different piston displacements with various T_{cc}. When T_{cc} is set to 2 ms, the FDV has been closed before the piston travels to the BDC. It can be found from pictures (c) and (d) that the FDV flow descends to zero at the last of the suction stage, and the flow through the auxiliary check valve is the highest correspondingly. Picture (b) shows that there is a sudden pressure reduction in the piston chamber due to the too-early closing of the FDV. The reason for this appearance is that the pressure difference with only the auxiliary check valve working is greater than the one of the simultaneous operation of the two valves. Moreover, we can see that the pressure rise of the piston chamber with T_{cc} being −2ms lags behind the other. The pressure variation in the piston chamber is identical except for −2 ms because the auxiliary check valve can compensate for the shutting off of the FDV.

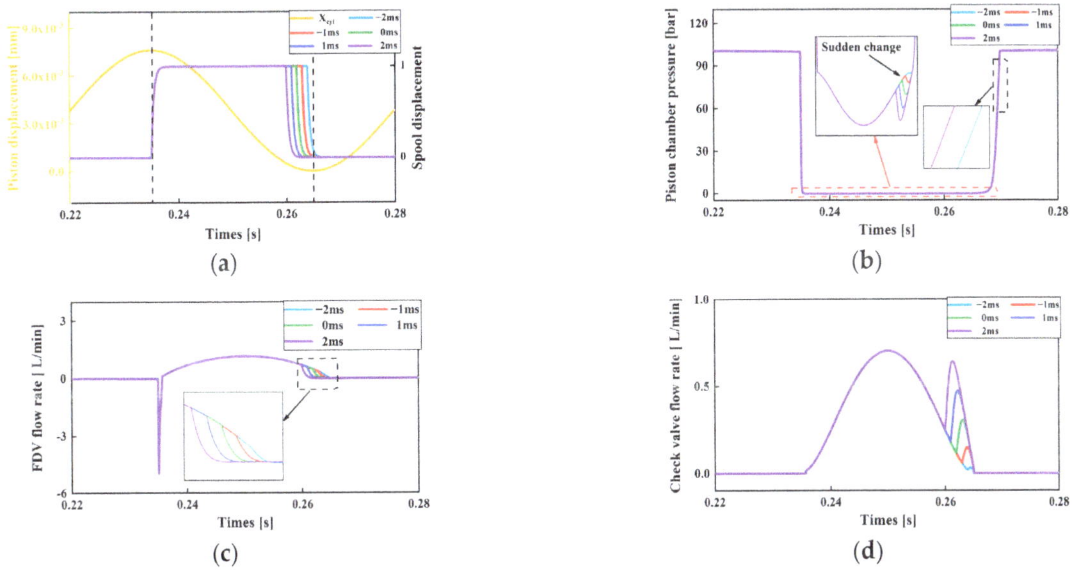

Figure 6. Characteristics with different T_{cc}: (**a**) spool displacement; (**b**) piston chamber pressure; (**c**) FDV flow rate; (**d**) auxiliary check valve flow rate.

5. Adaptive Control of the FDV

5.1. Control Principle of Adaptive Control of the FDV

From the viewpoint of the producing mechanism of the backflow and the volumetric efficiency decrease, we propose an adaptive control method for the FDV of a digital pump, which detects the rotation angle of the digital pump shaft and the pressure difference between the inlet and outlet of the FDV. The flowchart of the adaptive control method is demonstrated in Figure 7.

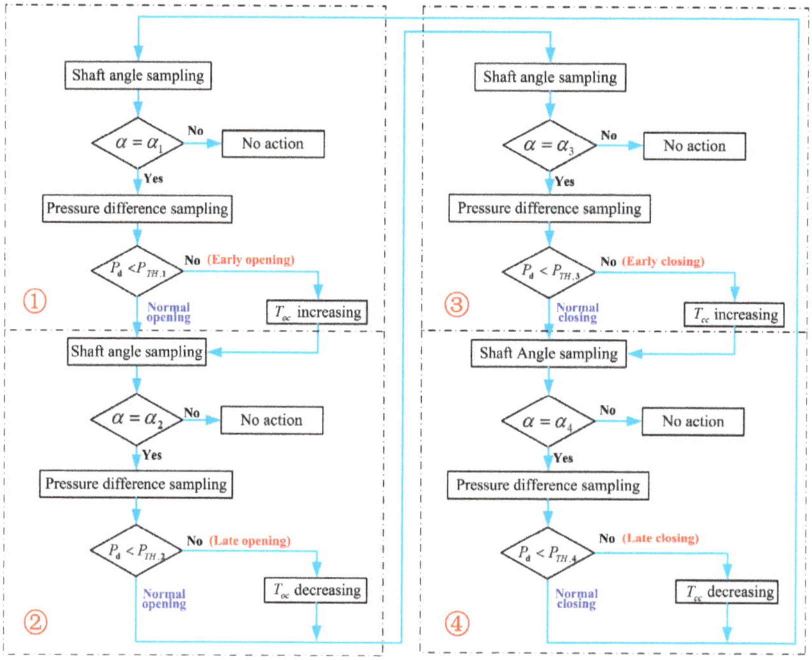

Figure 7. The flow chart for adaptive control of the FDV.

As can be seen from Figure 7, the whole adaptive control of the FDV is divided into four parts that are as follows: ① early opening handling module, ② late opening handling module, ③ early closing handling module, and ④ late closing handling module. These four modules are executed once in order when the pump shaft rotates by 360 degrees, which keeps the adaptive control operation of the FDV running and achieves the adaptive adjustment of the T_{oc} and T_{cc}. The adaptive control of the FDV has four important shaft angles α_1 , α_2, α_3, and α_4, at which the early opening, late opening, early closing, and late closing are inspected, respectively. The locations of these four critical angles are represented in Figure 8.

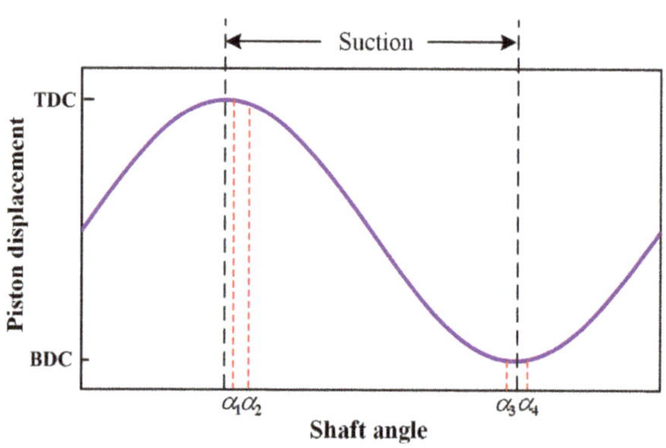

Figure 8. The diagram of the inspection angles of flow distribution.

The detailed control strategy of adaptive control of the FDV is described as follows.

1. In the early opening handling module, when the shaft angle collected by the rotary encoder is not equal to the early opening assessment angle α_1. Nothing special has to be performed. Conversely, the pressure difference between the inlet and outlet of the FDV is sampled and compared with the set threshold $P_{TH,1}$. If the pressure difference is greater than the threshold, the FDV is opening too early and the T_{oc} is raised. According to Figure 5b, when the FDV is opening in advance, the pressure of the piston chamber is not as great as expected because the high-pressure oil in the piston chamber will be discharged through the FDV; on the contrary, the control of the FDV is normal, and the next module is subsequently called.
2. When the shaft of the digital pump is late opening assessment angle α_2, the late opening of the FDV is detected and treated. If the FDV is not activated at α_2, the oil enters the piston chamber only via the auxiliary check valve, and the pressure difference of the FDV is smaller than the one of the two valves opening. The pressure difference comparison of the FDV closing and the two valves opening is displayed in Figure 9. If the late opening of the FDV is perceived, the decrease in T_{oc} is implemented.
3. The operating principle of adaptive control of the FDV early closing is similar to the late opening; the T_{cc} decreases when the FDV is prematurely deactivated. The comparison between P_d and $P_{TH,3}$ and T_{cc} adjustment are both carried out with the shaft angle equaling to α_3.
4. The solution to the late closing of the FDV is entirely consistent with the early opening handling module. If the FDV is not closed in time, the piston chamber is connected to the tank, and the pressure of the piston chamber can not rise to the threshold $P_{TH,4}$ at the shaft angle α_4. In this case, the T_{cc} is gradually increased to the early opening of the FDV.

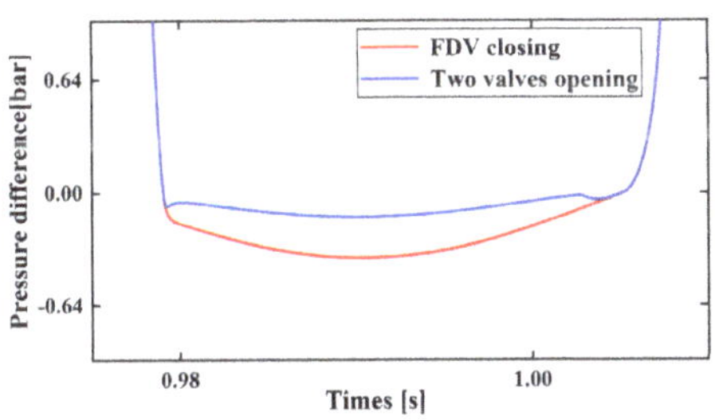

Figure 9. Pressure difference comparison of FDV closing and two valves opening.

5.2. Simulation Analysis of Adaptive Control of the FDV

5.2.1. The Results with Motor Speed Changes

The step changes in moto speed are used to analyze and verify the effectiveness of the adaptive control method proposed in this paper. The flow rate of the FDV is shown in Figure 10a with motor speed fluctuations. It can be seen that the flow rate of the FDV is proportional to the motor speed, but the backflow of the FDV emerges after the motor speed variation. This phenomenon is weak when the motor speed step increases but stronger when the speed step decreases. The increase in motor speed means that the activation and inactivation of the FDV are performed; hence, there is a small amount of backflow at the moment of the FDV closing after 2 s. From pictures (b), (c), and (d), the T_{oc} and T_{cc} are automatically elevated from -12 to -6 and from -22 to -32 under adaptive control and kept in the optimum values related to the motor speed.

When the motor speed reduces by 500 r/min at 4 and 6 ms, a violent flow backward appears in the opening process of the FDV. In the timeline, the moment of action of the FDV does not change, but the movement of the plunger was delayed. It is obvious that the sudden reduction in the motor speed is equivalent to the early opening and closing of the FDV. According to the module 1 and 3 in Figure 7, the T_{oc} and T_{cc} should be diminished with the premature motion of the FDV, which can be proved by the curves reducing from

−6 to −16 and from −32 to −21 between 4 and 8 ms in Figure 10b,c. Additionally, it can be seen from Figure 10d that this method can improve the situation of excessive pressure difference during FDV action. As a result, the adaptive control approach is better able to cope with the motor speed change, and the backflow phenomenon is eliminated.

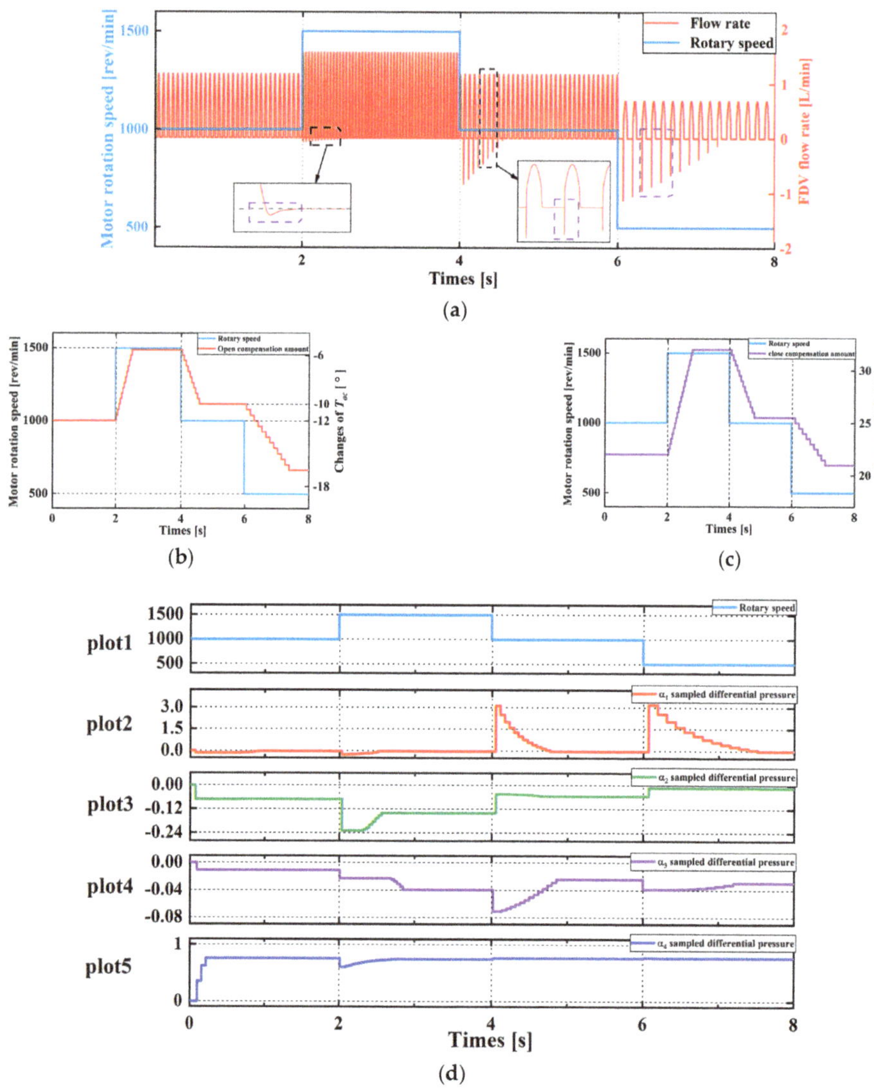

Figure 10. The results with different motor speeds: (**a**) flow rate of FDV; (**b**) T_{oc} curve; (**c**) T_{cc} curve; (**d**) plot 1: Motor speed; plot 2–4: differential pressure at α_1, α_2, α_3, α_4.

5.2.2. The Results with the Different Dynamic Responses of the FDV

It is found from Figure 2 that the dynamic response of the FDV contains the delay and spool moving two parts. In this paper, the spool moving time of the FDV is altered to ascertain the availability of the adaptive control, the motor speed is set to 1000 rpm, and the delay time of the FDV is still held at the primitive value 2 ms. As shown in Figure 11a, the time spent by the spool moving varies from 1 ms to 9 ms. The initial phase is set so that the FDV is already in a perfect flow distribution when the spool starts moving in the opening stage. The time the spool starts to open and close has no change because the delay time and control signal of the FDV is invariant, which prevents the early or late opening of the FDV from happening. Hence, the value of T_{oc} shown in Figure 11e is equivalent. The longer spool moving time may cause the complete closing of the FDV at the pumping stage of the digital pump, making more backflow, which can be observed in Figure 11b. Figure 11c shows the flow of the FDV closing process utilizing the adaptive control, and the backflow is completely removed. The backflow contrast curves of the FDV before and after optimization are unlocked in Figure 11d. When the closing moving time of the FDV is greater than 3 ms, the late closing trouble takes place and is solved by means of increasing T_{cc}, which can be confirmed by the increase, respectively, from 22 to 29, 38, and 42 in Figure 11f.

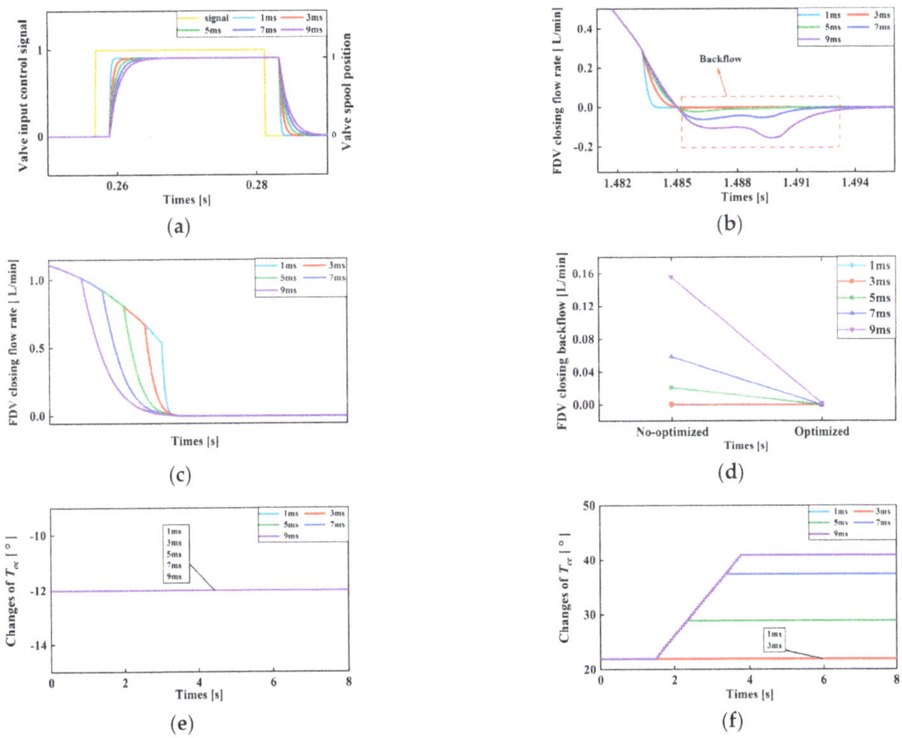

Figure 11. The results of different dynamic characteristics of FDV: (**a**) spool displacement; (**b**) flow rate before optimization; (**c**) optimized flow rate; (**d**) comparison backflow of FDV closing process; (**e**) T_{oc} curve; (**f**) T_{cc} curve.

6. Conclusions

In this paper, the self-adaption flow distribution of the digital pump with FDV and an auxiliary check valve is studied. Inappropriate control signal T_{oc} and T_{cc} generated by the changes in the FDV dynamic properties and motor speed will bring the early or late working of the FDV and lead to the backflow of oil in the piston chamber and more volumetric efficiency losses. As a consequence, T_{oc} and T_{cc} of the FDV must be corrected in real-time to avoid flow distribution errors.

A new adaptive control method for the FDV is proposed to automatically adjust the T_{oc} and T_{cc} and compensate for the inevitable undulation of the motor speed and the variation in FDV response. The approach suggested is completed by conducting four compares between the pressure difference sampled at the shaft angle α_1, α_2, α_3, α_4 with the pre-set thresholds.

The simulation results reveal that when the adaptive control is implemented, the T_{oc} and T_{cc} can be autonomously revised with the aforementioned changes, and the undesired backflow is cleaned up. The work in the future is the test rig development of the digital pump by utilizing the semi-physical simulation system in our laboratory and validating the availability of this solution through the experiment.

Author Contributions: Conceptualization, Y.L. and L.W.; methodology, L.W.; software, D.Y. and X.Z.; validation, D.Y., X.Z. and Z.L.; formal analysis, Y.L. and L.W.; investigation, Z.L. and Y.Z.; resources, L.W.; data curation, D.Y. and X.Z.; writing—original draft preparation, D.Y. and X.Z.; writing—review and editing, L.W. and Z.L.; visualization, X.Z.; supervision, Z.L.; project administration, Z.L.; funding acquisition, L.W. and D.Y. All authors have read and agreed to the published version of the manuscript.

Funding: This research was funded by the National Key R&D Program of China (Grant No. 2020YFB2009803).

Data Availability Statement: Not applicable.

Conflicts of Interest: The authors declared no potential conflict of interest with respect to the research, authorship, and publication of this article.

References

1. Zhang, Q.W.; Kong, X.D.; Yu, B.; Ba, K.X.; Jin, Z.G.; Kang, Y. Review and Development Trend of Digital Hydraulic Technology. *Appl. Sci.* **2020**, *10*, 579. [CrossRef]
2. Danfoss. Data Sheet-Digital Displacement®Pump DDP096 and DPC12. 2022. Available online: https://www.bibus.sk/fileadmin/editors/countries/bibsk/Suppliers/Danfoss/documents/displacement-pumps/danfoss-DOC332065464017.pdf (accessed on 30 December 2019).
3. Zhao, J.; Wang, M.; Wang, Z.; Grekhov, L.; Qiu, T.; Ma, X.J.A.T.E. Different boost voltage effects on the dynamic response and energy losses of high-speed solenoid valves. *Appl. Therm. Eng.* **2017**, *123*, 1494–1503. [CrossRef]
4. Zhao, J.; Yue, P.; Grekhov, L.; Ma, X.J.A.T.E. Hold current effects on the power losses of high-speed solenoid valve for common-rail injector. *Appl. Therm. Eng.* **2018**, *128*, 1579–1587. [CrossRef]
5. Su, M. Investigation on Control Characteristics and Method of Solenoid High Speed on/off Valve. Ph.D. Thesis, Guizhou University, Guiyang, China, 2010.
6. Gao, Q.; Zhu, Y.; Luo, Z.; Bruno, N. Investigation on adaptive pulse width modulation control for high speed on/off valve. *J. Mech. Sci. Technol.* **2020**, *34*, 1711–1722. [CrossRef]
7. Zhong, Q. Research on Control Algorithm for High-speed on/off Valves that Adaptive to Supply Pressure Changes. *J. Mech. Eng.* **2021**, *57*, 224–235.
8. Zhong, Q. Performance Analysis of High Speed on/off Valve by Multi-voltages Compound Excitation. *J. Mech. Eng.* **2021**, *57*, 191–201.
9. Chapple, P.; Lindholdt, P.N.; Larsen, H.B. An Approach to Digital Distributor Valves in Low Speed Pumps and Motors. In Proceedings of the ASME/BATH 2014 Symposium on Fluid Power and Motion Control, Bath, United Kingdom, 10–12 September 2014.
10. Yue, D.; Li, L.; Wei, L.; Liu, Z.; Liu, C.; Zuo, X. Effects of Pulse Voltage Duration on Open–Close Dynamic Characteristics of Solenoid Screw-In Cartridge Valves. *Processes* **2021**, *9*, 1722. [CrossRef]
11. Liu, Z.; Li, L.; Yue, D.; Wei, L.; Liu, C.; Zuo, X. Dynamic Performance Improvement of Solenoid Screw-In Cartridge Valve Using a New Hybrid Voltage Control. *Machines* **2022**, *10*, 106. [CrossRef]
12. Pan, J.; Shi, G. Distribution Characteristics of Hydraulic Motor with Digital Distribution. *Trans. Chin. Soc. Agric. Mach.* **2011**, *42*, 203–208.
13. Pedersen, N.H.; Johansen, P.; Andersen, T.O. Feedback Control of Pulse-Density-Modulated Digital Displacement Transmission Using a Continuous Approximation. *IEEE/ASME Trans. Mechatron.* **2020**, *25*, 2472–2482. [CrossRef]
14. Pedersen, N.H.; Johansen, P.; Andersen, T.O. Four Quadrant Hybrid Control Oriented Dynamical System Model of Digital Displacement (R) Units. In *Fluid Power Systems Technology*; American Society of Mechanical Engineers: New York, NY, USA, 2018.
15. Pedersen, N.H.; Johansen, P.; Andersen, T.O. Model Predictive Control and Discrete Analysis of Partial Stroke Operated Digital Displacement Unit. In *2018 Global Fluid Power Society Phd Symposium (Gfps)*; IEEE: Piscataway, NJ, USA, 2018.
16. Pedersen, N.H.; Johansen, P.; Andersen, T.O. Event-Driven Control of a Speed Varying Digital Displacement Machine. In Proceedings of the Asme/Bath Symposium on Fluid Power and Motion Control, Sarasota, FL, USA, 16–19 October 2017.
17. El-Breidi, F. *Investigation of Digital Pump/Motor Control Strategies*; Purdue University: West Lafayette, Indiana, 2016.
18. Breidi, F.; Garrity, J.; Lumkes, J., Jr. Investigation of a Real-Time Pressure Based Valve Timing Correction Algorithm. In Proceedings of the ASME/BATH 2017 Symposium on Fluid Power and Motion Control, Sarasota, FL, USA, 16–19 October 2017.
19. Nordas, S.; Beck, M.M.; Ebbesen, M.K.; Andersen, T.O. Dynamic Response of a Digital Displacement Motor Operating with Various Displacement Strategies. *Energies* **2019**, *12*, 1737. [CrossRef]

20. Tian, H.; Li, P.Y.; Van de Ven, J.D. Valve Timing Control for a Digital Displacement Hydraulic Motor Using an Angle-Domain Repetitive Controller. *IEEE/ASME Trans. Mechatron.* **2019**, *24*, 1306–1315. [CrossRef]
21. Darnet, J.; Bideaux, É. State-of-the-art of Variable Displacement Technologies for Radial Piston Hydraulic Machines. In Proceedings of the Asme/Bath Symposium on Fluid Power and Motion Control, Bath, UK, 14–16 September 2022.
22. Pedersen, N.H.; Johansen, P.; Andersen, T.O. Lqr Feedback Control Development for Wind Turbines Featuring a Digital Fluid Power Transmission System. In Proceedings of the 9th Fpni Ph.D Symposium on Fluid Power, Florianópolis, SC, Brazil, 26–28 October 2016.

machines

MDPI

Article

Multibody Modeling of a New Wheel/Track Reconfigurable Locomotion System for a Small Farming Vehicle

Andrea Grazioso [1], Enrico di Maria [2], Nicola Ivan Giannoccaro [1,*] and Kazuo Ishii [2]

[1] Dipartimento di Ingegneria dell'Innovazione, Università del Salento, 73100 Lecce, Italy
[2] Graduate School of Life Science and Systems Engineering, Kyushu Institute of Technology, Kitakyushu 808-196, Japan
* Correspondence: ivan.giannoccaro@unisalento.it; Tel.: +39-0832297813

Abstract: Tracks and wheels are the most widely used running gear for the locomotion of agricultural vehicles. The main difference between the two systems is the contact area with the ground and, consequently, the pressure distribution. Evaluating the pressure distribution on the ground is important because soil damage and vehicle performance depend on it. This analysis is especially difficult for tracked vehicles, owing to their complexity compared with wheeled systems. In this paper, we describe a multibody model of a flexible track to evaluate the pressure distribution upon contact with soft terrain. The track considered in this study is part of a reconfigurable locomotion system of a small farming vehicle, which can vary the pressure distribution by switching from a wheeled vehicle to a half-tracked vehicle. The aim of such a vehicle is to minimize soil damage in addition to optimizing its performance. The model is used to characterize this vehicle and evaluate the pressure distribution with varying characteristic parameters, such as track tension, the position of the vehicle's center of gravity, the weight distribution on the track itself, and the stiffness of the suspension system.

Keywords: multibody analysis; reconfigurable locomotion system; flexible track; terramechanics; farming vehicle; ground pressure distribution

1. Introduction

Tracked systems are among the most common running gear used for various off-road vehicles, from military applications to the agricultural field. Tracked vehicles can be divided into two main categories: rigid and flexible tracks [1]. In the first case, the track belt structure consists of a solid chain track made of steel plates, whereas in the second case, the belt is made of rubber reinforced with steel wires along the longitudinal length. The track belt is forced onto the ground by a series of wheels called rollers, and it is moved by an additional motored wheel called a sprocket, the position of which depends on the system layout. Although the two types of tracks (rigid and flexible) lead to differing performances of the vehicles, a common aspect is the capability to deform upon contact with the soil and adapt to it to some extent. An optimal design of these systems requires taking into account this flexibility. This makes the analysis of tracked vehicles challenging compared with wheeled vehicles. The vehicle's performance depends on the forces exerted on the ground as a function of the normal and shear stress at the contact patch. Accordingly, it is important to have a reliable tool to quantify the stress distribution at the point of contact between the running gear and soil in order to analyze and design tracked vehicles.

Several methods have been proposed to carry out such analysis. The first adopted strategy uses empirical methods [2], whereby simple soil measurements and some main vehicle characteristics are used to assess the performance. Another strategy involves using analytical models, among which the method developed by Bekker is the most famous [2]. These strategies have the drawback of relying on simplified assumptions, and in the case of analytical models, the pressure distribution at the contact point with the ground is assumed beforehand. Common pressure distributions are constant, trapezoidal, and multipeak sinusoidal. In this last case, the pressure peaks are related to the rollers. The real pressure distribution is unlikely to be of the functions mentioned above. It is affected by the position of the vehicle's center of gravity, the tension of the track belt, the geometry, the disposition of the rollers, and the suspension system forcing the rollers onto the belt. Whereas parametric models of increasing complexity have been proposed to analyze tracked systems [3–6], the high computational capacity of modern computers enabled the development of numerical methods to quantify the vehicle performance. An example is NTVPM [2], which can model the interaction between the track and soil in detail.

Multibody systems have proven to be promising tools for the modeling of tracks. In [7], a multibody model was developed, assuming that the track can be approximated as a set of imaginary wheels. In [8], a multibody model was developed, employing a nonlinear finite-element representation that can simulate both flexible rubber tracks and rigid tracks.

In this work, we propose a step-by-step implementation of a multibody model to study the pressure distribution of a reconfigurable wheeled/tracked vehicle on soft terrain. The vehicle consists of a rear-wheel axle and a wheel/track reconfigurable front axle, which can adjust the contact area with the ground. A tracked configuration can better distribute pressure and causes less compaction than a wheel. On the other hand, tracks cause more soil distortion [9] and are less efficient during steering maneuvers than wheels. The vehicle is intended to adjust its contact area based on the soil conditions to minimize soil damage while considering the rolling resistance and traction performance. To design such a system, it is important to quantify the pressure distribution at the contact point with the soil; this is the purpose of the multibody model developed in this work. Although a similar task could be accomplished by relying on dedicated plugins of commercial software [10], adopting a more diffuse and general-purpose tool such as MATLAB® enables the development of a model from scratch, facilitating a deeper understanding and a better compromise between accuracy and computational effort. Moreover, a step-by-step implementation permits customization of the model based on future research needs.

In this paper, we describe the strategy pursued in modeling the system consisting of half the robot frame, the reconfigurable front axle, the rear wheeled axle, and the ground.

2. Materials and Methods

A 3D model of the system in the simulation environment is shown in Figure 1. A detailed model of the tracks and the reconfiguration mechanism was combined with a simplified body representing the chassis and the rear-axle wheel.

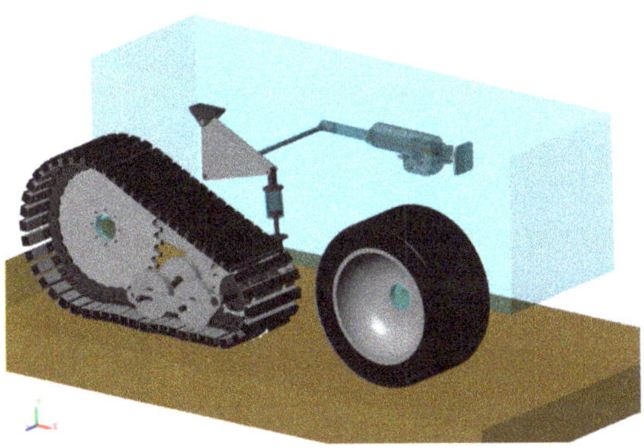

Figure 1. 3D view of the multibody model of the farming robot.

A functional sketch of the model is shown in Figure 2, in which the main components are represented:

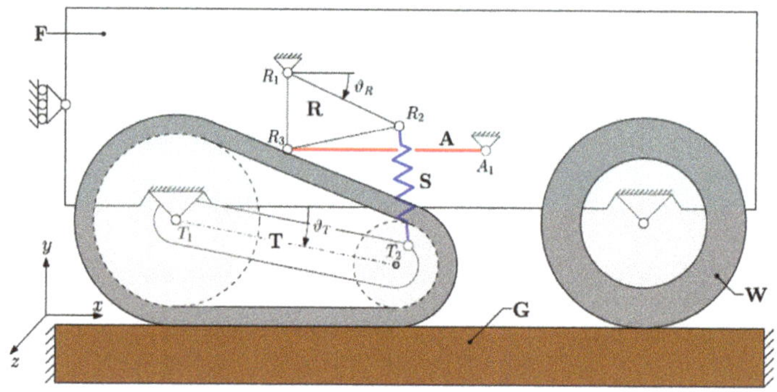

Figure 2. Scheme of the main components of the model.

Ground, indicated with the letter G, represents the soil;

Frame, indicated with the letter F, represents the vehicle chassis;

Track module, indicated with the letter T, is a set of bodies assembled to form the system shown in Figure 3. The assembly in Figure 4 is hinged to the vehicle chassis (F) through a revolute joint located at point T_1;

Figure 3. Track module.

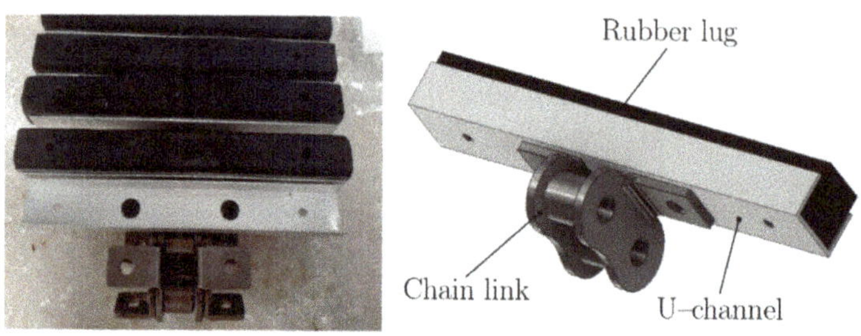

Figure 4. Detail the track belt (**left**) and the track link (**right**) components.

Shock absorber, indicated with the letter S (blue line). According to one of the requirements discussed above, this element is inserted to endow the mechanism with the capability of letting the track better adapt to bumpy, uneven ground. The bottom end of the shock absorber is hinged to the track frame (T) through a revolute joint;

Rocker, indicated with the letter R, represents the key element of the switching mechanism, deputed to transfer the motion from the actuator to the track frame. It is a ternary link that is often used in passive [11] and active adjustable suspension systems [12,13]. It is hinged to the vehicle chassis (F) through a revolute joint at point R_1. The upper end of the shock absorber (S) is connected to the rocker through a revolute joint located at point R_2;

Actuator, indicated with the letter A (red line), is the element that drives the mechanism. One end of the actuator is hinged to the frame (F) through a revolute joint located at point A_1, whereas the other end is linked to the rocker (R) through a revolute joint at point R_3; and

Wheel, indicated with the letter W, represents the wheel of the robot's rear axle.

The track module is a set of bodies assembled together. This module comprises the sprocket; the idler; the track frame; the undercarriage; and the chain, which constitutes the track belt. In order to capture the functioning of this module, the aforementioned elements, especially the links making up the chain, cannot be rigidly connected to one another, as they experience multiple contacts, the implementation of which is detailed in Section 2. The shock absorber is composed of two main bodies—the cylinder and the piston—connected by a prismatic joint. The multibody model of the actuator is composed of three blocks: the main body, the leadscrew, and the plunger (also called piston or rod. The relative motion between the actuator body and the plunger is dictated by a prismatic joint, allowing only translation along the axis of the leadscrew; the position along the longitudinal axis of the cylinder varies from 0 mm when the actuator is fully retracted to 100 mm (equal to the stroke) when the actuator is fully extended. The wheel is modeled as a rigid body composed of two parts, the tire and rim, which are clearly visible on the right side of Figure 1.

The functioning of the proposed switching mechanism is extremely simple; when the actuator (A) retracts (distance between points A_1 and R_3 decreases), it causes the rocker (R) to rotate about point R_1 in a counterclockwise direction; consequently, the track module (T) is forced through the shock absorber (S) linked to the rocker (R) to rotate about point T_1 in a counterclockwise direction, thus lifting the idler section of the track from the ground. Conversely, if the actuator (A) extends (the distance between points A_1 and R_3 increases), the rotations of the rocker R and the track module T occur in a clockwise direction, and the reconfigurable section of the track is lowered to the ground. The motion imposed by the actuator is transferred to the track module in an effective yet simple way, thus providing for the reconfiguration of the system.

2.1. Track Module

The track module shown in Figure 3 was designed, realized, and tested at the Ishii Laboratory of the Kyushu Institute of Technology [14]. A roller chain track was chosen among the various possible mechanical solutions to build a custom track. The belt consists of an attachment chain (also known as a conveyor chain) with built-in flanges on which treads can be easily attached to form a track.

As shown in Figure 3, the roller chain track is made up of a chain engaging with a toothed drive sprocket on one side (hereafter referred to as a sprocket) and a toothed idler sprocket on the other (hereafter referred to as an idler). The sprocket moves the track along, and the idler keeps the track straight. The sprocket has 33 teeth and an outer diameter of 281 mm, whereas the idler has 15 teeth and an outer diameter of 135 mm, with pitch diameters of 267.21 mm and 122.17 mm, respectively.

On each one of the chain attachments, an aluminum U channel is fastened by two bolts, as shown in Figure 5.

The links of the track belt that gradually enter the lower branch, specifically in the section separating the sprocket from the idler, must be pressed against the ground to effectively redistribute the weight of the vehicle.

This important function is accomplished by the undercarriage shown in Figure 6. It is composed of two rows carrying two rollers each. The shafts about which the rollers are free to rotate are supported by two L–shaped pieces, which are connected on one end to the idler shaft and on the other end to the arm between the sprocket and idler shafts.

The arm connecting the sprocket and idler shaft is represented in Figure 7, hereafter referred to as the track frame. It is an ensemble of three telescopic squared aluminum frames: a central frame, a solid block, and a frame that can slide inside the other two, which are hollow and connected by a hexagonal nut. This nut acts as a tensioner for the track belt; one of the ball joints at the end of the nut is right-threaded, whereas the other has a left thread so that the rotation of the nut can increase or decrease the distance between the idler and sprocket, thus tensioning the chain.

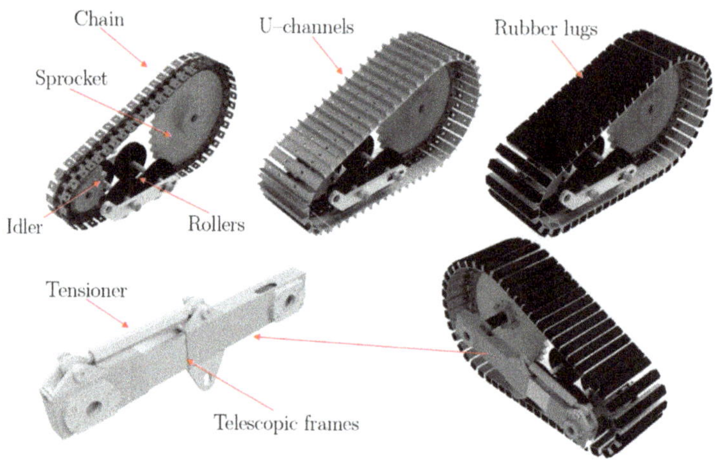

Figure 5. Components of the track module and final assembly.

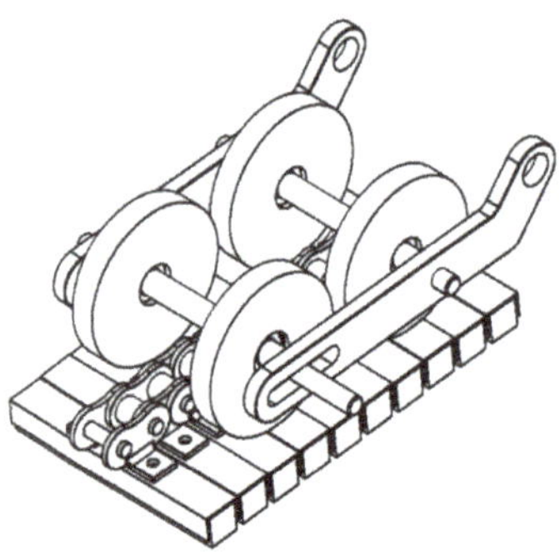

Figure 6. The undercarriage of the track module: rollers and L–shaped supports.

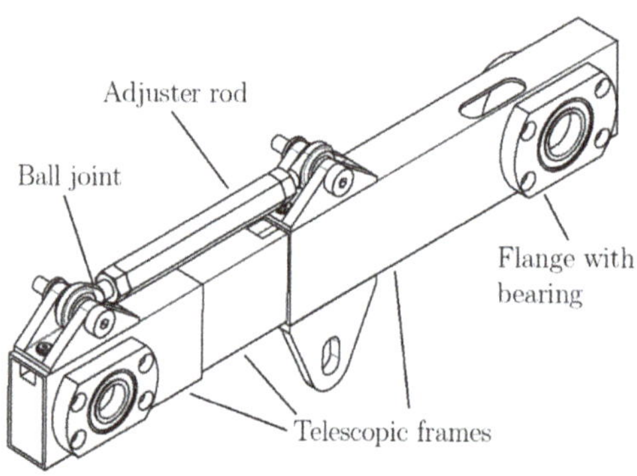

Figure 7. Track frame connecting the sprocket shaft and the idler shaft.

2.1.1. Track Frame

The track frame shown in Figure 7 is the element that connects the sprocket shaft and the idler shaft. It is composed of two telescopic frames that can slide relative to each other; a tensioner nut with a right thread on one side and a left thread on the other side provides a simple way to tension the chain that constitutes the track belt. On the idler side of the track frame, the shock absorber has to be mounted to allow the switching mechanism raise and lower part of the track; therefore, as shown in Figure 8, the attachment bracket of the tensioning system is equipped with an additional hole to house the lower eyelet of the shock absorber. To endow the model with the possibility of specifying an arbitrary tension for the track chain, thus improving its fidelity, the two parts composing the track frame that rotate about the sprocket shaft and carry the idler shaft are connected by a prismatic joint with base (T_{s1}) and follower (T_{s2}) (see Figure 9). This joint may be assigned a null displacement or a certain force. In the first case, a null displacement between T_{s1} and T_{s2} is imposed, and the chain is loose, i.e., the tension is zero; in the second case, an arbitrary force (F_{ts}) oriented from T_{s1} to T_{s2} is applied to the joint, and consequently, the track chain experiences a tension (T_{ch}) as expressed by (1),

$$T_{ch} = \frac{F_{ts}}{2\cos(\alpha)} \tag{1}$$

where α is the chain angle shown in Figure 9. In light of such implementation, it is possible to simulate the model with differing values of chain tension, which are usually expressed as fractions of the vehicle weight [1].

Figure 8. Track frame connecting the sprocket shaft (in blue) and the idler shaft.

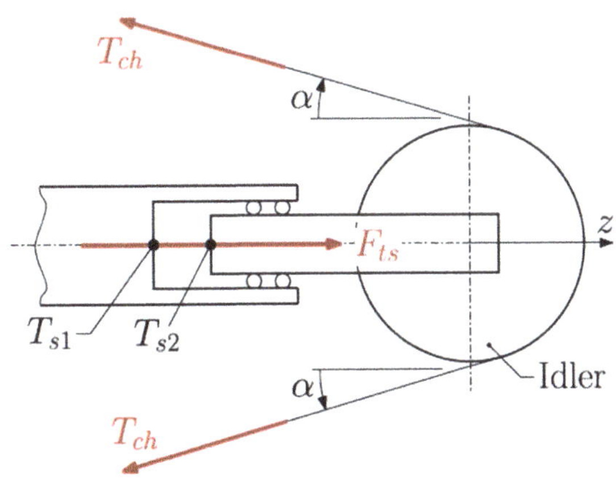

Figure 9. Scheme of the track belt tensioner.

2.1.2. Track Belt

The track belt comprises a chain with attachments (see Figure 10). Although the chain is composed of articulated links forming a closed kinematic loop, the belt can be modeled by connecting all the links to form a unique rigid body made up of the chain, the sprocket, the idler, the undercarriage, and the track frame. Such a modeling strategy, although simple and fast, does not capture the flexibility of the track belt, thus leading to an unrealistic, almost uniform pressure distribution over the ground contact area, which is equivalent to assuming a rigid footing, as is often the case in the literature on rolling resistance evaluation. To fulfill the goal of assessing how the pressure originated by the track–terrain interaction distributes over the contact patch, it was necessary to model the chain more accurately.

Figure 10. Detailed photo of the track belt.

The chain shows the alternation of external links (26 pin links such as that on the left in Figure 11a) and internal links (26 bushing links such as that on the right in Figure 11a). Each link is connected to the two adjacent links using revolute joints to permit relative rotations, as shown in Figure 11a. The total of 52 links are divided into 13 groups of 4. Figure 11a,b indicate the contact surfaces, highlighted in different colors:

- In red, the surfaces of the bushings of the inner link that come into contact with the two gears of the track module;
- In green, the portions of the upper surface of U channels that come in contact with the undercarriage rollers; and
- In blue, the surface of the rubber lug that comes in contact with the ground.
- The contacts mentioned above were analyzed and implemented in the proposed model.

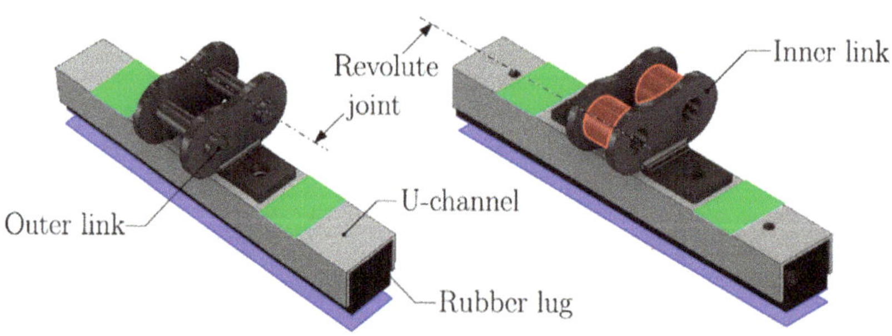

(**a**)

Figure 11. *Cont.*

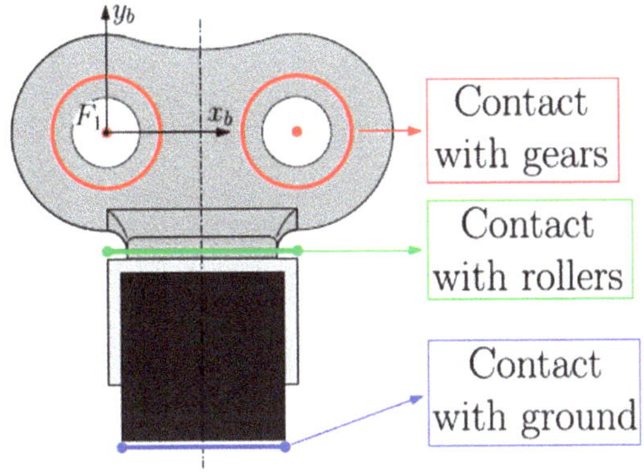

(**b**)

Figure 11. Links of the chain that constitutes the track belt: (**a**) outer and inner links, (**b**) contact surfaces.

2.2. Contact Model

As previously mentioned, the bodies of the multibody model, especially those comprising the track module shown in Figure 11, in addition to coming into contact with the ground (the rubber lugs), collide with each other, so a precise and accurate representation of impact or contact is needed [15–17]. The contact of the track links with the gears, and the track module's rollers are modeled using a linear spring–damper model. According to this method, the local deformations of the contacting bodies in the contact region are simplified, i.e., allowing the bodies to penetrate each other to a limited extent. The amount of penetration, denoted by δ, is equal to the sum of local deformations undergone by the two contacting bodies. The penetration causes a pair of resistive contact forces to act on the two bodies in opposite directions.

The contact force can be decomposed into the normal and tangential components, which act along unit vectors n and t. The normal contact force that A exerts on B (F_{nAB}) shown in Figure 12 is computed as if it were generated by a linear spring–damper system placed between points A and B so that [18]

$$F_{nAB} = -\sigma\left(\delta_{BA}\right)\cdot\left(K\cdot\delta_{BA} + \gamma\left(\dot{\delta}_{BA}\right)D\dot{\delta}_{BA}\right)n_A \quad (2)$$

where:

- K is the contact spring stiffness and represents the resistance opposing the geometric penetration of the two colliding bodies. The larger the value of the spring stiffness, the harder the contact. The value $K = 10^6$ N/m was set based on an iterative tuning;
- D is the damping coefficient representing the resistance of the contact damper opposed to motion when the bodies penetrate. The larger the damping coefficient value, the more energy is lost during the collision and the faster the contact vibrations are dampened. The value $D = 10^3$ Ns/m was chosen based on iterative tuning; and
- δ_{BA} is the penetration depth of B into A, defined by (3):

$$\delta_{BA} = (r_b - r_a)\cdot n = r_{BA}\cdot n \quad (3)$$

where r_A and r_B are the position vectors of point A and point B, respectively; r_{BA} is the position of B relative to A; and $\dot{\delta}_{BA}$ is the penetration velocity of B into A, as expressed by (4):

$$\dot{\delta}_{BA} = (v_b - v_a)\cdot n = v_{BA}\cdot n \quad (4)$$

where v_A and v_B are the velocity vectors of points A and B, respectively; v_{BA} is the velocity of B relative to A (i.e., as demonstrated by a reference frame attached to A);

- $\sigma(\delta_{BA})$ is the contact force activation factor, defined as in (5),

$$\sigma(\delta_{BA}) = \begin{cases} 0 & if\ \delta_{BA} \geq 0 \\ \tanh(\frac{\ln 199}{2 \cdot \delta_{TRW}} \cdot |\delta_{BA}| & if \delta_{BA} < 0 \end{cases} \quad (5)$$

where δ_{TRW} is the transition region width, which specifies the region over which the spring–damper force increases to its full value. The hyperbolic tangent that appears in the definition of the function $\sigma\ (\delta)$ smoothly ramps up the force, whereas the penetration depth moves through the transition region ($0 < \delta \leq \delta_{TRW}$). When $\delta = \delta_{TRW}$, at the end of the transition region, the contact force reaches a value equal to 99% of the full stiffness and damping. Similarly, on the rebound, the elastic and viscous components of the force are smoothly decreased back to zero. Reducing the transition region width makes the onset of contact sharper and the model more accurate; however, the solver is forced to take smaller time steps to deal with a discontinuous problem, and the simulation speed decreases. The value δ_{TRW} = 0.1 mm was chosen, as it proved to be a good compromise between speed and accuracy.
- $\gamma\left(\dot{\delta}_{BA}\right)$ is the damping-enabling factor, defined as in (6)

$$\gamma(\dot{\delta}_{BA}) = \begin{cases} 0 & if\ \dot{\delta}_{BA} \geq 0 \\ 1 & if\ \dot{\delta}_{BA} < 0 \end{cases} \quad (6)$$

which deactivates the damping component of the contact force on the rebound.
1. $-n_A$ is the unit normal vector pointing outwards from the surface of A at the point of contact.

The force that B exerts on A (F_{nBA}) is equal and opposite to F_{nAB}, i.e., $F_{nBA} = -F_{nAB}$. Owing to contact friction, the tangential component was neglected in the present analysis. It does not play a considerable role with respect to the switching phases and causes the solver to incur an increased computational burden. This is the reason why the subscript n is hereafter dropped from the symbol of the contact force.

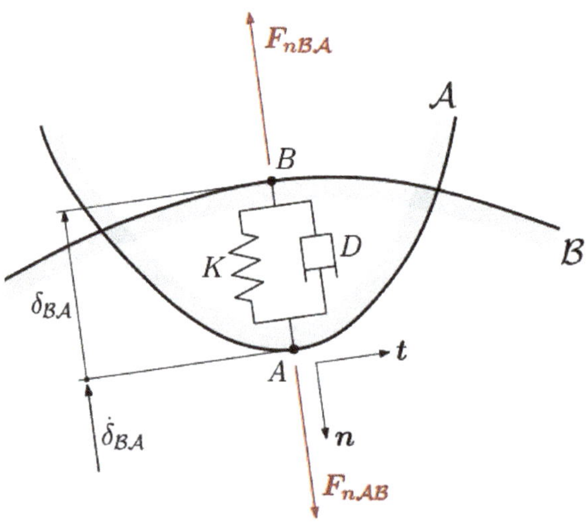

Figure 12. Linear spring–damper for frictionless contact modeling.

2.2.1. Contact between Track Links and Gears

The alternation of the internal links (bushing links) and external links (pin links) shown in Figure 11a forms the roller chain track of the robot. It is evident that to simulate the switching phases, it is necessary to model the contact between the chain bushing/rollers and the teeth of both the sprocket and the idler [19–21]. First, an assumption was made about the motion of the links, i.e., that they are constrained through a planar joint to perform a planar motion on the plane of the two gears so that the model accounts for the contact between the bushing cylinder (shown in red in Figure 11) and the surface of the sprocket teeth, whereas any side contact is excluded. A reference frame is attached to the sprocket as follows:

- The origin is placed at the center of the sprocket denoted by the symbol O_s;
- Axes x_s and y_s are configured as in Figure 13; and
- The z_s axis is perpendicular to the plane of the illustration in Figure 13.

The profile of the surface of each tooth of the sprocket is quite complex and would require detailed geometric features, such as an x–y spline curve, to be faithfully represented.

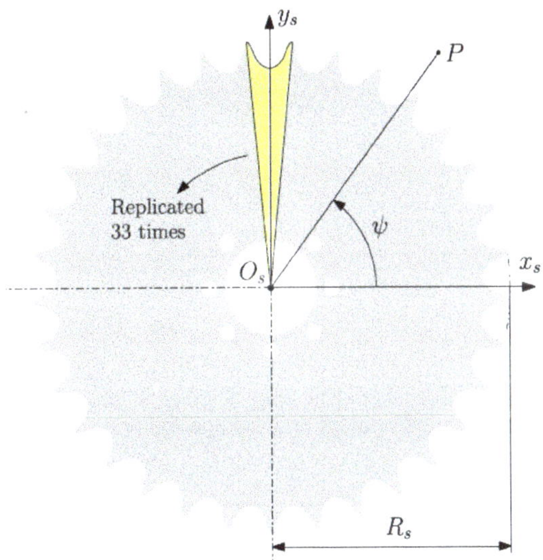

Figure 13. Sprocket gear.

In the present analysis, to simplify the model and reduce the computation time, for each tooth of the sprocket, six points were selected relative to the sprocket center in such a way as to construct an approximated piecewise linear tooth profile sufficiently close to the real profile, as shown in Figure 14. The coordinates ($x^s{}_{i,1}$, $y^s{}_{i,1}$) expressed in the reference system (O_s; x_s, y_s) of the point $P_{i,1}$ $i = 1, \ldots, 6$ used to approximate the first tooth are summarized in Table 1. The six points identify five lines that represent as many surfaces of possible contact; the nth of these lines, with $n = 1, \ldots, 5$, has a tangent unit vector given by (7).

Table 1. Coordinates of the profile points of the first sprocket tooth.

Point	x (mm)	y (mm)
$P_{1,1}$	13.4	140.0
$P_{2,1}$	12.0	140.1
$P_{3,1}$	5.2	125.6
$P_{4,1}$	−5.2	125.6
$P_{5,1}$	−12.0	140.1
$P_{6,1}$	−13.4	140.0

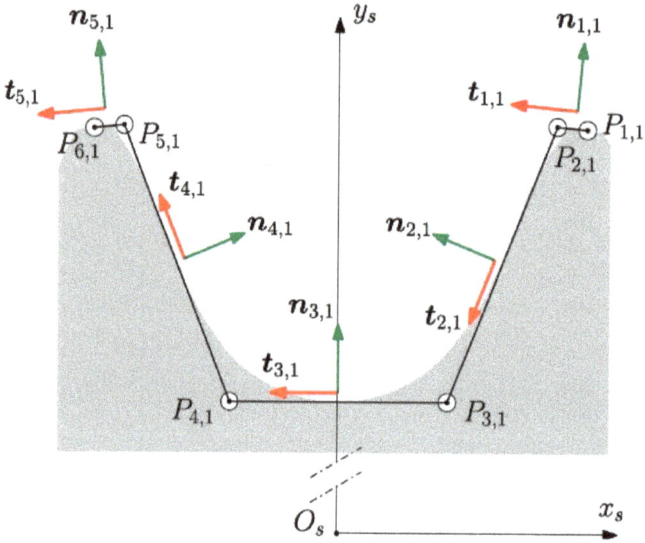

Figure 14. Approximation of the sprocket tooth profile using five straight-line segments.

$$t_{n,1} = \frac{x^s_{n+1,i} - x^s_{n,i}}{\sqrt{\left(x^s_{n+1,i} - x^s_{n,i}\right)^2 + \left(y^s_{n+1,i} - y^s_{n,i}\right)^2}} \cdot x_s + \frac{y^s_{n+1,i} - y^s_{n,i}}{\sqrt{\left(x^s_{n+1,i} - x^s_{n,i}\right)^2 + \left(y^s_{n+1,i} - y^s_{n,i}\right)^2}} \cdot y_s \quad (7)$$

where x_s and y_s are the unit vectors of the x_s and y_s axes, respectively. The unit vector of the tooth profile can be obtained through a 90 degree clockwise rotation of t_n, i.e., in (8).

$$n_{n,1} = \frac{y^s_{n+1,i} - y^s_{n,i}}{\sqrt{\left(x^s_{n+1,i} - x^s_{n,i}\right)^2 + \left(y^s_{n+1,i} - y^s_{n,i}\right)^2}} \cdot x_s - \frac{x^s_{n+1,i} - x^s_{n,i}}{\sqrt{\left(x^s_{n+1,i} - x^s_{n,i}\right)^2 + \left(y^s_{n+1,i} - y^s_{n,i}\right)^2}} \cdot y_s \quad (8)$$

The first tooth profile that is the set composed of the points $P_{i,1}$, $i = 1 \ldots, 6$ and the vectors $t_{n,1}$ and $n_{n,1}$, $n = 1, \ldots, 5$ is replicated around the z_s axis 33 times according to the number of sprocket teeth (see Figure 13), yielding 6× (33 teeth) = 198 points, 5× (33 teeth) = 165 tangent vectors, and as many normal vectors. The ith point of the profile of the kth sprocket tooth, with $i = 1, \ldots, 6$, and $k = 1, \ldots, 33$, is denoted by the symbol $P_{i,k}$. Similarly, the vectors tangent and normal to the nth profile segment of the kth tooth, with $n = 1, \ldots, 5$, are indicated by $t_{n,1}$ and $n_{n,1}$, respectively. Because the coordinates of the characteristic points of the tooth profile remain constant for the sprocket reference system, the tangent and normal vectors, given by Equations (7) and (8), respectively, are computed before the simulation and stored for later repeated calculations to reduce the running time. Figure 15 shows the geometry of the contact between a chain bushing and one of the sprocket teeth. Each inner link of the chain has two bushings (highlighted in red in Figure 11) that come into contact with the toothed surface of the sprocket; as shown in Figure 11b, a reference frame is attached to the single bushing, with the origin placed at the center point (F_1) and axes x_b, y_b, and z_b, with the latter perpendicular to the plane of the diagram.

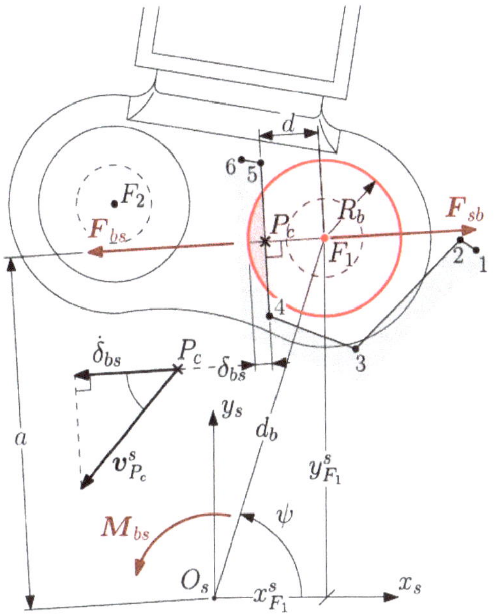

Figure 15. Illustration of the contact between a chain bushing and a sprocket tooth.

The algorithm that calculates forces and torques originated by such interactions is articulated through the following steps:

1. The distance between the sprocket center (O_s) and the bushing center (F_1) is calculated by (9),

$$d_b = \sqrt{{x^s_{F_1}}^2 + {y^s_{F_1}}^2} \tag{9}$$

   where $x^s{}_{F1}$ and $y^s{}_{F1}$ are the x y coordinates of point F_1 with respect to the reference system (O_s; x_s, y_s, z_s).

2. A check is performed to verify whether the bushing in question can collide with the sprocket. Contact can take place if and only if the distance (d_b) in (9) is less than or equal to a prespecified proximity range, i.e., (10):

$$d_b \leq \delta^s_{prox} = R_s + P_{ch} \tag{10}$$

The proximity range ($\delta^s{}_{prox}$) was set equal to the sum of the outer sprocket radius (R_s) and the pitch of the chain (P_{ch}). Inequality (10) is a necessary but not sufficient condition for the sprocket and the bushing to come into contact; if it is verified, the algorithm continues through the next steps; otherwise, the algorithm does not continue. The check in question avoids needless calculations for links that are excessively distant from the sprocket.

3. The sprocket tooth with which the considered bushing is likely to collide is identified based on the angle (ψ) between the vector from O_s to F_1 and the x_s axis of the sprocket reference frame, i.e., (11)

$$\Psi = \arctan(\frac{y^s_{F_1}}{x^s_{F_1}}) \tag{11}$$

Knowing the value of ψ, it is possible to select the index (k^*) of the tooth candidate for contact, along with relevant information, such the tangent vectors ($t^{k*}{}_n$) and the previously determined normal vectors ($n^{k*}{}_n$, n = 1, . . . , 5) (see the Equations (7) and (8)). Thanks to this step, the calculations exposed in the next points are performed only for the candidate tooth, thus preventing the algorithm from cycling over all the sprocket teeth.

4. The positions of the profile points of the $k^*{}_{th}$ tooth with respect to the bushing center (F_1) are computed as (12)

$$L_{i,k*} = \left(x^s_{F_1} - x^s_{i,k*}\right) \cdot x_s + \left(y^s_{F_1} - y^s_{i,k*}\right) \cdot y_s, \qquad i = 1, \ldots, 6 \tag{12}$$

where $x^s{}_{i,k}{}^*$ and $y^s{}_{i,k}{}^*$ are the x and y coordinate, respectively, of the point ($P_{i,k}{}^*$) expressed in the sprocket reference system (O_s; x_s, y_s).

5. For each segment composing the *k*th* tooth profile, a check is performed to verify whether contact could occur with the circle representing the bushing. Contact can occur between the bushing circle and the profile line segment if the orthogonal projection of the center of the circle (F_1) onto the line of the profile falls between the two ends of the segment (see Figure 15). Mathematically, this translates into the following inequality (13),

$$(t_{n,k*} \cdot L_{n,k*})(t_{n,k*} \cdot L_{n+1,k*}) \leq 0 \qquad n = 1, \ldots, 5 \tag{13}$$

where represents dot product, and the vectors $L_{n,k}$ are given by (12). The profile segments for which condition (13) is true are those probably in contact with the bushing, whereas the others attribute no contribution to the interaction in question. Here, for clarity and brevity of the explanation of the following steps, we assume that only the *n*th* line of the tooth profile satisfies the inequality (13); the code specifically written for the task at hand is capable of dealing with completely generic circumstances in which the bushing contacts more than one side of the tooth profile at the same time.

6. The penetration of the bushing (b) into the sprocket (s) is defined in (14):

$$\delta_{bs} = d - R_b \tag{14}$$

where R_b is the bushing outer radius, and d is the distance from the center of the bushing (point F_1) to the n*th line of the k*th tooth profile given by (15)

$$d = \frac{\| (P_{n*+1,k*} - P_{n*,k*}) \wedge L_{n*,k*} \|}{\| (P_{n*+1,k*} - P_{n*,k*}) \|} \tag{15}$$

where $P_{n^*,k^*} = x^s{}_{n^*,k^*}x_s + y^s{}_{n^*,k^*}y_s$ and $P_{n^*+1,k^*} = x^s{}_{n^*+1,k^*}x_s + y^s{}_{n^*+1,k^*}y_s$ are the position vectors of the n*th and the (n* + 1)th point of the sprocket k*th tooth profile, respectively, expressed in the reference frame (O_s; x_s, y_s).

7. The velocity with which the bushing penetrates into the sprocket is calculated as in (16):

$$\dot{\delta}_{bs} = (v^s_b - d\omega^s_b t_{n*,k*}) \cdot n_{n*,k*} \tag{16}$$

where $v^s{}_b$ is the velocity vector of the bushing center (F_1) as demonstrated by the sprocket reference system (O_s; x_s, y_s); and $\omega^s{}_b$ is the angular velocity of the frame (F_1, x_b, y_b) attached to the bushing, as demonstrated by the sprocket reference system (O_s; x_s, y_s). The quantity in parentheses in Equation (16) represents the velocity of the point of contact (Pc) shown in Figure 15, the position vector of which in the coordinate system (O_s; x_s, y_s) is given by (17).

$$P_c = \left(x^s_{F_1} x_s + y^s_{F_1} y_s\right) - d n_{n*,k*} \tag{17}$$

8. The force exerted by the sprocket tooth on the chain bushing (F_{sb}) is computed (18) by substituting the penetration δ_{bs} from (14) and the penetration velocity $\dot{\delta}_{bs}$ from (16) into Equation (1)

$$F^s_{sb} = -\sigma(\delta_{bs})\left(K\delta_{bs} + \gamma\left(\dot{\delta}_{bs}\right) D\dot{\delta}_{bs}\right) n_{n*,k*} \tag{18}$$

where the superscript s indicates that the force vector components calculated through Equation (18) are expressed in the reference system integral with the sprocket; therefore, to apply the force to the bushing, it is necessary to express its components in the coordinate system attached to the bushing itself. This can be achieved using the relative rotation matrix $[R]_{sb}$, as in (19).

$$F^s_{sb} = \begin{bmatrix} R_{xx} & R_{xy} & R_{xz} \\ R_{yx} & R_{yy} & R_{yz} \\ R_{zx} & R_{zy} & R_{zz} \end{bmatrix}^T_{sb} \begin{Bmatrix} F^S_{sb} \cdot x_s \\ F^S_{sb} \cdot y_s \\ F^S_{sb} \cdot z_s \end{Bmatrix} \quad (19)$$

$F^b{}_{sb}$ is then applied to the bushing.

9. The force exerted by the chain bushing on the sprocket tooth is determined by simply inverting the sign in (19), i.e., $F^s{}_{bs} = -F^s{}_{sb}$. The force ($F^s{}_{bs}$) also causes a torque ($M^s{}_{bs}$) with respect to the sprocket center (O_s) (arm denoted by an in Figure 15), given by (20).

$$M^s_{bs} = P_c \wedge F^s_{bs} \quad (20)$$

Both the force ($F^s{}_{sb}$) and the moment ($M^s{}_{bs}$) are applied to the sprocket.

At each time step, the implemented model carries out the calculations related to the contact of the two gears of the track module with each of the 52 chain bushings. Much of the algorithm outlined above, except the first two points, is implemented through a specially written MATLAB® function. The contact of the chain links with the idler is dealt with using a procedure similar to that described for the sprocket.

2.2.2. Contact between Track Links and Rollers

The rollers placed in the undercarriage of the track module shown in Figure 2 are essential to evenly distribute the vehicle's weight over the ground contact area. Therefore the contacts between the rollers and every link composing the chain must be modeled to obtain reliable results. Because the chain links are constrained to perform a planar motion in the plane of the two gears, the contact in question ultimately reduces to the problem of the interaction between a finite line (representing the upper surface of the U channel highlighted in red in Figure 11) and a circle (representing the outer diameter of the roller), as shown in Figure 16. A reference frame is placed on the U channel of each chain link, with the origin positioned at the midpoint (F_3) of the contact surface, the x_u axis and y_u axis parallel and normal, respectively, to the contact surface, and the z_u axis perpendicular to the plane of the illustration in Figure 16. A coordinate system is also associated with the roller, originating at the center point (O_r) and axes x_r, y_r, and z_r. The x and y coordinates of the roller center's reference frame (F_3; x_u, y_u, z_u) are denoted by the symbols $x^u{}_{Or}$ and $y^u{}_{Or}$, respectively. Because the total length of the contact line is 22 mm, three possible contact situations may arise:

1. $x^u{}_{Or} > 11$ mm, in which case the projection of the roller center onto the line of contact falls "beyond" the second edge (E_2), as represented in Figure 16b. Contact takes place at point E_2, and the penetration occurs along the direction of the following unit vector (21),

$$n_u = \frac{x^u_{Or} - x^u_{E2}}{d} x_u + \frac{y^u_{Or}}{d} y_u \quad (21)$$

where $x^u{}_{E2} = 11$ mm is the x coordinate of point E_2 in the U-channel frame (F_3; x_u, y_u, z_u), and d is the distance from the edge point (E_2) to the roller center (O_r) given by (22).

$$d = \sqrt{\left(x^u_{Or} - x^u_{E2}\right)^2 + {y^u_{Or}}^2} \quad (22)$$

The penetration of the roller (r) into the U channel (u) is computed as in (23),

$$\delta_{ru} = d - R_r \quad (23)$$

where R_r = 50 mm is the outer radius of the roller. The penetration velocity of the roller into the U channel is given by (24),

$$\dot{\delta}_{ru} = (v^u_{Or} - d\omega^u_r z_u \Lambda n_u) \cdot n_u \tag{24}$$

where $v^u{}_{Or}$ is the velocity vector of the roller center (O_r), and $\omega^u{}_r$ is the angular velocity of the roller from the perspective of the reference frame attached to the U channel (F_3; x_u, y_u, z_u).

The quantity in parentheses in Equation (24) is the velocity vector of the point of contact. The contact force that the U channel exerts on the roller can be computed by substituting (23) and (24) into Equation (2), as in (25),

$$F^u_{ur} = -\sigma(\delta_{ru})\left(K\delta_{ru} + \gamma\left(\dot{\delta}_{ru}\right)D\dot{\delta}_{ru}\right)n_u \tag{25}$$

where the superscript u indicates that the components of the vector are expressed in the reference frame (F_3; x_u, y_u, z_u), which is why (25) must be premultiplied by the transpose of the relative rotation matrix ($[R]_{ur}$) before being applied to the roller. Accordingly, the U channel undergoes a force ($F^u{}_{ru} = -F_u{}^{ur}$), which also causes a torque about point F_3, given by (26).

$$M^u_{ru} = (x^u_{E2} x_u) \wedge F^u_{ru} \tag{26}$$

2. -11 mm $\leq x^u{}_{Or} \leq 11$ mm, in which case the projection of the roller center onto the line of contact falls between the edges E_1 and E_2, as represented in Figure 16a. The penetration of the roller into the U channel is defined in (27).

$$\delta_{ru} = y^u_{Or} - R_r \tag{27}$$

where R_r = 50 mm is the outer radius of the roller. The velocity with which the roller penetrates the U channel is given by (28),

$$\dot{\delta}_{ru} = (v^u_{Or} + y^u_{Or}\omega^u_r x_u) \cdot y_u \tag{28}$$

where the quantity in parentheses is the velocity vector of the contact point P_c. The contact force that the U channel exerts on the roller can be computed by substituting (27) and (28) into Equation (2), as in (29).

$$F^u_{ur} = -\sigma(\delta_{ru})\left(K\dot{\delta}_{ru} + \gamma\left(\dot{\delta}_{ru}\right)D\dot{\delta}_{ru}\right)y_u \tag{29}$$

where the superscript u indicates that the components of the vector are expressed in the reference frame (F_3; x_u, y_u, z_u), which is why (29) must be premultiplied by the transpose of the relative rotation matrix ($[R]_{ur}$) before being applied to the roller (19). Accordingly, the U channel undergoes a force ($F^u{}_{ru} = -F^u{}_{ur}$), which also causes a torque about point F_3, given by (30).

$$M^u_{ru} = (x^u_{Or} x_u) \wedge F^u_{ru} \tag{30}$$

3. $x^u{}_{Or} < -11$ mm, i.e., the projection of the roller center onto the line of contact, falls "behind" the first edge (E_1). This third case is analogous to the first, the only difference being that the line of action of the force passes through point E_1 rather than E_2; hence, no further explanation is needed.

The interaction between every chain link and the rollers, governed by the three cases listed above, is implemented using a specially written MATLAB® function.

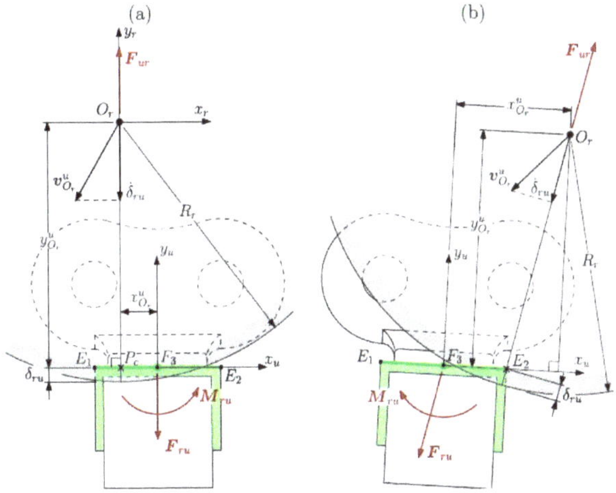

Figure 16. Illustration of the contact between a sprocket tooth and a roller pin. (**a**) the projection of the roller center onto the line of contact falls between the edges E_1 and E_2. (**b**) the projection of the roller center onto the line of contact falls "beyond" the second edge E_2.

2.3. Ground Modeling

2.3.1. Soil Model

To estimate the sinkage and the normal pressure distribution at the vehicle–terrain interface, a model capable of describing the soil behavior when compressed in a normal direction is needed. M. G. Bekker developed the most famous and widely used model, which combines on-field measurements taken with a bevameter (a measuring tool designed by Bekker himself) with an analytical formulation to provide the following pressure–sinkage relationship (31), (also known as the Bekker equation) [22]

$$p = \left(\frac{k_c}{b} + k_\phi\right) y^n \tag{31}$$

where p is the normal pressure exerted by soil, expressed in kPa; y is the vertical sinkage coordinate, expressed in m; b is the smaller dimension of the contact patch, expressed in m; and n, k_c, and k_ϕ are the pressure–sinkage parameters.

Equation (31) is essentially an empirical equation, according to which the parameters k_c and k_ϕ have variable dimensions, depending on the value of the exponent n (equal to kN/m^{n+1} and kN/m^{n+2}, respectively). The exponent n is dimensionless.

Based on a more rigorous approach to soil mechanics and experimental evidence, Reece proposed the following pressure–sinkage relationship (32) for homogeneous soil [23],

$$p = \left(ck_c' + \gamma_s b k_\Phi'\right)\left(\frac{y}{b}\right)^n \tag{32}$$

where c is the cohesion of the terrain, expressed in kPa; γ_s is the weight density of the terrain, expressed in kg/m^3; and n, k'_c, and k'_Φ are the pressure–sinkage parameters for the Reece equation. The constants c, γ_s, k'_c, and k'_ϕ in Equation (32) can be combined as $k''_c = ck'_c$ and $k''_\phi = \gamma_s k'_\phi$ to yield (33).

$$p = \left(k_c'' + bk_\Phi''\right)\left(\frac{y}{b}\right)^n \tag{33}$$

The parameters n, k''_c, and k''_ϕ are derived from a procedure for fitting pressure–sinkage data with the Reece Equation (33); their mean values for various mineral terrains are tabulated in [2]. Because the parameters k''_c and k''_ϕ, unlike what happens with Equation (31), have constant dimensions (kN/m^2 and kN/m^3, respectively). Equation (33) was chosen to characterize the pressure–sinkage relationship of the modeled soil. Furthermore, a viscous linear term was introduced in Equation (33) to account for energy dissipation, yielding (34).

$$p = \left(k_c'' + bk_\Phi''\right)\left(\frac{y}{b}\right)^n + D_g\dot{y} \tag{34}$$

where $\dot{y}$ is the sinkage velocity (expressed in m/s), and D_g is the soil damping coefficient [24], the value of which was set to 500 kPa·s/m.

2.3.2. Contact between Track Links and Ground

The soil is represented as a brick body fixed to the world frame.

As shown in Figure 17, a coordinate system is associated with the ground body, with an origin at point O_g and:

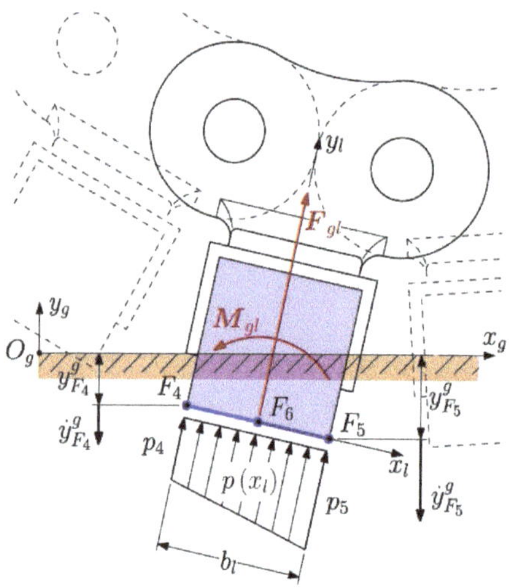

Figure 17. Illustration of the contact between rubber lugs and the ground.

- The x_g axis lying on the upper plane of the brick solid representing the soil and oriented from left to right (fixed horizontal direction);
- The y_g axis normal to the upper plane of the brick solid representing the soil and oriented upwards (fixed vertical direction); and
- The z_g axis emerging from the plane of the diagram in Figure 17 so that the right-hand rule is satisfied.

The dimensions of the body representing the ground are 1.5 m, 0.1 m, and 0.7 m in the x_g, y_g, and z_g directions, respectively. The vertical sinkages of the links that make up the track belt are evaluated with respect to the undeformed upper surface of the ground solid, i.e., the $x_g - z_g$ plane of the soil reference system. As described, each track link has a rubber lug meant to come in contact and exchange forces with the ground. To quantify the sinkage of the lower surfaces of the lugs, two points (F_4 and F_5) are attached to each one of these elements, as represented in the diagram in Figure 17. A reference frame is placed at the midpoint (F_6) of segment F_4F_5, with:

- the x_l axis parallel to the contact surface (shown in blue in Figure 17) and oriented from point F_4 to point F_5;
- The y_l axis normal to the contact surface and oriented towards the inside of the rubber lug; and
- The z_l axis emerging from the plane of the diagram in Figure 17 so that the right-hand rule is satisfied.

The reference frame (F_6; x_l, y_l, z_l) is fixed to the rubber lug.

The mathematical model of deformable soil defined by the non-linear elastic relationship (34) was integrated into the MATLAB® Simscape™ environment to correctly simulate the contact and the behavior of the track on agricultural terrain. In particular, at each time step of the simulation of the multibody model:

1. The vertical sinkages $y^g{}_{F4}$ and $y^g{}_{F5}$, together with the sinkage velocities $\dot{y}^g_{F4}$ and $\dot{y}^g_{F5}$, of the points F_4 and F_5, respectively, are evaluated with respect to the ground reference system (O_g; x_g, y_g, z_g);
2. The normal pressures p_4 and p_5, acting at points F_4 and F_5, respectively, are calculated according to Equation (34), yielding (35) and (36),

$$p_4 = \sigma_4 \left[(k_c'' + bk_\Phi'')\left(\frac{\left|y^g_{F4}\right|}{b_l}\right)^n + \gamma_4 D_g \dot{y}^g_{F4} \right] \quad (35)$$

$$p_5 = \sigma_5 \left[(k_c'' + bk_\Phi'')\left(\frac{\left|y^g_{F5}\right|}{b_l}\right)^n + \gamma_5 D_g \dot{y}^g_{F5} \right] \quad (36)$$

where b_l is the length of the segment F_4F_5, which represents the smaller dimension of the contact patch, equal to 25 mm; σ_4 and σ_5 are the contact-enabling factors, defined as in (37); and γ_4 and γ_5 are the damping-enabling factors, defined as in (38).

$$\sigma_4 = \begin{cases} 1 & if\ y^g_{F4} < 0 \\ 0 & if\ y^g_{F4} \geq 0 \end{cases} \qquad \sigma_5 = \begin{cases} 1 & if\ y^g_{F5} < 0 \\ 0 & if\ y^g_{F5} \geq 0 \end{cases} \quad (37)$$

$$\gamma_4 = \begin{cases} 1 & if\ \dot{y}^g_{F4} < 0 \\ 0 & if\ \dot{y}^g_{F4} \geq 0 \end{cases} \qquad \gamma_5 = \begin{cases} 1 & if\ \dot{y}^g_{F5} < 0 \\ 0 & if\ \dot{y}^g_{F5} \geq 0 \end{cases} \quad (38)$$

The factors σ_i, i = 4; 5, in (37) ensure that the pressure is zero when no contact occurs between the considered lug and the ground. Similarly, the factors γ_i, i = 4; 5, in (38) are essential to activating the damping term of Equation (34) only when the point of the rubber lug, either F_4 or F_5, is moving towards the inside of the ground body;

3. The force (F_{gl}) and the moment (M_{gl}) exerted by the ground on the rubber lug are calculated assuming a linear distribution of the pressure over the contact area (a simplification that has proven reasonable), as expressed in (39).

$$p(x_l) = \frac{p_5 - p_4}{b} x_l + \frac{p_4 x^l_{F5} - p_5 x^l_{F4}}{b} \quad (39)$$

where $x^l{}_{F4} = -bl/2 = -12.5$ mm and $x^l{}_{F5} = bl/2 = 12.5$ mm are the (constant) coordinates of points F_4 and F_5, expressed in the reference frame (F_6; x_l; y_l; z_l) attached to the lug. If the pressure (39) is supposed to be constant along the direction normal to the plane of the diagram in Figure 17, integrating over the contact area yields (40) and (41),

$$F_{gl} = w_l \int_{x^l_{F4}}^{x^l_{F5}} p(x_l) dx_l \mathbf{y}_l = \frac{w_l b_l}{2}(p_4 + p_5)\mathbf{y}_l \quad (40)$$

$$M_{gl} = w_l \int_{x^l_{F4}}^{x^l_{F5}} p(x_l) x_l dx_l \mathbf{z}_l = \frac{w_l {b_l}^2}{12}(p_5 - p_4)\mathbf{z}_l \quad (41)$$

where w_l is the width of the track, equal to 160 mm, i.e., the dimension of the rubber lug along the direction of the z_l axis; and y_l and z_l are unit vectors of the y_l z_l axis of the reference frame attached to the lug, respectively. The force lies along the direction normal to the contact surface, whereas the torque acts in the longitudinal plane of the track module, as represented in Figure 17. The previously calculated force (F_{gl}) and moment (M_{gl}) are applied to the considered rubber lug. The steps listed above are repeated for every link composing the track belt in each simulation time step. The ground experiences the force ($F_{lg} = -F_{gl}$) and the moment ($M_{lg} = -M_{gl}$) as a consequence of the contact with the single lug; however, these produce no effect, as the soil is rigidly connected to the fixed world frame.

2.3.3. Contact between Rear Wheels and Ground

The purpose of modeling half the robot necessarily requires considering the wheeled rear axle's presence and the reconfigurable module on the front. Because the vehicle does not move during the reconfiguration phases, the wheel is rigidly attached to the frame. The method adopted for the implementation of the contact between the wheel and the ground is that proposed by Bekker [25], which is part of the mechanics of a rigid wheel over unprepared terrain and relies on three major assumptions:

- The terrain reaction at all points on the contact patch is purely radial;
- The terrain reaction is equal to the normal pressure beneath a horizontal sinkage plate at the same depth in a pressure–sinkage test, as expressed by Equation (35); and
- No shear stress acts on the wheel–terrain interface.

Figure 18 represents a diagram of the simplified wheel–soil interaction model described here. The center point of the wheel body is denoted by the symbol O_w; it is the origin of the right-hand reference frame integral with the wheel, with x_w, y_w, and z_w axes. The right-hand coordinate system (O_g; x_g, y_g, z_g) is associated with the ground body. To implement the force exchanged between the wheel and, consequently, the vehicle frame and the body representing the ground, the following steps are taken:

1. The maximum sinkage (y_0) is calculated as in (42):

$$y_0 = \sigma_w \left(R_w - \left| y^g_{Ow} \right| \right) \tag{42}$$

where $y^g{}_{Ow}$ is the y coordinate of the wheel center (O_w) in the ground reference (O_g; x_g, y_g, z_g), R_w is the outer wheel radius (equal to 0.17 m), and σ_w is the contact-enabling factor, defined as in (43):

$$\sigma_w = \begin{cases} 1 & if \left| y^g_{Ow} \right| \leq R_w \\ 0 & if \left| y^g_{Ow} \right| > R_w \end{cases} \tag{43}$$

This ensures that the maximum sinkage is zero when there is no contact between the wheel and soil. Furthermore, the y component of the wheel center velocity expressed in the ground reference frame (O_g; x_g, y_g, z_g), $\dot{y}^g{}_{Ow}$, is used to calculate the sinkage velocity ($\dot{y}_0$) as in (44),

$$\dot{y}_0 = \sigma_w \gamma_w \left| \dot{y}^g_{Ow} \right|) \tag{44}$$

where γ_w, similarly to (38), is a factor needed to activate damping only when the wheel moves towards the inside of the ground body.

2. The normal pressure distribution over the contact area is computed according to Equation (34) by (45),

$$p(\theta_w) = (k_c'' + b_w k_\Phi'')(\frac{y(\theta_w)}{b_w})^n + D_g \dot{y}(\theta_w) \tag{45}$$

where θ_w is the contact angle of the wheel, ranging between zero and the maximum value (θ_0) given by (46); b_w is the width of the wheel measured in the direction normal to the plane of the diagram in Figure 18, equal to 0.16 m; and $y(\theta_w)$, $\dot{y}(\theta_w)$ are the sinkage and sinkage velocity functions of the contact angle as expressed by (47).

$$\theta_o = \arccos(\frac{R_w - y_0}{R_w}) \tag{46}$$

$$y(\theta_w) = R_w(cos\theta_w - 1) + y_0\dot{y}(\theta_w) = (cos\theta_w)\dot{y_0} \tag{47}$$

3. The force exerted by the ground on the wheel (F_{gw}) is determined by taking the integral of the pressure distribution (45) over the contact area, as in (48),

$$F_{gw} = 2b_w R_w \int_0^{\theta_0} p(\theta_w)\cos\theta_w d\theta_w y_w \tag{48}$$

where y_w is the unit vector of the y_w axis of the reference frame attached to the wheel. In deriving the force (48), the pressure (p) is assumed to be uniform along the direction of the z_w axis, i.e., normal to the plane of the diagram in Figure 18. The integral in (48) is solved numerically at each simulation time step. The force (F_{gw}) is then applied to the wheel. Because the shear stress at the wheel–terrain interface is supposed to be null, and the pressure (45) is symmetrical with respect to the wheel centerline, no torque contribution is included in this case.

Once again, the reaction force ($F_{wg} = -F_{gw}$) of the wheel on the ground has no effect.

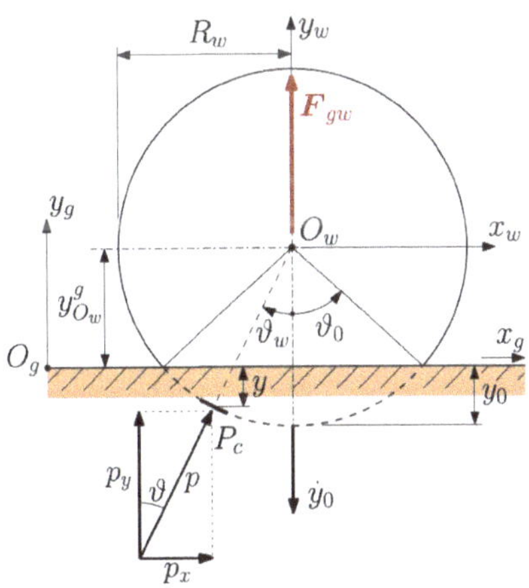

Figure 18. Illustration of the contact between the rear wheel and the ground.

3. Results and Discussion

The normal pressure distribution over the contact area between the track module and the ground is one of the most relevant, if not the principal, outcomes of the analysis and simulation of the implemented model (see Figures 19 and 20).

The reasons why it is important to predict how pressure distributes at the track–ground interface are diverse. First, the resistance opposed by the ground to the motion of the vehicle due to terrain compaction (R_c) is proportional to the work done in compacting the terrain [25], as shown in (49),

$$Work = b\int_0^l \int_0^{y(x)} p(y(x))dydx \tag{49}$$

where b is the width of the track contact area (equal to the width of the rut made by the track), l is the length of the track contact area (equal to the length of the rut made by the track), x is the distance from the front of the contact area, and $y(x)$ is the sinkage of the track in the point of coordinate x along the contact area. Because $R_c l$ = *Work*, Equation (49) yields (50),

$$R_c = \frac{b}{l}\int_0^l \int_0^{y(x)} p(y(x))dydx \tag{50}$$

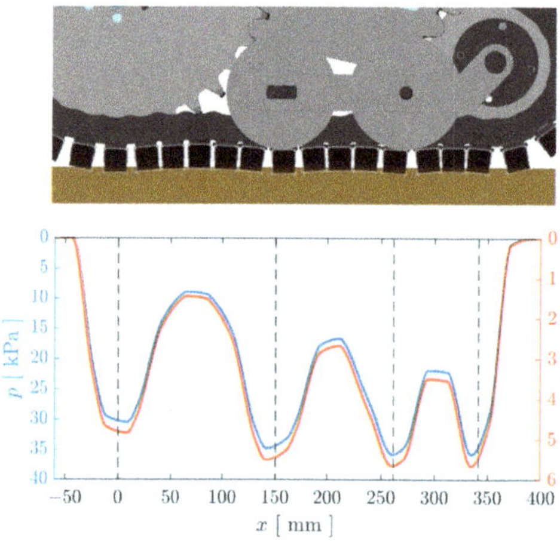

Figure 19. Normal pressure distribution (left y axis) and sinkage (right y axis) of the reconfigurable module in track mode.

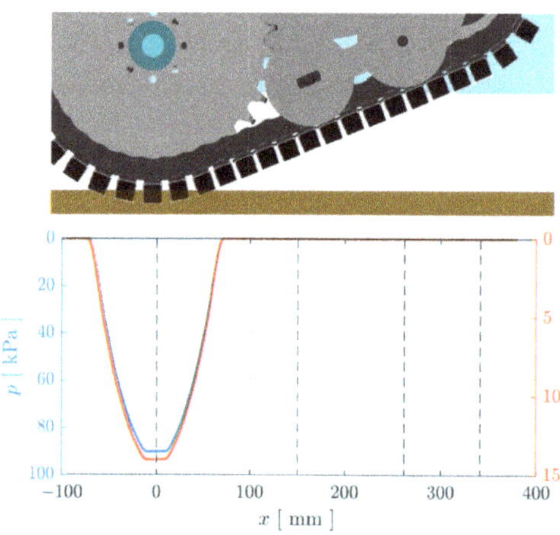

Figure 20. Normal pressure distribution (left y axis) and sinkage (right y axis) of the reconfigurable module in wheel mode.

From which it is evident that the motion resistance depends on the pressure distribution between the track and the ground. Usually, the uniform contact pressure is assumed in the first place so that the integral in (50) can be calculated and the resistance can be assessed. Other pressure distributions, such as trapezoidal or sinusoidal distributions, can be used to obtain R_c from (50); however, all of these lead to first approximation estimates. As stated in [14], assuming a more uniform pressure distribution than the real distribution leads to underestimated results, as in this case, as numerically verified by applying Equation (50) to calculate the rolling resistance both with the pressure distribution obtained from the simulation and a uniform distribution deduced by dividing a quarter of the vehicle weight by the track contact area (see Table 2).

Table 2. Motion resistance for three soil types: values calculated with a uniform pressure and with simulated pressure.

Soil Type (Moisture Content %)	Rolling Resistance (N)	
	Uniform Pressure	Simulated Pressure
Grenville loam (18.2%)	5.86	9.28
Upland sandy loam (44.3%)	3.58	5.10
North Gower clayey loam (52.0%)	11.42	16.83

The shearing of the terrain produces the tractive effort developed by the track or wheel. The torque applied to the tire or the track sprocket initiates a shearing action over the vehicle running gear–soil interface.

To predict vehicle thrust and associated slip, the relation between shear stress and the shear displacement for the considered terrain is needed, as highlighted by (51)

$$F = b\int_0^l \tau(j)dx \tag{51}$$

where F is the tractive effort, and τ and j are the shear stress and displacement, respectively. With the sole objective of qualitatively demonstrating the influence of normal pressure distribution on the generation of tractive effort, among the three most observed types of shear stress–shear displacement semiempirical relationship, we select the exponential function of the form proposed by Janosi and Hanamoto [26] so that Equation (51) becomes (52),

$$F = b\int_0^l (c + ptan\phi)\left(1 - e^{-\frac{j}{k}}\right)dx \tag{52}$$

where c and ϕ are the cohesion and angle of internal shearing resistance of the terrain, respectively, and K is the shear deformation modulus, representing a measure of the magnitude of the required shear displacement.

Equation (52) demonstrates how the tractive effort is affected, among other factors, by the normal pressure distribution (p) over the contact area. Although traction is not subject to the present work, it is notably influenced by the pressure distribution, the shape of which is therefore a matter of absolute interest.

Another reason for being interested in assessing the pressure distribution over the contact area is dictated by the peculiarity of the proposed system itself. Because wheel/track reconfiguration plays a leading role, it is required to monitor the development of contact pressure in response to what the actuator is controlled to do. In other words, it is important to predict the shape of the pressure curve, particularly its peak value, as the target imposed by the control system varies. This aspect is closely associated with the ultimate purpose for which the system is controlled, as the goal is to achieve a track configuration in which the peak pressure and, consequently, soil compaction, are minimized. In Figure 21, a side view of the front module captured during the track mode interval of the simulation cycle (from t_2 = 20 s up to t_3 = 25 s) is shown. As evidenced by the deflection of the chain due to the applied load and consequent deformation of the ground, the accurate modeling detailed in Section 2 has made it possible to achieve a plausible sinkage profile of the track. The rigid links making up the chain that constitutes the track belt can be divided into two groups: those that are in contact both with the terrain and the roadwheels (with this name indifferently denoting the sprocket, the idler, or the rollers) and the others in contact with the terrain only. The shape of the track segments in contact with the roadwheels stems from the shape of the roadwheels themselves.

In contrast, the sinkage of the track in contact with the terrain only is determined by the roadwheel spacing and the chain tension, as well as the assumed pressure–sinkage relationship. The observed sinkage profile resembles the normal pressure graphed in Figure 21; as expected, the distribution is far from uniform, as it presents with peaks under the sprocket, the two rows of rollers, and the idler. The lowest pressure value occurs where the spacing is greatest, i.e., in the section between the sprocket and the first row of rollers.

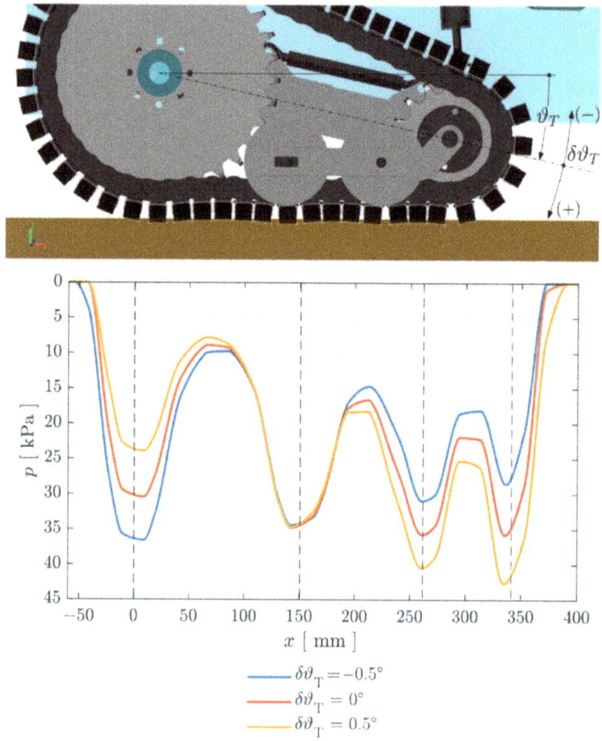

Figure 21. Normal pressure distribution under the reconfigurable module in track mode: effect of control target deviation ($\delta\vartheta_T$).

The system must be capable of moving from an initial posture to a final assigned posture, so the extremal points of the motion path must be defined. The wheel mode is achieved when the idler wheel is sufficiently distant from the ground, i.e., when the inclination angle of the track frame reaches a value of $\vartheta_{T,\text{wheel}} = 10°$. Regarding the track mode, it is difficult, if not impossible, to know a priori what position the mechanism has to assume to produce an optimal pressure distribution over the contact area. Therefore, the definition of the inclination angle ($\vartheta_{T,track}$) corresponding to the track configuration is more formal than substantial, as there is no way to select a single, specific value; instead a range in the proximity of a "geometric" reference position must be examined (53):

$$\theta_{T,track} = \sin^{-1}\left(\frac{R_i - R_s}{l_{1T}}\right) - \delta\theta_T \tag{53}$$

where l_{1T} is the length of the first arm of the track frame (segment T_1T_2 in Figure 2), and R_i and R_s are the outer radii of the idler and the sprocket, respectively. The first term in Equation (53) corresponds to the posture in which, supposing an ideal, perfectly flat, rigid ground, the lower branch of the track belt would be horizontal, and the entire track area would contact the soil. The second term in Equation (53) ($\delta\vartheta_T$), later referred to as control target deviation, can take either positive or negative values close to zero; its effect is demonstrated by the pressure curves in Figure 21:

- The blue curve corresponds to a control target deviation of $\delta\vartheta_T = -0.5°$, which yields $\vartheta_{T,track} = -11.61°$. In this case, the actuator is controlled to reach an overall length of 338.8 mm (the plunger is driven to travel 82.8 mm along the available stroke).
- The orange curve corresponds to a control target deviation of $\delta\vartheta_T = 0°$, yielding $\vartheta_{T,track} = -12.11°$. In this case, the actuator is controlled to reach an overall length of 341 mm (the plunger is driven to travel 85.0 mm along the available stroke).
- The yellow curve corresponds to a control target deviation of $\delta\vartheta_T = 0.5°$, yielding $\vartheta_{T,track} = -12.61°$. In this case, the actuator is controlled to reach an overall length of 343.3 mm (the plunger is driven to travel 87.3 mm along the available stroke).

Observation of the three curves mentioned above indicates the significance of the effect of the control deviation ($\delta\vartheta_T$) on the pressure distribution. According to Equation (53), the higher $\delta\vartheta_T$ is, the harder the control system forces the actuator to push the track frame against the ground; as a consequence, the peak of pressure beneath the idler (and partly that beneath the second row of rollers) increases at the expense of that under the sprocket. This is easily understood by considering that as $\delta\vartheta_T$ increases from $-0.5°$ to $0.5°$, as shown in Figure 21 (from the blue curve to the yellow curve), the actuator drives the rocker slightly further and causes the shock absorber to exert an increased compressing action on the track frame.

As an extreme consequence of this reasoning, if $\delta\vartheta_T$ exceeds a threshold limit, an undesired situation may occur in which the peak pressure beneath the sprocket vanishes; this means that the sprocket is detached from the ground and the vehicle front axle rests on the idler only. If $\delta\vartheta_T$ were too low instead, the part of the track belt beneath the rollers and the idler would barely touch the ground, providing insufficient contact pressure. Accordingly a value or a narrow range of values can be assigned to $\delta\vartheta_T$ to generate a distribution characterized by minimum peak pressure. The highlighted relationship between $\delta\vartheta_T$ and the peak of pressure beneath the idler is difficult to assess, as it vitally depends on knowledge of the exact geometry of the parts involved, as well as the exact value of all the elastic parameters. Therefore, a small mistake in any procedural step could cause the reconfiguration to fail, leading to a lack of contact or the generation of excessive force. The switching system, especially during wheel-to-track reconfiguration must exert a definite force on the external environment rather than reaching a certain position. Therefore, the task for which the system is intended should be approached with a force–control or impedance–control strategy [27,28]. Nonetheless, the position control scheme adopted here is a possible way to approach the question, even if only for the practicality of implementing position feedback in a first embodiment of the system, as opposed to the necessity of obtaining a measured and controlled force. With the implemented model, it is possible to evaluate the effect of variation of the control target, together with other important parameters of the system, on the contact pressure peak between the track and the ground.

3.1. *Effect of Chain Tension*

In Figure 22, the effect of the track belt tension on the contact peak pressure is examined for a control target deviation ($\delta\vartheta_T$) in the range of $-1°$ to $1°$. Three curves are shown:

- The blue curve corresponds to null tension, i.e., the chain is loose;
- The orange curve corresponds to a tension equal to 25% of the vehicle weight, i.e., 0.981 kN; and
- The yellow curve corresponds to a tension equal to 50% of the vehicle weight, i.e., 1.962 kN.

As the track belt tension increases, the contact peak pressure beneath the track decreases. This is beneficial effect, as the tension stiffens the system and therefore helps to redistribute the pressure more evenly. From the point of view of the control target deviation ($\delta\vartheta_T$), the three curves show no substantial difference.

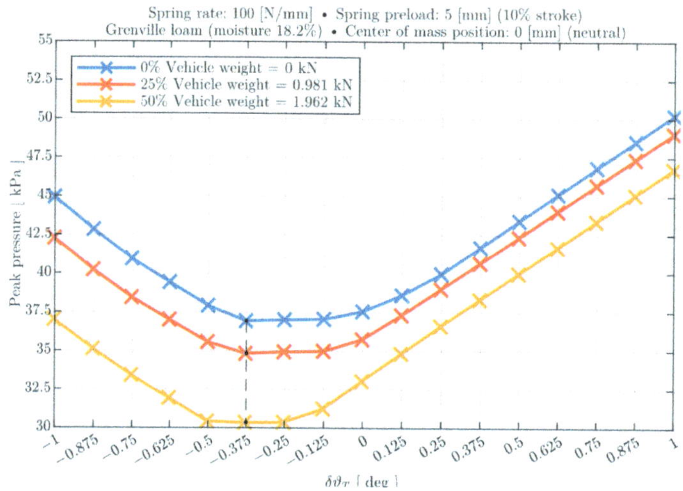

Figure 22. Peak pressure beneath the track module as a function of $\delta\vartheta_T$: chain tension effect.

3.2. *Effect of Shock Absorber Spring Rate and Preload*

Figure 23 shows the effect of the stiffness of the shock absorber spring on the contact peak pressure for a control target deviation ($\delta\vartheta_T$) in the range of $-1°$ to $1°$. Three curves are shown:

- The blue curve corresponds to a shock absorber with a spring rate equal to 40 N/mm;
- The orange curve corresponds to a shock absorber with a spring rate equal to 100 N/mm; and
- The yellow curve corresponds to a shock absorber with a spring rate of 160 N/mm.

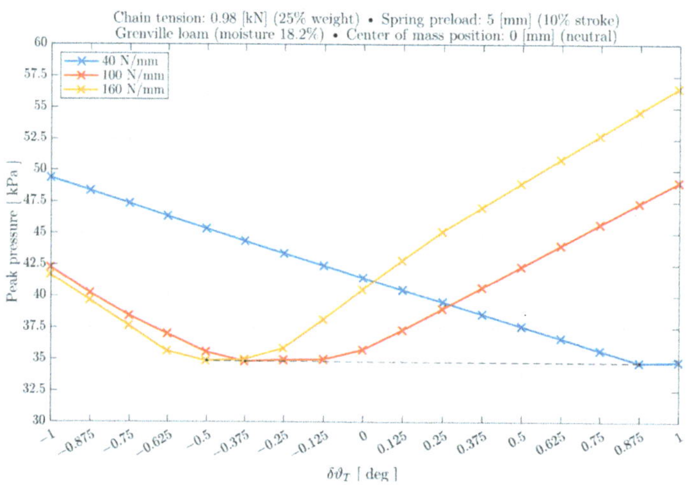

Figure 23. Peak pressure beneath the track module as a function of $\delta\vartheta_T$: spring rate effect.

The above-listed alternatives cover a wide range of elasticity, from a soft spring to the hardest one. All three curves exhibit a minimum point corresponding to different $\delta\vartheta_T$ values. The blue curve reaches the minimum at $\delta\vartheta_T = 0.875°$, whereas for the orange and the yellow curves, the minimum peak pressure is obtained at $\delta\vartheta_T = -0.375°$ and δϑT = $-0.5°$, respectively. Such differentiations are easily understood by considering that the shock absorber and the ground can be thought of as two springs in series, with stiffnesses of k_s and k_g, respectively, so the overall stiffness is defined as in (54)

$$k = \frac{k_s k_g}{k_s + k_g} = \frac{k_g}{1 + \frac{k_g}{k_s}} \tag{54}$$

Equation (52) shows that if a certain force is to be developed on a given terrain, the softer the spring (the lower k_s), the further the control system has to push the idler to achieve a minimum peak pressure (the higher the $\delta\vartheta_T$). An comparable result is obtained by varying the preload of the spring, as shown in Figure 24.

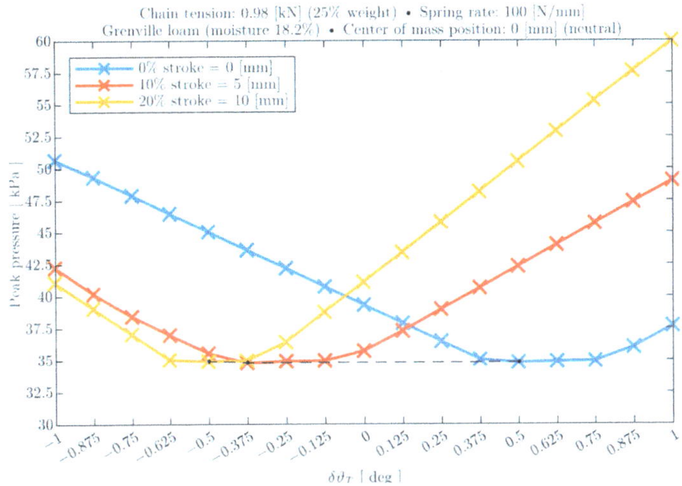

Figure 24. Peak pressure beneath the track module as a function of $\delta\vartheta_T$: spring preload effect.

Figure 25 shows the effect of the position of the center of gravity of the vehicle frame on the contact peak pressure.

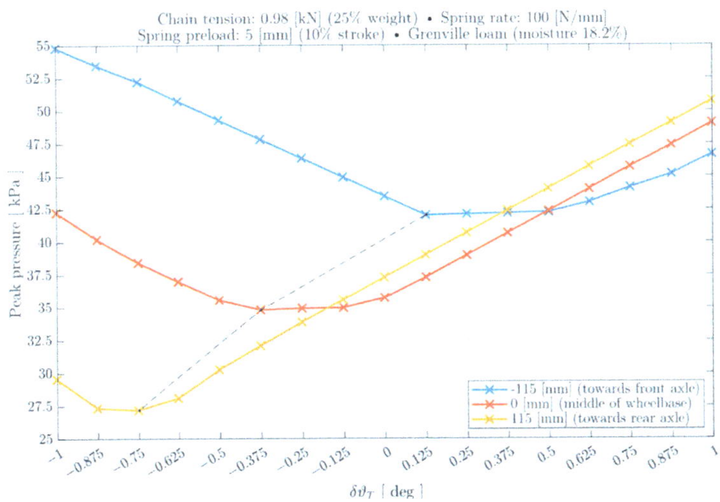

Figure 25. Peak pressure beneath the track module as a function of $\delta\vartheta_T$: center of gravity position effect.

Such effect is important because it is reasonable to expect that the distribution of mass on the vehicle (specifically the payload) will likely change during normal operation. The control target deviation ($\delta\vartheta_T$) is given for the x axis and varies from $-1°$ to $1°$, whereas for the y axis, it is the peak of the pressure distribution between track and ground. Three curves are shown:

- The blue curve corresponds to a frame with the center of gravity displaced by 115 mm towards the front axle with respect to the middle point of the wheelbase;
- The orange curve corresponds to a frame with the center of gravity in the middle of the vehicle wheelbase; and
- The yellow curve corresponds to a frame with a center of gravity 115 mm towards the rear axle with respect to the middle point of the wheelbase.

The vehicle wheelbase is 768 mm. The distance separating the frame center of mass from the middle of the wheelbase for the blue and yellow curves is set as equal to 15% of the wheelbase. First, a substantial increase in the minimum peak pressure is observed when passing from the yellow to the blue curve (from about 27 kPa to 43 kPa); this is easily explained by the increased fraction of the vehicle weight supported by the front wheel/track reconfigurable axle. For the same reason, as the center moves towards the front axle, the control target deviation necessary to achieve the minimum peak pressure increases: equal to $-0.75°$ for the yellow curve, $-0.375°$ for the orange curve, and $0.125°$ for the blue curve.

4. Conclusions

The development of a multibody model to study the pressure distribution of a reconfigurable wheeled/tracked vehicle on soft terrain is reported herein. The main aim of this research was the construction of a reconfigurable wheel/track locomotion system to be installed on board a grape-transporting robot to switch between a half-tracked configuration and a wheeled configuration depending on the soil type and conditions. The objective was to realize a running gear capable of actively adapting to the environment to attain improved performance while preserving the porous structure of the soil, which is often damaged by the intensive use of heavy machinery, to the greatest extent possible.

The scope of the presented multibody model is to simulate the wheel-to-track and track-to-wheel switching phases. As detailed in Section 2, the goal in question was achieved through the accurate modeling of the multiple interactions between each of the links of the chain that constitute the track belt and the various elements that are part of the system, i.e., the sprocket wheel, the idler wheel, the undercarriage rollers, the soil, etc.. The chain that constitutes the track is modeled as groups of rigid links constrained by revolute joints to account for the flexibility of the track itself. A specifically designed algorithm was implemented through MATLAB® functions to compute the contact forces applied to the track elements. The model used to characterize the soil behavior when compressed in a normal direction is based on Reece's non-linear elastic pressure–sinkage relationship. Furthermore, the contact between the rear wheel and the ground was accounted for to ensure the possibility of assessing the effect of the position of the center of mass of the vehicle frame on the pressure distribution over the ground contact patch.

As mentioned in Section 3, the proposed multibody model allows for realistic simulation of the vehicle reconfiguration from half-tracked mode to wheeled mode and vice versa and to obtain useful information regarding both the controlled actuation unit and the contact of the track module with the ground. The main outcome is the normal pressure distribution between the track module and the ground; when in tracked configuration, it presents a non-uniform shape with peaks directly beneath the sprocket wheel, the idler wheel, and the rollers. In particular, the pressure distribution in track mode is strongly influenced by the action exerted on the system by the controlled actuator; the further the plunger is pushed forward, the more the pressure peak under the idler increases. Numerous simulations of the system prove that there exists a value, or at least a range of values, of the control target that yields the optimal ground pressure distribution characterized by the minimum peak. Herein, we demonstrated that such an optimum range is highly affected by the spring rate and preload of the shock absorber, as well as the position of the vehicle center of mass, with a much more minimal influence of the chain tension.

Author Contributions: Conceptualization A.G. and E.d.M.; methodology, A.G.; software, A.G. validation, A.G., E.d.M., N.I.G. and K.I.; formal analysis, A.G. and E.d.M.; investigation, A.G. and E.d.M.; resources, N.I.G. and K.I.; data curation, A.G.; writing—original draft preparation, A.G. and N.I.G.; writing—review and editing, A.G., E.d.M. and N.I.G.; visualization, A.G. and E.d.M.; supervision, N.I.G. and K.I.; project administration, E.d.M.; funding acquisition, N.I.G. All authors have read and agreed to the published version of the manuscript.

Funding: This research received no external funding.

Conflicts of Interest: The authors declare no conflict of interest.

References

1. Muro, T.; O'Brien, J. *Terramechanics: Land Locomotion Mechanics*; A. A. Balkema Publishers: London, U.K., 2005.
2. Wong, J.Y. *Terramechanics and Off-Road Vehicle Engineering: Terrain Behaviour, Off-Road Vehicle Performance and Design*; Butterworth-Heinemann: Oxford, UK, 2010.
3. Nicolini, A.; Mocera, F.; Somà, A. Multibody simulation of a tracked vehicle with deformable ground contact model. *Proc. Inst. Mech. Eng. Part K J. Multi-Body Dyn.* **2019**, *233*, 152–162. [CrossRef]
4. Slättengren, J. Utilization of ADAMS to Predict Tracked Vehicle Performance. SAE Transactions. 2000; Volume 109, pp. 1–7. JSTOR. Available online: http://www.jstor.org/stable/44650730 (accessed on 1 January 2022).
5. Solis, J.M.; Longoria, R.G. Modeling track–terrain interaction for transient robotic vehicle maneuvers. *J. Terramech.* **2008**, *45*, 65–78. [CrossRef]
6. Mocera, F.; Somà, A.; Nicolini, A. Grousers Effect in Tracked Vehicle Multibody Dynamics with Deformable Terrain Contact Model. *Appl. Sci.* **2020**, *10*, 6581. [CrossRef]
7. Mahalingam, I.; Padmanabhan, C. A novel alternate multi-body model for the longitudinal and ride dynamics of a tracked vehicle. *Veh. Syst. Dyn.* **2021**, *59*, 433–457. [CrossRef]
8. Ma, Z.D.; Perkins, N.C. A Super-Element of Track-Wheel-Terrain Interaction for Dynamic Simulation of Tracked Vehicles. *Multibody Syst Dyn.* **2006**, *15*, 347–368. [CrossRef]
9. Lamandé, M.; Greve, M.H.; Schjønning, P. Risk assessment of soil compaction in Europe—Rubber tracks or wheels on machinery. *Catena* **2018**, *167*, 353–362. [CrossRef]
10. Grazioso, A.; Galati, R.; Mantriota, G.; Reina, G. Multibody Simulation of a Novel Tracked Robot with Innovative Passive Suspension. In *Mechanisms and Machine Science*; Niola, V., Gasparetto, A., Quaglia, G., Carbone, G., Eds.; Advances in Italian Mechanism Science. IFToMM Italy 2022; Springer: Cham, Switzerland, 2022; Volume 122. [CrossRef]
11. Grazioso, A.; Di Maria, E.; Margheriti, S.; Gallone, R.; Prete, G. Modellazione della Sospensione Formula SAE SRT15 in MSC Adams/Car. November 2014. Available online: https://www.mscsoftware.com/it/page/modellazione-della-sospensioneformula-sae-srt15-msc-adamscar (accessed on 1 January 2022).
12. Visconte, C.; Cavallone, P.; Carbonari, L.; Botta, A.; Quaglia, G. Design of a Mechanism with Embedded Suspension to Reconfigure the Agri_q Locomotion Layout. *Robotics* **2021**, *10*, 15. [CrossRef]
13. Jiang, H.; Xu, G.; Zeng, W.; Gao, F.; Chong, K. Lateral Stability of a Mobile Robot Utilizing an Active Adjustable Suspension. *Appl. Sci.* **2019**, *9*, 4410. [CrossRef]
14. di Maria, E. Running Gear Size Selection of a Wheel/Track Reconfigurable Grape Transporting Vehicle by FEM Analysis. Ph.D. Thesis, Graduate School of Life Science and Systems Engineering, Kyushu Institute of Technology, Fukuoka, Japan, 2021.
15. Cheli, F.; Pennestrì, E. Cinematica e Dinamica dei Sistemi Multibody. Casa Editrice Ambrosiana: Milano, Italy, 2009; Volume 1.
16. Flores, P.; Lankarani, H.M. *Contact Force Models for Multibody Dynamics*; Springer: Berlin/Heidelberg, Germany, 2016. [CrossRef]
17. Lankarani, H.M.; Nikravesh, P.E. Continuous contact force models for impact analysis in multi-body systems. *Nonlinear Dyn.* **1994**, *5*, 193–207. [CrossRef]
18. Nikravesh, P.E. *Planar Multibody Dynamics: Formulation, Programming with MATLAB®, and Applications*; CRC Press: Boca Raton, FL, USA, 2018.
19. Omar, M.A. Chain Drive Simulation Using Spatial Multibody Dynamics. *Adv. Mech. Eng.* **2014**, *6*, 378030. [CrossRef]
20. Pedersen, S. Simulation and Analysis of Roller Chain Drive Systems. Ph.D. Thesis, Department of Mechanical Engineering, Technical University of Denmark, Lyngby, Denmark, 2004.
21. Pedersen, S. Model of contact between rollers and sprockets in chain-drive systems. *Arch. Appl. Mech.* **2005**, *74*, 489–508. [CrossRef]
22. Bekker, M.G. *Introduction to Terrain-Vehicle Systems*; University of Michigan Press: Ann Arbor, MI, USA, 1969.
23. Reece, A.R. Principles of Soil-Vehicle Mechanics. *Proc. Inst. Mech. Eng. Automob. Div.* **1965**, *180*, 45–66. [CrossRef]
24. Rubinstein, D.; Coppock, J.L. A detailed single-link track model for multi-body dynamic simulation of crawlers. *J. Terramechanics* **2007**, *44*, 355–364. [CrossRef]

25. Wong, J.Y. *Theory of Ground Vehicles*; John Wiley & Sons: Hoboken, NJ, USA, 2008.
26. Janosi, Z.; Hanamoto, B. The analytical determination of drawbar-pull as a function of slip for tracked vehicles in deformable soils. In Proceedings of the 1st International Conference on the Mechanics of Soil-Vehicle Systems, Torino, Italy, 12–16 June 1961; pp. 707–736.
27. Villani, L.; De Schutter, J. Force Control. In *Springer Handbook of Robotics*; Siciliano, B., Khatib, O., Eds.; Springer Handbooks; Springer: Cham, Switzerland, 2016. [CrossRef]
28. De Schutter, J.; Van Brussel, H. Compliant Robot Motion II. A Control Approach Based on External Control Loops. *Int. J. Robot. Res.* **1988**, *7*, 18–33. [CrossRef]

machines

MDPI

Review

Conventional, Heuristic and Learning-Based Robot Motion Planning: Reviewing Frameworks of Current Practical Significance

Fatemeh Noroozi [1,*], Morteza Daneshmand [1,†] and Paolo Fiorini [2]

[1] Norwegian Institute of Bioeconomy Research (NIBIO), 1431 Ås, Norway; morteza.daneshmand@nibio.no
[2] Department of Engineering for Innovation Medicine, University of Verona, 37134 Verona, Italy; paolo.fiorini@univr.it
* Correspondence: fatemeh.noroozi@nibio.no
† M.D.'s affiliation is previously with Institute of Technology, University of Tartu, 50411 Tartu, Estonia.

Abstract: Motion planning algorithms have seen considerable progress and expansion across various domains of science and technology during the last few decades, where rapid advancements in path planning and trajectory optimization approaches have been made possible by the conspicuous enhancements brought, among others, by sampling-based methods and convex optimization strategies. Although they have been investigated from various perspectives in the existing literature, recent developments aimed at integrating robots into social, healthcare, industrial, and educational contexts have attributed greater importance to additional concepts that would allow them to communicate, cooperate, and collaborate with each other, as well as with human beings, in a meaningful and efficient manner. Therefore, in this survey, in addition to a brief overview of some of the essential aspects of motion planning algorithms, a few vital considerations required for assimilating robots into real-world applications, including certain instances of social, urban, and industrial environments, are introduced, followed by a critical discussion of a set of outstanding issues worthy of further investigation and development in future scientific studies.

Keywords: motion planning; mobile robots; social robots; self-driving cars; humanoid robots

1. Introduction

Artificial Intelligence (AI)-based utilities and techniques have their most prevalent applications in robotics, where agents possessing computing capabilities gather, analyze, and use information from their environment to make and act upon informed decisions. Despite the usual lack of sufficiently reliable information about their environment, a robotic agent has to make rational decisions or at least be capable of autonomously making a sequence of moves that optimize the expected values of indicators of certain functional criteria to an acceptable extent, while minimizing the costs incurred. The environment may change during the course of executing a motion plan, which motivates the incorporation of the capability for modifying the plan based on the aforementioned variations. Taking into account the fact that it is not feasible to compute reactions to all possible scenarios in advance, it is common to instead perform replanning computations and optimizations based on the specific scenario at hand. This may include various factors, such as model and tracking errors, as well as different types of constraints, according to the sensor feedback and user inputs [1,2].

An autonomous (e.g., robotic) system might involve multiple rigid parts connected using numerous types of links, hinges, and joints. They may be either dependent or independent in terms of their movement [3]. The navigation of an autonomous system requires localization, perception and cognition, and motion planning [4], with the latter constituting the central focus of this review. A schematic representation of the role of motion planning within the workflow of AI-based systems is shown in Figure 1, together with more specific illustrations of its place in the structure governing the performance of robotic systems and self-driving cars. The figure was inspired by the definitions and representations provided in [1,5].

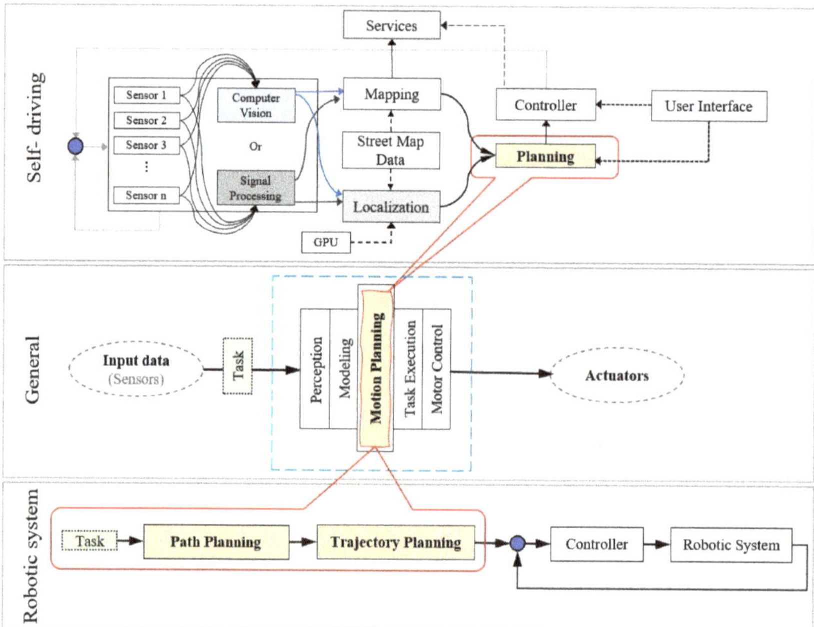

Figure 1. A schematic representation of the place of motion planning within a general AI-based pipeline (**middle**), with more specific illustrations of self-driving cars (**top**) and robotics systems (**bottom**). The figure was inspired by the conceptions and diagrams reported in [1,5].

Motion planning in cluttered environments is important in various applications relying on autonomous systems. A basic motion planning problem for an autonomous system with the shape known from the outset deals with a 2D or 3D environment, which may include obstacles with known shapes and positions, where the aim is to determine a collision-free route for the autonomous system from an initial position and orientation (pose) to the final one, if a route exists [3].The pose needs to be estimated throughout the operation using, e.g., image processing techniques, to rectify the path, and correspondingly the control commands, based on the actual situation. Various types of robotic manipulators, such as spatial and tentacle-like, require motion planning, with considerable uncertainties arising from size limitations, compliance and collision-avoidance requirements, and sensing unreliability [6]. Thus, finding a suitable path depends on a variety of experimental factors, such as the presence of static or dynamic obstacles and other types of uncertainties, and must be handled using relevant control strategies, e.g., Dynamic Programming (DP) [7].

Perception of the surrounding environment by the robot may be achieved using various sensory modalities, such as sound signals. In this configuration, a set of audition capabilities enable the robot to locate a sound source, among other functionalities. For example, a microphone array installed on a robot may be utilized to measure the angle of arrival, thereby estimating the location of the sound source, where the front–back ambiguity may be alleviated based on the robot's motion, even in noisy and reverberant environments, potentially with moving sources, using, e.g., Kalman filtering [8].

The results include a set of locations, i.e., via points, at which the autonomous system may be positioned, as well as rule sets intended to constrain the movements of the autonomous system between them [9]. Generally, path planning aims to find the intermediate points, such that the final pose can be reached within the shortest possible period, while taking into account an environmental and spatial model [10].

Path planning may be performed either in the joint space or in the operating space. It is limited to kinematic planning, and the criteria for evaluating performance mostly concern the computational costs, meaning that it does not account for the physics-based requirements of the problem [11]. Therefore, subsequently, trajectory planning needs to be performed, which incorporates the time information into the plan, to determine the dynamic aspects of the motion, as well as the time at which each of the via points are passed. The resulting system is often called a Kinodynamic motion Planning (KP) algorithm.

The importance of such concepts stems from the fact that inertial forces and torques are affected by the robot's acceleration, where jerk, which is mathematically represented as the derivative of acceleration, has a direct impact on mechanical vibrations. Thus, a trajectory planning method may concentrate on optimizing one or several factors, including execution time, energy consumption, jerk, route length, safety, computation time, and smoothness [12], and this can be achieved using, e.g., an artificial bee colony algorithm for a local search, as well as evolutionary programming for refining the path [13]. While planning a trajectory, the relevant constraints, including temporal, physical, geometric, and friction-related, as well as joint limits and equilibrium requirements, where applicable, should also be satisfied [2,10,14]. Mathematically, a broad category of classical problems concern motion planning under differential constraints [15].

Motion planning considering dynamics can be significantly challenging; since, due to the lack of a local planner, the only primitive toward the state space comes from the controls being forward-propagated. This has been alleviated in the literature through employing propagation of random controls in each iteration, utilizing tree Sampling-Based Motion Planning (SBMP) techniques, leading to asymptotic optimality, i.e., ultimate optimality in terms of the cost of the path, as the number of points grows toward infinity. Nevertheless, the convergence of the resulting approaches to suitable trajectories may be unacceptably slow [16,17].

Adjustable links and joints can be considered to model trajectories for improved flexibility against unexpected dynamic obstacles, such that the trajectory can be conveniently modified by creating multiple obstacle trajectories based on the relationships between them and using queries along the link chains [18].

Despite the above geometric perspectives, practical use cases may entail technical complexities of different sorts, including functional uncertainties and noise during sensing and control, probabilistic incompleteness, nonholonomic constraints, optimality requirements such as time or vibration minimization [19], dynamic constraints, error recovery concerns, route caching preconditions, and a lack of essential information about the scene and obstacles. Therefore, during the last four decades, researchers from several communities, such as computational geometry and robotics, have been engaged in developing suitable frameworks for this purpose [3].

A certain aspect of the problem concerns robot autonomy, which is necessary for enhancing the robot's safety and its performance, in terms of delivering insightful information about the scene it is supposed to navigate through [20]. This, in turn, gives rise to extra factors and requirements regarding the fact that the robot has to be capable of capturing and processing some or all of the information for determining its actions.

Another challenge underlying the task of motion planning arises from the intricacy of achieving onboard implementation of the procedures required for handling the dynamics of using the robot's computational resources [21], as well as the lack of sensory capabilities that are required for a consistent performance across different situations, i.e., where the robot needs to make inferences about possible valid configurations [22]. Additionally, different configuration parameters may affect the quality of the trajectory returned by a motion planning algorithm, and these are usually adjusted beforehand [23].

As a common practice, the problem of motion planning is converted from the Cartesian space (workspace) to the higher-dimensional configuration space, which is often referred to as the C-space. The goal is to reduce the problem of checking for 3D collisions between spatial objects to that of simpler point-like tests. Nevertheless, making the required map in the C-space may become intractable computationally and memory-wise with the higher dimensions of the C-space. This may stem from factors such as high numbers of Degrees of Freedom (DoF), or from the time-dependency of the environment, e.g., because of obstacles dynamically appearing, disappearing, or moving, and requirements for being incorporated online into the C-space [24].

Other challenges may also be presented by the necessity of finding a probabilistic compromise between safety and speed in collision checking at the edges of the sensing field [25]. It should, however, be noted that forcing a robot to perform as fast as possible may undermine its accuracy and repeatability. Therefore, it is vital to plan trajectories that can be performed quickly and highly efficiently, and which are smooth enough not to require excessive acceleration or extreme actions, which would otherwise damage mechanical parts such as actuators or control modules, because of, e.g., extreme vibrations [12].

Once a trajectory has been planned, a low-level controller needs to be employed so the autonomous system follows it, while maintaining the safety and compensating for tracking errors using, e.g., sum-of-squares programming for upper-bound limits of the tracking error [26].

Furthermore, completeness and exactness are considered essential properties of any motion planning algorithm [27]. Probabilistic completeness refers to the property of an algorithm, where the probability of it being unable to return a solution if at least one exists decays to zero as the number of samples grows to infinity [28]. Similarly, a motion plan needs to be singularity-free, meaning that the end-effector should not reach regions of the workspace where it is not able to move along one or more of its DoF or where it could move irrespective of the input joint positions. It is worth noting that avoiding local coordinates may be useful for tackling singularities or ambiguities [29].

Generally, a navigation package consists of two main modules: a global planner that finds the optimal path according to prior knowledge of the environment and the static obstacles, and a local planner that refines the path, to avoid collisions with dynamic obstacles [30].

Motion planning algorithms involve loosely coupled multilayered designs that aim at dynamically feasible solutions, which are found quickly enough to be suitable for online planning. Lower-dimensional projection spaces are used for approximating a lead path from the start to the goal configuration. The results are then employed as initial guesses for the second planner, considering the dynamic constraints and possible cases of Inevitable Collision State (ICS). This is also fundamental when it comes to leveraging the computational cost by performing calculations up to a reasonable horizon at each interval, and focusing on regions where high-quality solutions are deemed most likely to be found [21].

SBMP, which will be further discussed in the upcoming sections, is highly reliable, in terms of avoiding obstacles, and is therefore useful in cases where the task must be performed while inferring and trying to satisfy a set of dynamic constraints. Instead of specifically and numerically programming a robot, planning its motions and performing a certain task may be accomplished through learning the actions and constraints. For example, expert demonstrations may be used within a Demonstration-Guided Motion Planning (DGMP) framework for tasks involving, e.g., holding an object in an upright or horizontal position while moving it or cleaning a surface [31].

Motion planning algorithms have been extended and refined to improve their efficiency, reliability, energy and time consumption, computational cost, physical feasibility, robustness, and stability. However, most types of robot are not yet capable of working in close contact, while cooperating, and collaborating with each other and with human beings. Thus, efficiently incorporating them into practical contexts involves unresolved challenges. For example, social robots should behave seamlessly, smoothly, and similarly to humans. This necessitates additional performance indicators measuring their aptness for social interactions.

Motion planning is of great value in, e.g., search and rescue, Simultaneous Localization And Mapping (SLAM) exploration, teleoperation, inspection, construction, and architecture. For example, it could be utilized for searching collapsed buildings that are considered either hazardous or unreachable. Applications of robots in home and office environments also abound. Given the fact that teleoperation may become temporarily unavailable, it is essential to equip a robot with autonomy, where distances are predicted using regression and ranking with a steer function, and the two-point boundary value problem is solved using, e.g., non-linear parametric models benefiting from constant-time inferences. Each motion planning problem aimed at optimizing a robot's performance needs to be set up based on the mission's representative indicators. For example, when it comes to SLAM, the goal is to determine a sequence of motions that yield a set of viewpoints suitable for making reasonably accurate models, suffering the least from known obstacles, such as holes [32,33].

The remainder of this review is organized as follows: First, in Section 2, relevant studies and taxonomies are reviewed, to clarify the differences and advantages of the present survey in comparison with similar surveys. Then, a brief overview of a set of motion planning techniques widely applied to robots in the existing scientific and technical literature is presented in Section 3. Next, we review some of the most fundamental aspects of their practical applicability to different types of robot in Section 4. Finally, a list of concluding remarks and hints for possible future research and development directions are outlined in Section 5.

Overall, the contributions of the present review are as follows:

- By analyzing the frequency of use of conventional, heuristic, and learning-based algorithms, we demonstrate the current transition, whereby more and more studies have recently embraced learning-based models instead of older heuristic approaches, while the conventional techniques are still being utilized at a similar ratio as before;
- We provide two different categorizations of these algorithms, one based on being conventional or heuristic, and the other based on being global or local. To the authors' knowledge, our categorizations are the most comprehensive proposed in the literature to date, thereby enabling a sound and quick judgment of the role, importance, and relevance of each technique for a given application;
- By identifying and studying the most common motion planning pipelines of current practical significance, we exclude algorithms that are no longer being actively employed as of the time of writing or that have not yet proven useful in real-world applications. Considering the large amount of literature being published on the topic of motion planning every year, the materials presented throughout this survey are of essential value for readers whose purpose is to grasp an overall understanding of the field, as opposed to deeply analyzing the mathematical and theoretical backbones.

2. Related Work and Taxonomy

The taxonomies developed to classify motion planning techniques include different types. In this section, a selection of the most recent reviews will be briefly discussed, to provide a broad insight into the taxonomies, as well as their relevance and applicability.

From the broadest perspective, collision-avoiding motion planning techniques belong to two main categories; namely, SBMP methods and trajectory optimization approaches [34]. Historically, each motion planning method can considered classical or heuristic, the former being represented by RoadMap (RM), cell decomposition, subgoal networks, and potential fields, as popular examples. These are deemed to be capable of handling simplified motions and environments only [17]. It is widely perceived that the classical methods are simple, but their main disadvantages are being computationally expensive or intractable, and they may fail to tackle uncertainties appropriately [4].

Motion planning has always been dependent on computer science, aiming at devising solutions that ensure a good level of reliability, accuracy, and precision. Nevertheless, one of the predicaments in motion control is that, in modeling, different frameworks are required for motion planning and obstacle avoidance [27].

Assuming that proper safety considerations are incorporated, robots may improve various factors, such as mobility, transportation convenience, and efficiency in people's daily lives. This applies to numerous types of robots and self-driving vehicles. More clearly, a system needs to be planned to work smoothly, safely, conveniently, and efficiently, e.g., in terms of energy consumption, within a dynamic environment that may include objects, agents, and live beings. Actions are either decided or rethought based on the feedback received about the present state. For an informative review of motion planning and control techniques aimed at urban environments, see [35]. By discussing and contrasting the associated motion models and the extent of applicability of the methods for certain types of environments and computational resources, we provide insights into possibilities for improving the system design.

Onboard sensors and networks communicating information between the agents and the environment, as well as between the agents themselves, can play essential roles in automating the functionalities of intelligent vehicles. This gives rise to the possibility, and in a broader context, the necessity, of having agents cooperate for the application to be viable within complex real-world environments involving Vulnerable Road User (VRU)s. For a comparative study of motion planning frameworks for intelligent vehicles, see [36].

A survey of 3D path planning techniques for robots was provided in [10]. Algorithmic motion planning techniques dealing with simple problems, as opposed to heuristic practical techniques, were reviewed in [3], which offers a theoretical analysis of the methods from the point of view of computational and combinatorial geometry, as applied to surfaces and curves, as well as the worst-case asymptotic bounds.

From another perspective, the problem of motion planning can be viewed focusing on whether the autonomous system targeted is holonomic or nonholonomic. In simple words, if the number of controllable DoF an autonomous system possesses is the same as its overall number of DoF, then it is considered holonomic, and if not, it is nonholonomic. Due to the extremely high level of complications arising in nonholonomic motion planning, it has attracted significantly more attention. A survey of motion planning and obstacle avoidance techniques for nonholonomic mobile robots was reported in [27].

Due to the recent increasing interest in autonomous driving, numerous studies have surveyed the state of the art of self-driving vehicles. An informative review of motion planning techniques for on-road intelligent vehicles and transportation systems was provided in [1]. It first explored the related concepts from the point of view of motion planning for mobile robots and then clarified the nuances and necessary considerations for employing the inferences in the context of autonomous vehicles.

Another survey of motion planning and control techniques for intelligent vehicles in urban environments was reported in [35], where different strategies toward mobility model and various environment structures were discussed, along with the relevant implications regarding the computational aspects, aiming at providing a more insightful perspective on the role of choices made at the system design level.

In [36], a review of motion planning for self-driving cars was presented, covering overtaking maneuvers. A study of the feasibility and optimality of different methods under various environmental conditions was provided in [37].

Randomized path planning algorithms may be either single-query or multiple-query. The methods falling under the former category try to solve a single path planning problem quickly, without performing or requiring preprocessing, while the methods belonging to the latter intend to solve multiple path planning problems in the same environment. Thus, in the case of multiple-query path planning methods, data structures containing information from preprocessing could help achieve a faster performance [38]. The basic Rapidly-exploring Random Trees (RRT) construction procedure is summarized in Algorithm 1, where the EXTEND operation is illustrated in Figure 2. Moreover, the RRT-connect is shown in Algorithm 2.

Algorithm 1: Basic RRT construction. This algorithm was taken from [38].

```
Function BUILD_RRT(q_init)
    T.init(q_init);
    for k = 1 to K do
        q_rand ←RANDOM_CONFIG();
        EXTEND(T, q_rand);
    end
    return T
────────────────────────────
Function EXTEND(T, q)
    q_near ←NEAREST_NEIGHBOR(q, T);
    if NEW_CONFIG(q, q_near, q_new) then
        T.add_vertex(q_new);
        T.add_edge(q_near, q_new);
        if q_new = q then
            return Reached
        else
            return Advanced
        end
    return Trapped
```

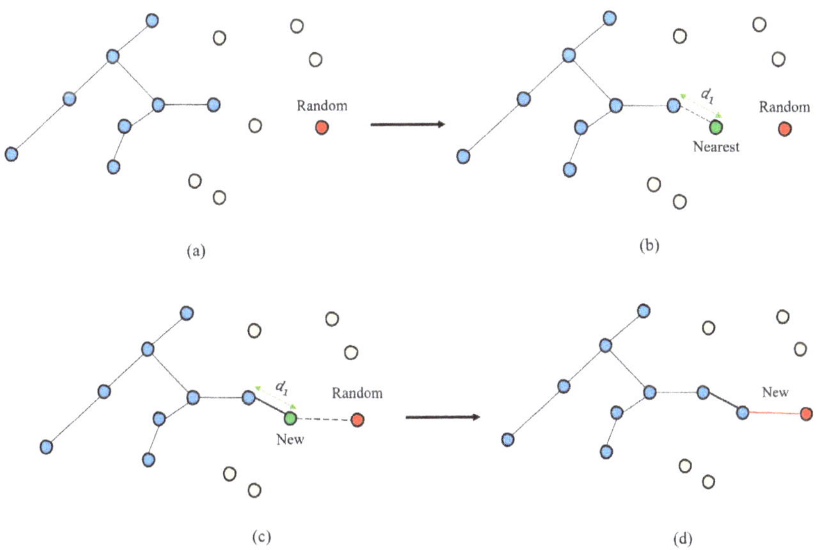

Figure 2. The four steps of the EXTEND operation. The search tree is shown in its original form, i.e., at the time of starting to find a path to a random point (**a**). Then the nearest neighbors are determined (**b**), followed by making a new node (**c**), and finally, finding the random point (**d**).

Algorithm 2: The RRT-Connect algorithm, which was taken from [38].

```
Function CONNECT(T, q)
    repeat
        S ←EXTEND(T, q);
    until (S ≠ Advanced);
    return S
---------------------------------------------
Function RRT_CONNECT_PLANNER(q_init, q_goal)
    T_a.init(q_init); T_b.init(q_goal);
    for k = 1 to K do
        q_rand ←RANDOM_CONFIG();
        if EXTEND(T_a, q_rand) ≠ Trapped then
            if CONNECT(T_b, q_new) = Reached then
                return PATH(T_a, T_b);
        SWAP(T_a, T_b);
    end
    return Failure
```

Optimality guarantees for SBMP algorithms under differential constraints were theoretically evaluated and compared in [15], concentrating on drift-less control-affine dynamical models, where the concept of converging with probability has been introduced, resulting in a higher flexibility and convergence rate bounds, in contrast to sure converging.

The nature-inspired approaches, fuzzy logic, Neural Network (NN)s, and hybrid methods utilized in heuristic-based path planning were reviewed in [4]. Another general survey of motion planning was provided in [12].

A review of the 3D motion planning techniques widely utilized for underwater, ground, and aerial robots was conducted in [10], where the exploration mechanisms were divided into five categories and compared in terms of their implementable areas and time efficiencies.

The advantages and disadvantages of path planning techniques from the point of view of special robotic operations were reviewed in [39]. Several approaches to analyzing the safety and feasibility in path planning were investigated and compared, considering their role in contexts involving underwater vehicles, Unmanned Aerial Vehicle (UAV)s, Autonomous Guided Vehicle (AGV)s, and industrial robots, with applications in mountainous areas, warehouses, and production lines. Examples concerning AGVs in medical and industrial scenarios were specifically discussed, followed by a detailed analysis of a certain robotic spray painting problem.

As aforementioned, the technical aspects of motion planning problems and pipelines have been well studied from various perspectives in the existing literature, which seems to have overlooked the requirements associated with the practical assimilation of these technologies and strategies. For example, much more could be achieved in the sense of fast, safe, and reliable communication, cooperation, and collaboration with other agents and human beings present in the same environment. The taxonomy employed in the present survey, therefore, concentrates on the application side, aiming at shedding some light on additional considerations for, e.g., social, mobile, and humanoid robots, as well as real-world frameworks involving, e.g., self-driving, object manipulation, and multi-robot teams working in cooperation. From a technical viewpoint, on top of the path planning and trajectory optimization components of motion planning algorithms, typical online replanning capabilities are reviewed, along with other topics with a functional impact, including Demonstration-Based Learning (DBL), scene uncertainties and dynamics, stability, and computational feasibility.

As the focus of the present review is on incorporating motion planning techniques into practical contexts, the articles' suitability and importance were judged according to their significance in the field, as measured based on their number of citations, which reflects their impact on how the relevant applications are evolving. Therefore, the basic list of studies for this survey consisted of 200 articles, being compiled by searching the keyword "motion planning" on Google Scholar, to ensure that no prior bias or subjective preference would affect the type of motion planning algorithms being picked up and explored.

The period of publication for the above list was set to cover the period from 2015 to 2019, so that each article could be deemed to have had enough time to collect citations from the time of publication, but, at the same time, would not be too old to present notions with current practical implications. Nevertheless, some of the papers were excluded from the basic list, due to their lack of direct relevance to present applications with real-world usage.

However, some older studies have also been included where necessary for presenting theoretical foundations. On the other hand, newer studies have also been covered where a similar search for articles published after 2019 indicated significantly high numbers of citations within a short period. More clearly, these papers were considered to have made a considerable contribution, in terms of either introducing a new approach or showing great potential to emerge as one.

A summary of the motion planning algorithms reviewed in this survey, along with their important characteristics, is provided in Table 1, which also lists selected references for each approach, together with the frequency of the articles covered based on the publication year. Similarly, a graphical overview is shown in Figure 3. Moreover, the most significant criteria involved in judging the suitability of each motion planning algorithm are schematically listed in Figure 4.

Figure 3. A graphical summary of the motion planning algorithms reviewed in the present survey.

Table 1. A summary of the most essential properties of motion planning algorithms discussed throughout this survey, and their selected references, along with the number of articles covered for each publication year. The criteria used while judging the advantages and disadvantages of each algorithm are illustrated in Figure 4.

Category	Description	Advantages	Disadvantages	Selected References
		Converntional		
SBMP	Analyzing alternative trajectories based on path length, by low-dispersion or random sampling, using, e.g., RRT, RMP, FMT, FMT*, RRT*, RRT#, RRT^X, RRT-Connect, RRT*-Connect, RRG, CBF-RRT, LBT-RRT, RRT2.0, BTT, BIT*, or MC-RRM	Not requiring a model of the environment, high reliability in avoiding obstacles, as well as inferring and satisfying dynamic constraints, the possibility of achieving asymptotic optimality and probabilistic completeness, simplicity, and quick performance	Slow convergence, unreliability in cases involving complex motions or environments, runtime computational heaviness and intractability, especially in the presence of obstacles, and poor safety because of randomness	[6,8,15,17,28,32, 38,40–67]
RM	Multi-query search using a pose graph to find feasible motions based on path length or accumulated pose uncertainty, using, e.g., cell decomposition, a sub-goal network, PRM, PRM*, visibility graph, or spanners	Simplicity, probabilistic completeness, asymptotic optimality, the possibility of parallelization of collision detection over the RM edges, and reliability in tackling dynamic obstacles and constraints	Poor performance for complex motions or in unstructured environments, high runtime computational cost and intractability, weak safety due to randomness, and the possibility of excessive growth or failure in the presence of uncertainties	[68–71]
Potential field	Guiding the motion based on repulsion from obstacles and attraction toward the goal, with respect to the distances, thereby handling dynamics, as well as utilizing velocity and orientation information	Real-time performance, simplicity, high safety, asymptotic optimally	Probabilistic incompleteness, high computational cost and intractability, unreliability against uncertainties, narrow paths, and dynamic environments, and the possibility of local minima	[72,73]
		Heuristic		
Bio-inspired	Using multi-query search algorithms, such as bee colony, GA, PSO, BBO, or the bat algorithm	Handling discrete functions and achieving convergence for complex problems, and ease of combining with other methods	Probabilistic incompleteness, high computation cost, slowness, weakness against dynamics and probabilities, and the possibility of local minima	[13,74–82]
NN	Making search trees by predicting the cost, based on dynamics, heuristics, and obstacles, using ANNs, or LSTM	High generalizability	Challenging dataset preparation and the possibility of becoming stuck in local minima	[16,83–86]
		Learning-based		
DBL	Using human requests based on a kinetic model to learn near-optimal heuristics representing the operator's choice from path planning, and making cost maps and features, possibly combined with RL or IRL	Handling unseen or unstructured environments by generalizing from primitive motions to more sophisticated ones, without explicit task constraints, and predicting human motions for improved speed	Requiring large datasets and powerful online adaptation	[31,87–96]
DL	Automatic selective sampling using, e.g., OracleNet, DeepSMP, MPNet, GANs, or CoMPNetX	Theoretical worst-case guarantees, real-time performance, and handling higher dimensionalities, as well as cluttered or unfamiliar environments	Challenging requirements for training datasets	[97–101]
RL	Sequential decision-making through LPV state-space representations, using, e.g., PPO, TD3, EASE, SAC-based methods, DDPG, or other dual architectures	Actively exploring the domain, handling unfamiliar environments and high-dimensionalities, and tackling dynamic obstacles accurately by identifying their boundaries and distances to the robot	Requiring high numbers of samples, training difficulties due to convergence and robustness issues caused by ambiguities between Cartesian and joint spaces, continuous workspaces, and redundant DoF	[102–114]

					Frequency by year					
<2015	2015	2016	2017	2018	2019	2020	2021	2022	2023	Total
17	29	37	40	28	31	7	11	8	5	213

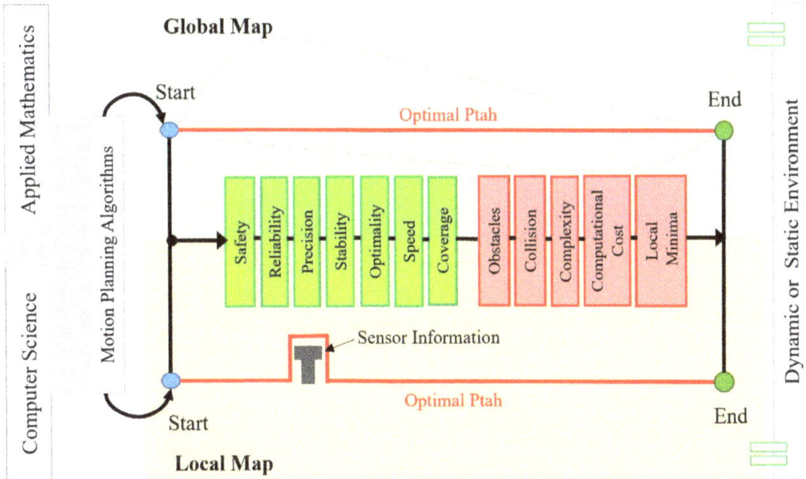

Figure 4. A schematic list of the main criteria considered while evaluating the suitability of motion planning algorithms, where green and red boxes indicate positive and negative factors, respectively. Computer science and applied mathematics constitute the theoretical backbones of motion planning, where the and the goal is to find the optimal path from the starting point to the goal using a local or global planner, within a dynamic or static environment.

3. Motion Planning Pipelines

Motion planning is part of a more general problem referred to as Task and Motion Planning (TMP), which typically takes place under partial observability. A symbolic decision tree is employed, which may undergo additional branching based on new observations. Independent optimization of the symbolic trajectories is performed using approximate path costs, followed by picking up the best policy and optimizing a joint trajectory tree [115,116]. In what follows, the essential aspects of the most common approaches to motion planning in robotics will be concisely discussed.

3.1. SBMP

Not requiring an explicit model of the obstacles present in environments, SBMP algorithms utilize a collision-checking module, which analyzes the practicability of alternative trajectories. Through connecting a set of collision-free configurations in space, they build a graph of feasible trajectories [40]. On the other hand, the set of connected edges thought to lead to the shortest path from the initial to the final configuration is called the shortest-path-to-goal sub-tree [41]. The continuity of the algorithm depends on the configuration space's topological complexity [117].

A fundamental strength of data-driven SBMP algorithms arises from their relatively low reliance on explicit mathematical models, which are not available from the outset in most practical scenarios [42].

RRT is a well-known example of this type of motion planning algorithm. It has engaged hundreds of researchers, each of whom has tried to verify and demonstrate its applicability to an application of a certain kind. Consequently, investigating the performance of RRT has been the subject of a substantial number of studies, where making modifications aiming at improving the speed, fluency, accuracy, and stability were among the main factors in evaluating the resulting functionality [43].

SBMP methods mainly depend on uninformed path sampling procedures, whose examples include low-dispersion and random sampling. Thus, the information required for collision testing needs to be accessible to the planner. However, this might not be feasible in many scenarios [44].

Occupancy maps are commonly associated with SBMP methods. Nevertheless, they suffer from the deficiency that the resulting cost estimation is limited to the search heuristic utilized over the unknown environment. Thus, only intermediate objectives along the frontiers can be considered [118].

Multi-Component Rapidly-exploring RoadMap (MC-RRM) is a SBMP motion planning method that aims to optimize a learned cost metric. It is useful for incrementally computing a motion plan according to DBL [31], which will be further discussed in Section 3.5.

RRT* and Rapidly-exploring Random Graph (RRG) are examples of algorithms that present slight differences with RRT. They lead to asymptotic optimality but are slower than RRT. Performing continuous interpolation between the fast RRT and its asymptotically optimal alternatives, with a cost function representing the path length, brings the ability to quickly solve path planning problems, while maintaining asymptotic optimality. Taking advantage of a sub-graph of the RM returned by RRG and an auxiliary lower-bound graph, a parameter may be utilized and adjusted to specify how similar to RRG or RRT* the behavior of the algorithm needs to be.

For example, Lower Bound Tree (LBT)-RRT is an efficient single-query SBMP method taking advantage of the above strategy. It offers an asymptotically near-optimal solution within an approximation factor $1 + \epsilon$ of the optimal solution. It makes use of an approximation factor, which if set to 1 or infinity, makes the algorithm behave like RRG or RRT, respectively, and for any other value, yields a compromise between the speed of RRT and the path quality, i.e., asymptotic optimality, of RRT*. The LBT-RRT algorithm applies to problems of DoF ranging from 3 to 12 [45]. Moreover, probabilistic belief networks resulting from pose SLAM could be utilized as belief RM for optimal collision-free trajectory planning by searching through the pose graph, regardless of the map reference frame, and considering the accumulated robot pose uncertainty as one of the criteria [46,68].

On the other hand, running multiple independent RRTs could be considered in a context where the aim is to achieve asymptotic convergence to a minimal Collision Probability (CP) [6]. For complex motion planning problems that are in high-dimensional configuration spaces or for problems that involve a high number of obstacles for which it is computationally expensive to check collisions, lazy dynamic programming recursion [47] on a prescribed number of samples that have been probabilistically determined helps deal with the problem, through dividing it into smaller sub-problems, as well as allowing skipping collision-checks when evaluating local connections, thereby making the problem more tractable. Moreover, to further improve efficiency, the incremental growth of the tree of candidate paths can be one-pass, i.e., it could take advantage of a heapsort technique to determine the appropriate sample point systematically, and consequently, grow only in the outward direction, thereby avoiding backtracking over sample points that have already been evaluated.

To improve mathematical flexibility, aiming at making it possible to impose a lower bound on the convergence rate, instead of almost sure convergence, the concept of convergence in probability can be considered. The Fast Marching Tree (FMT)* algorithm is a viable example of such a strategy, which can offer asymptotic optimality, while still converging on the optimal solution faster than alternatives such as Probabilistic RoadMap (PRM)* and RRT*. Under a set of considerations for the configuration spaces and tuning parameters, a lower bound of the order $O\left(n^{-1/d+\rho}\right)$ can be achieved for the convergence rate, with n, d, and ρ being the number of sampled points, the dimension of the configuration space, and an arbitrarily small constant, respectively. It has been shown that the FMT* algorithm is asymptotically optimal, even in cases where a non-uniform sampling of the configuration space is taken into account, a general cost function is used instead of the prevalent path length, or the connections are determined according to the number of nearest neighbors, e.g., by using the k-Nearest-Neighbor (kNN) algorithm, instead of considering a predetermined connection radius [28].

Probabilistic SBMP algorithms, such as PRM and RRT, have proven extremely successful, mainly because of their favorable mathematical advantages, including probabilistic completeness and asymptotic optimality, as well as their practical usefulness. Their property of being probabilistic stems from the notion that they find a path through connecting independently and identically distributed (i.i.d.) random points within the robot's configuration space. However, they also entail drawbacks. For example, they may not be able to guarantee safety and do not allow offline computations, which could otherwise help alleviate the runtime computational intensiveness.

To counteract the above phenomena caused by the randomness of probabilistic SBMP methods, one may think of ways to incorporate deterministic characteristics into the system. Using deterministic low-dispersion sampling sequences, along with a suitable configuration of tuning parameters, PRM leads to deterministic asymptotic optimality. The convergence rate, which is represented by a sub-optimality factor, may be controlled through an upper bound, in terms of the $l2$-dispersion of the sampling sequence and PRM's connection radius. It is worth mentioning that $l2$-dispersion stands for how well a set of points covers the space. Geometrically, this represents the largest Euclidean ball that does not touch any of the points.

The above framework is associated with a space and computational complexity that could be arbitrarily close to $O(n)$, which is the theoretical minimum, where n is the number of points in the sequence. It is possible to achieve similar performances, i.e., with reasonable success rates and path costs, on differentially-constrained problems [48].

In dynamic environments, performing offline computations to obtain prior information may be impossible. Therefore, the ability to carry out fast replanning, while maintaining asymptotic optimality under a single-query framework, is crucial. As soon as it is found that the appearance or motion of an obstacle will make it collide with the robot while following the shortest-path-to-goal sub-tree, replanning needs to be performed to modify the graph, such that collisions are avoided. Older single-query replanning algorithms performed the task in sch a way that the branches disconnected as a consequence of this operation are pruned away, followed by regrowing parts or all of the graph.

However, being able to transfer information quickly is essential for a timely reaction in dynamic environments involving unpredictable changes in the presence or location of obstacles. Thus, the computation time could be considerably enhanced by trying to modify and repair the existing graph instead. This may be achieved by making use of a fast graph rewiring cascade, which incorporates new information and accordingly updates the shortest-path-to-goal sub-tree.

Moreover, the graph and the shortest-path-to-goal sub-tree could be built in the state-space of the robot itself, to ensure that the motion complies with the kinematics of the robot, and gradually improve throughout the navigation. The RRT^X algorithm uses this strategy. It offers probabilistic completeness. RRT^X does not perform global and local planning separately, but it is still suitable for real-time performance in dynamic environments, due to its capability of quickly reacting to the emergence and movement of obstacles.

A major factor affecting the competence of an algorithm to quickly react to changes in obstacles is the information transfer time, which is the amount of time the algorithm needs to transfer information regarding a decrease in the cost associated with a node to the relevant parts of the graph. The time required to transfer information regarding an ϵ-cost reduction to a graph of size n is $O(n \log n)$, $\Omega\left(n\left(\frac{n}{\log n}\right)^{1/D}\right)$, and $\omega\left(n \log^2 n\right)$ for RRT^X, RRT* and RRT$^\#$, respectively. These are all asymptotically optimal single-query algorithms.

RRT and RRT* have a $O(\log n)$ amortized iteration time in static environments. This is $\omega\left(\log^2 n\right)$ in the case of RRT$^\#$. For RRT^X to achieve a $O(\log n)$ amortized iteration time in static environments, each node can bear a set of $O(\log n)$ expected neighbors and the graph may retain ϵ-consistency for a prescribed ϵ [41].

Combining the virtues of some of the aforementioned randomized algorithms, new ones such as RRT*-connect have been devised for solving single-query problems using a bidirectional search. They are faster than RRT* but , despite RRT-Connect, are capable of converging to the theoretical optimum, i.e., offering asymptotic optimality. Such a strategy is particularly useful for a faster performance in applications such as autonomous driving in complex environments [49].

Asymptotic optimality or near-optimality of a broad category of SBMP algorithms depends on the connection radius around a configuration q, which is expected to be connected to all configurations inside the ball. To accelerate the functionality, in contrast to algorithms operating based on connecting to the k nearest neighbors, the foregoing notion could be adapted to non-Euclidean spaces. This is often seen in practical applications [50].

Alternatively, motion planning may be performed using the Monte Carlo algorithm, based on criteria such as Shannon entropy or the standard deviation of the estimated belief for the source location [8].

Despite the RRT algorithm, RRT2.0 is asymptotically optimal. Avoiding Boundary Value Problem (BVP), which is required to establish connections between the nodes within a tree for algorithms such as RRT*, FMT*, and Batch-Informed Trees (BIT)*, it utilizes forward propagation, where every point in the configuration space is represented in an augmented fashion through supplying the associated cost-to-come, which represents the aggregation of the costs incurred by the edges constituting the path. RRT2.0 offers optimality under the assumption that the objective function and the dynamics are Lipschitz-continuous, where a trajectory with positive clearance from the obstacles and a piecewise-constant control function can be approximated [51].

When it comes to SBMP algorithms in the Euclidean space with uniform random sampling, e.g., in the case of methods such as BoTtleneck Tree (BTT), FMT*, and RRG, as well as PRM-based methods, given n samples and d dimensions, connection radiuses less than a critical value proportional to $\Theta\left(n^{-1/d}\right)$ lead to failure of the algorithm in returning asymptotically (near) optimal results. For greater values, the probability of success will be at least $1 - O(n^{-1})$. Additionally, instead of $\Theta(\log n)$ induced by a radius of order $\left(\frac{\log n}{n}\right)^{1/d}$, only $\Theta(1)$ connections will be required [52].

Among notable KP algorithms, Kinodynamic motion Planning by Interior-Exterior Cell Exploration (KPIECE) has a relatively high success rate in finding time-optimal solutions, and Synergistic Combination of Layers of Planning (SyCLoP) yields comparatively high levels of power-optimality [11].

Although numerous SBMP strategies offer asymptotic optimality, they may suffer from a slow convergence, which could be alleviated through intelligence contributed by certain heuristics aimed at faster exploration. More clearly, once a tentative solution is found, further sample points can be taken from a collection of points commonly referred to as the "informed set". The heuristic needs to be chosen paying due attention to providing a reasonable estimate of the cost associated with the solution. Further focusing the search is possible based on metrics such as cost-to-come, thereby determining more limited point sets such as the "relevant region" proposed in [53].

3.2. Trajectory Optimization

Although SBMP algorithms may be able to offer feasible solutions to the problem under high-dimensional configuration spaces, the resulting solutions might entail undesired properties, such as unnecessary movements or jerk, which is further worsened in the presence of tight navigation constraints or sparsely scattered obstacles. As a remedy for the foregoing shortcomings, by analyzing the problem from a probabilistic viewpoint, smooth continuous-time trajectories can be obtained as samples from a Gaussian Process (GP) [119].

Trajectory optimization may target criteria such as the time required to follow the trajectory, velocity, jerk, curvature, or a combination of these, while the step-time is optimized simultaneously [46].

Complex motion planning tasks for many-DoF robots can be handled using functional gradient algorithms, which are capable of finding an optimal, smooth trajectory within a search space, while also taking into account geometric requirements such as smoothness. They rely on a finite parameterization of the trajectories represented by a list of waypoints.

It is worth noting that the performance may suffer from drawbacks such as an extremely small step size or a high number of iterations, which can be alleviated in adaptively lower dimensions through avoiding waypoints by representing trajectories as linear combinations of kernel functions within Reproducing Kernel Hilbert Spaces (RKHSs).

In the above context, optimization may be performed using different kernels such as Basis splines (B-splines), as well as Laplacian or Gaussian Radial Basis Functions (RBFs) [120]. Splines are useful for reducing a problem's dimensions and the number of constraints, where solving the problem with a receding horizon helps handle modeling errors and environmental uncertainties [121]. The occupancy gradients may then be optimized considering an update rule defined based on a stochastic process [34,122]. As a relevant point, an online adaptation of the aforementioned plans based on experience requires differentiation [23].

Numerous adaptive bio-inspired strategies such as the bat algorithm [74] also exist, which may be utilized for optimal motion planning.

In higher-dimensional state spaces, trajectory optimization can also be combined with other approaches to locomotion, such as graph search, which leads to high flexibility, dynamical feasibility, and probabilistic optimality [123].

Piecewise linear paths may be further improved in terms of curvature, tangential or acceleration continuity, and boundedness, which are important for avoiding undesired phenomena such as vibrations, as well as confined chord error, by utilizing interpolation and performing optimization for the Curvature Variation Energy (CVE). Furthermore, additional computational costs can be alleviated using mixed linear and quartic Bezier segments [124].

Various criteria may need to be considered and accounted for while formulating a trajectory optimization problem. These include path continuity, infinite-norm velocities, the joints' maximum angles and velocities, and the possible drifts caused by redundancies. For example, the constraints arising from the requirement that the speeds amount to zero at the end of a task may be dealt with using frameworks such as Infinity-Norm Velocity Minimization (INVM), Repetitive Motion Planning (RMP), or linear variational inequalities.

3.3. *Bio-Inspired Algorithms*

Biological structures found in nature have been the source of inspiration for various optimization algorithms. Examples include the movement of ants and the work of honey bees, who do not normally interfere with each other's motion [78–80].

Genetic Algorithm (GA)-based approaches have been widely utilized in the literature for route design problems, due to their capability for handling discrete functions. Particle Swarm Optimization (PSO) frameworks are also inspired by biological patterns derived from the social life of birds and their coordinated movements [75,82].

Different types of objective functions can be optimized using such bio-inspired methods. However, they are not capable of producing high-quality solutions in real time, because of their notoriously high computational cost. Furthermore, they cannot tackle dynamic environments and probabilities, because they typically return only one solution and are prone to becoming trapped in local minima [81].

Nevertheless, they were still the foremost category of optimization techniques utilized in motion planning at the time of the widespread deployment of Convolutional Neural Network (CNN)s, which have dominated the field since the year 2015, and more conspicuously, since 2020 [5,125]. This will be discussed in more detail in Section 3.5.

3.4. *Online Replanning*

Despite the prior considerations and predictions, actual performance is usually affected by various factors that prevent a robot from correctly following the trajectory planned. Fixing these issues requires performing execution-time replanning. On the other hand, there may be occasions where the newly captured sensory information renders the previous inferences invalid.

Other factors, such as uncertainties in the robot's kinematic model, sensor noise, and the obstacles' motion reinforce the necessity and importance of online replanning. A robot's success in carrying out its mission and the operation time are greatly dependent on its capability for replanning, especially in environments involving unstructured terrain or extreme occlusions.

The manner in which such problems are alleviated also depends on the source and cause of the uncertainties, among other factors. For example, a neural dynamics design may be employed through a pseudo-inverse-type formulation, to alleviate the impacts of time-varying noise [126]. Parallel instances of the same planner may be utilized to search through numerous heuristic weights and update them, thereby achieving fast replanning. High-Frequency Replanning (HFR) performs this task while letting the robot carry out the prior decided plan at the same time. Stochastic nonlinear dynamics are typically linearized, for the noise and uncertainties to become tractable. Assuming the availability of sufficient computational resources, asymptotic optimality may also be achieved [54,55].

3.5. DBL

In numerous motion planning tasks, the ideal behavior is supposed to be the that presented by a human expert. However, due to the complexity of deriving formal descriptions of the reward functions underlying expert behaviors systematically, a suitable strategy consists in employing nonlinear Inverse Reinforcement Learning (IRL) based on maximum entropy. In other words, it tries to extract approximate reward functions, which however may not match the inherent unknown one perfectly, leading to a similar behavior under similar conditions [87]. This can be accomplished by making cost maps from demonstrations. For example, expert driving behaviors in a complex urban environment can be learned based on a large number of demonstrations. This may be achieved using raw sensor data that are converted to cost maps and features directly, without requiring manual interference or processing. Using a Fully Convolutional neural Network (FCN) makes it possible to handle large datasets and complex behaviors efficiently [88].

Personal robots could also learn how to plan their motions to perform simple household tasks, such as cleaning the surface of a table or moving a spoon of sugar from a bowl to a cup, based on demonstrations [31]. Raw sensor data captured from expert demonstrations, e.g., 2D laser range findings, could be learned and directly mapped to robotic motion plans represented in the form of steering commands, in an end-to-end fashion. The foregoing procedure would need to be performed using supervised learning aimed at an existing motion planner, which may be employed for seen or unseen environments of a virtual or real nature, using e.g., the grid-based global approach [89].

In SBMP, task constraints aimed at specific missions, e.g., holding a glass of water upright to avoid spilling, need to be explicitly programmed. However, DBL provides the possibility of automatically deriving task constraints, which may be hard to express or determine manually, from the principal patterns representing the constraints, entailing less variation across different demonstration instances [31].

Despite these challenges, predicting human motion is feasible, at least under predefined task descriptions, such as single-arm reaching motions for manufacturing, and this is very beneficial for fast and efficient performance. Assumptions include the fact that human motion is optimal with respect to an unknown cost function and that online iterative replanning is capable of capturing the adaptation of a human's motion to that of a robot. Upon detecting sample motion trajectories, inverse optimal control may be utilized to learn the aforementioned cost function based on a human kinetic model using Stochastic Trajectory Optimizer for Motion Planning (STOMP) [90]. Safety requirements may be ensured by applying upper bounds to the collision probabilities based on the Gaussian distributions according to which the predicted motion is represented [91].

On the other hand, the operator's intentions can appear in the form of explicit requests, which need to be disambiguated and interpreted in terms of specific motion commands, through variable grounding based on the likelihoods of valid choices within a goal region [92]. For more complicated tasks, user preferences in terms of, e.g., the desired temporal or spatial constraints, may be learned and iteratively revised based on the choice of trajectory among the existing options [93].

In various contexts, a robot's performance may be affected by its capability for generalizing from basic motions and actions to more sophisticated ones. This would be of substantial benefit in terms of having the robot take on responsibilities it has not been specifically programmed to do. It would also make the robot adaptable, significantly beyond carrying out repetitive tasks, which is not sufficient for operating in unstructured environments. A promising approach to the foregoing task consists in combining kinesthetic user guidance with Reinforcement Learning (RL). More clearly, once the features of the primitive actions are captured based on the screw transformation of the end-effector, they are mapped and employed to compose the reward function for RL by training a new motion planning policy. The system developed in [96] takes advantage of this strategy and offers the option of requesting further learning of basic motions if the tasks available from the outset do not suffice for considering the new constraints. A more detailed discussion of the role of RL in motion planning will be presented in Section 3.7.

3.6. *Deep Learning*

As aforementioned, SBMP may incur a significantly high computational complexity, which grows excessive with increased problem dimensionality. One way to alleviate the foregoing shortcoming is to sample the configuration space selectively and focus only on regions that stand a higher chance of leading to the optimal path. Adaptive sampling may be performed using either manual heuristics or learning. Nevertheless, the former cannot handle higher dimensionalities and is not suitable for cluttered or unfamiliar environments [97].

Deep Learning (DL) networks, however, do not suffer from such drawbacks and may be trained to produce heuristics. They are capable of creating sample nodes. This enhances the computational efficiency and helps accommodate higher dimensionalities, where the resulting hybrid approach retains theoretical worst-case guarantees. DL can handle kinematic constraints and produce near-optimal trajectories in real time.

The stepping Recurrent Neural Network (RNN) OracleNet [99], Deep Sampling-based Motion Planner (DeepSMP) [97], Motion Planning Networks (MPNet) [98], Generative Adversarial Network (GAN)s [100], and the gradients-based Constrained Motion planning Networks x (CoMPNetX) [101] are examples of NNs that have been successfully utilized for DL-based motion planning.

For example, DeepSMP consists of a contractive autoencoder that uses point clouds to encode workspaces, followed by a stochastic deep feedforward NN with dropout that recursively creates node samples for an optimal end-to-end path based on the starting and goal configurations.

Nevertheless, preparing suitable datasets enabling the network to achieve this is typically not straightforward. This may be alleviated by decreasing the number of samples involved in training through active learning, where only certain samples chosen by the DL network itself are supplied in each round of training [5,97,98].

3.7. *Reinforcement Learning*

Similarly to DL, RL, which is aimed at sequential decision-making, can be employed for motion planning in unfamiliar environments. It can resolve high-dimensional problems involving dynamic obstacles by taking into account their location over a limited number of timestamps within the past horizon [108].

RL methods such as the Proximal Policy Optimization (PPO) algorithm [105] actively explore the domain. They are more accurate than supervised learning approaches, which lack data on the boundaries of obstacles [112]. However, RL suffers from the shortcoming that the number of samples needed for training may be too high. This undermines their practical usefulness. Nevertheless, an efficient reward function may help reduce the amount of data required [5,104].

Examples of the RL-based algorithms utilized in motion planning include the twin delayed deep deterministic policy gradient (TD3) [107] and Exploitation of Abstract Symmetry of Environments (EASE). The latter relies on locally adopting spatial symmetry abstractions obtained from naïvely trained agents [108].

Soft Actor Critic (SAC)-based methods constitute another category of such approaches, where optimizing several NNs based on maximum entropy [110] can be applied for an improved sampling efficiency. Moreover, Prioritized Experience Replay (PER) enables a more efficient use of data for training by changing the weights of samples based on Temporal-Difference (TD) error. Nonetheless, it should be noted that PER is computationally expensive, and inefficiently adjusting its hyperparameters may undermine the performance of RL [111,113].

Other schemes, such as Deep Deterministic Policy Gradient (DDPG), have also been successfully deployed [112]. Meta-learning can be accompanied by experience replay separation based on a success/failure discrimination within separate buffers randomly sampled at a ratio determined through NN-based learning [111]. Hindsight Experience Replay (HER) is another technique for utilizing training data more efficiently [106].

As a common approach to motion planning based on RL, a Linear Parameter Varying (LPV) state-space representation can be used to tackle dynamic obstacles. More specifically, a switching mechanism is embedded within a dual architecture, where the mode is determined based on the distance between the robot and the obstacles. The two modes may differ from each other in terms of, e.g., whether or not the joint positions are directly controlled, whereby the RL planner takes control when obstacles are perceived to have become too close to the robot [109].

As aforementioned, training RL models is challenging in terms of convergence and robustness. The reasons for this include ambiguities in the relationship between the Cartesian and joint spaces, continuous workspaces, and redundant DoF, which result in unnecessary explorations [107]. This could be alleviated using a NN to produce an initial policy for guiding the training of the RL framework [105].

3.8. Scene Uncertainties and Dynamics

Scene characteristics constitute one of the main factors influencing the performance of motion planning frameworks. This signifies the importance of finding out unknown or unexpected phenomena, and updating existing information regarding the environment.This may be acquired using an autonomous system's sensory equipment from the rest of the autonomous systems available within a reasonably close vicinity, through communication networks connecting them via Vehicle-to-Vehicle (V2V) protocols, or from management centers, via Vehicle-to-Infrastructure (V2I) protocols [36]. Doing so requires further incorporation of transportation engineering concepts [37].

A principal component of such information concerns obstacles and their properties, including their position and shape. These may be dynamic or even nonexistent from the point of view of the planner, until they appear. In cases where the constraints, e.g., the ones related to the obstacles, need to be captured and dealt with in real time, i.e., during the runtime, they may emerge in an even more complex form, especially when affected by sensing noise [127]. If the terrain is harsh or uneven or involves a clutter of static or dynamic obstacles, this can result in an unstructured environment [20].

The navigational complexity of a topological space may be evaluated based on the underlying homotopy [128]. Uneven or irregular surfaces are often encountered in the course of carrying out search and rescue tasks and conducting space missions. They require properties such as modularity, with which the robot adjusts its configuration according to the terrain properties. The terrain can be extremely rough and involve various obstacles, in which case, the area reachable by the robot is significantly limited, demanding the generation of a higher number of locomotion gates, e.g., walking or crawling, and thereby enabling the robot to switch between different motion primitives, considering the requirements arising from the number and properties of obstacles [56].

Motion planning for modular robots can greatly benefit from elementary motion primitives. This may be realized using locomotion generators, leading to flexibility in the shape of the robot, the environmental conditions, and the number of modules, where the necessity for reconfiguration is also obviated [56].

Terrain dynamics can be affected by media penetration, resulting in deformations appearing as, e.g., dirt or sand. This negatively affects the reliability of maneuvers involving walking or jumping. Closed-loop dynamics may be modeled using an added-mass description of the grain motions, taking into account the hydrodynamic and hydrostatic effects [129].

Foothold and motion planning could be coupled using a parameterized dynamic model, to efficiently analyze, at whole-body level, torque and kinematic limits, as well as the friction cones, at the same time as trajectory optimization. This enables considering the effect of topology based on terrain normals [130].

Irregular surfaces also result in complications such as stochastic bouncing and non-linear gravity, which may be handled by, e.g. solving Lambert's orbital boundary value problems, followed by propagating model and control uncertainties to compute landing distributions, which are, in turn, utilized to figure out an energy- or time-optimal hopping strategy based on a policy gradient. The final step consists in performing sequential planning through a Markov decision process using Least Squares Policy Iteration (LSPI), leading to an off-policy, off-line, and sample-efficient RL strategy [114].

Approximate Clearance Evaluation (ACE) is particularly useful for handling uneven terrain, where the terrain height bounds are utilized to find out the state bounds of articulated suspension systems, e.g., in the case of planetary rovers, thereby evaluating the worst possible safety indicators based on constraint violation levels. Although ACE may lead to extreme conservatism, a newer variant proposed in [131] has a favorable computational cost and alleviates the pessimism, while still retaining empirically estimated probabilistic safety guarantees based on the distributions calculated over the states.

Additionally, avoiding skidding and slippage requires proper modeling of the surface and producing sufficient amounts of torque to carry out curvilinear motions based on energy-efficient motion planning. Possible changes to the surface need to be accounted for and incorporated into the model using online learning through, e.g., NNs and Extended Kalman Filter (EKF) for the dynamic and kinematic aspects, respectively. The former has importance in estimating the pose, and the latter in determining the constraints on the turn radius and for producing estimates of energy consumption [20,83].

Alternatively, reasoning about the contact and motion may be handled simultaneously through Mixed-Integer Convex Programming (MICP), which is effective in tackling other complications such as uneven terrain, friction, and certain types of non-convexity [132]. Similarly, the motion within a free space can evolve toward motion while in contact by letting the tree grow in the combined space through factoring it into a belief over the configuration and the contact state [133].

Motion planning for legged robots is generally easier than for wheeled robots, due to the fact that they provide a wider range of mobility. Nevertheless, even in the case of legged robots, duly accounting for all the conditions and uncertainties arising from the terrain's unevenness is not straightforward, since it requires full-body motion plans made within a high-dimensional space based on the limited information available about the terrain [130].

It is also worth noting that collaboration with human operators may make a significant contribution in handling uncertainties, e.g., occlusions disrupting the robot's perception of the target object to be grasped. This can take place either by providing knowledge or taking part in the execution of higher-level tasks, where a tree-shaped set of feasible plans is created using geometric reasoning [134].

3.9. Stability

Uncertainties of various types, including disturbances, inaccuracies of parametric models, and environmental parameters, as well as complex geometry, may affect the performance of a motion planning algorithm [127]. Therefore, path planning and trajectory optimization may need to be followed by feedback control, to ensure stability and to overcome the nonlinear dynamics and uncertainties. The output of such a module is a control policy that stabilizes the system, from a bounded set of initial conditions to a goal state. As a result, a tree of trajectories could be created and stabilized using feedback.

By exploring a bounded set based on random samples, the possible necessity of deriving additional trajectories can be determined. Subsequently, funnels can be found iteratively, until they cover the bounded set, meaning that the control policies can stabilize the bounded set to the goal. The task of approximating the funnels can be performed using, e.g., sums-of-squares verification or through sampling followed by falsification based on simulation [57]. On the other hand, various post-processing procedures can be employed for smoothing and improving a trajectory, as well as replanning to avoid dynamic obstacles using model-based predictions [43]. For example, linear-quadratic-Gaussian controllers yield reliable reference-tracking functionalities [58].

Consideration of the dynamics of a problem can be realized by performing asymptotically optimal KP using two-point BVP-solvers based on optimization and numerical optimal control methods, such as Sequential Quadratic Programming (SQP) [46]. Artificial potential fields realized through, e.g., Improved Rapidly-exploring Random Trees (I-RRT)* constitute a viable alternative [135–137].

With information about the distance from obstacles, using the Monte Carlo (MC) algorithm, CP may be approximated with asymptotic correctness, which enhances the chance of finding a feasible plan.The CP can be found for each iteration for the estimated optimal path, followed by inflating or deflating the obstacles to make the CP reach a target value. Statistical variance-reduction techniques such as Control Variates (CV) and Importance Sampling (IS) contribute to a faster performance or real-time functionality through parallelization [58].

The results of CP approximation need to be transformed to the configuration space, to estimate the location of the points with the highest CP. This could be performed using a Newton method to determine the closest point that may cause a collision [6]. For safety-critical tasks, Control Barrier Function (CBF) can be used to incorporate obstacle-avoidance constraints through Quadratic Programming (QP) [59]. It is also possible to devise a random search tree reconstruction framework by predicting the cost using NNs. This results in asymptotic near-optimality [84].

Non-convexities in the cost functions and obstacles, as well as control and state constraints, are handled in energy-optimal motion planning by decomposing the decision variables and creating a convex representation of collision constraints. This results in two QP problems, accommodating the decoupled Linear Time-Invariant (LTI) dynamics in each axis. Assuming that an initial feasible trajectory exists, the foregoing problem may be solved iteratively using an Alternating Quadratic Programming (AQP) algorithm, which converges to a homotopic local optimum with respect to the initial guess faster than the MICP alternatives [138].

Applying splines to the resulting paths, to generate a possibly complex shape while maintaining parametric continuity, is another approach for improving the quality, in terms of, e.g., reduced lateral accelerations, enhanced robustness against disturbances, or improved tracking performance. In autonomous driving, additional benefits may include enhanced passenger comfort, as well as reduced tire wear and mechanical failure [17].

3.10. Computational Feasibility

Physics-based motion planning, state sampling, and state steering are computationally expensive, due to the costs of state propagation of real-time replanning performance [20], especially for high-DoF robots [119]. High-dimensional dynamics and external disturbances lead to extremely high computational costs, which are often achieved for efficient, real-time planning by sacrificing dynamical feasibility and safety. Various efforts have been made to come up with a compromise between safety and planning efficiency, resulting in, e.g., the modular Fast and Safe Tracking (FaSTrack) system [139], which can be used for tracking the multi-dimensional paths devised by different trajectory optimizers. On the other hand, using factor graphs and GP interpolation has been widely reported to improve the computation time required for making inferences [122].

Moreover, the computational load of creating a funnel of trajectories may be alleviated using convex optimization, through sums-of-squares programming [127]. For TMP algorithms, this could be leveraged by enabling add-and-remove constraints on motion feasibility at the task level, a successful example of which is the Iteratively Deepened Task and Motion Planning (IDTMP) proposed in [140], offering probabilistic completeness and scalability for plans involving several objects.

PRM* is known to minimize the computational cost of generating RMs. Nevertheless, it may be slow, and result in excessive growth, undermining online query resolution and storage. Incremental sparse sub-graph-based approaches, referred to as spanners, have alternatively been proposed with asymptotic near-optimality. They yield a lower path quality but alleviate the aforementioned shortcomings, and improve the RM density [69]. Examples of computationally heavy functions that increase overheads include checking for nearest neighbors and possible collisions in sampling, while using, e.g., RRT. This may be obviated by utilizing the CBF-RRT algorithm instead.

It has been stated that by discretizing the reachable state space and solving it numerically, the state sampling and nearest-neighbor search speeds can be improved. The former can be performed online and results in a reachable map, which reduces the number of states involved in sampling, as well as unsuccessful motion validity check queries. Moreover, states that are not reachable or can be reached only after a certain horizon are excluded from the nearest-neighbor search [60]. On the other hand, for problems such as motion-specific self-calibration, using EKF may accelerate the process of approximating model parameters or dynamics, to achieve the maximum belief informativeness based on a prescribed budget. This is beneficial in SBMP, which is required to react to certain circumstances promptly.

For a set of robots moving in parallel and with separability, assuming that each of them is expected to move at most d units from their initial location, the distance traveled by each robot is $O(d)$, i.e., constant stretch, meaning that a constant-factor approximation is possible for the optimization problem. Moreover, NP-hardness can be maintained, even if the robot positions are restricted to a regular grid. If separability is not guaranteed, e.g., for densely-packed disks that may not be well separated, the stretch factor increases to $\Omega\left(N^{1/4}\right)$, and with certain considerations, to $O\left(N^{1/2}\right)$ [141].

The distance metric is not necessarily Euclidean, and this choice greatly affects the computational cost of RRT-based motion planning. It also influences the path quality and space coverage. Determining this appropriately, e.g., based on learning, is essential for achieving a reasonable computational cost [32].

Exploiting data-driven approaches for predicting the map in an unknown environment, as opposed to relying on frontier selection heuristics, may help expedite planning [118]. In numerous applications requiring dense mapping of an unknown environment, e.g., for agile robots possessing visual sensors and constrained embedded computational capabilities, leveraging the computational cost may be accomplished by using only parts of the existing sensory information and combining the mapping, i.e., perception, and planning problems. In this manner, by switching between the two iteratively, each of them is updated based on the results of the other [142].

When it comes to problems involving human motion tracking, it is basically assumed that the subject moves according to a plan devised with a cost function. However, such algorithms are prone to computational intractability [143]. A remedy would be to utilize NNs learning near-optimal heuristics from path planning. For instance, the MPNet can learn continually and actively from expert demonstrations. This reduces the amount of learning data required. By recursively constructing connectable paths, the workspace is encoded based on point cloud measurements. Incorporating classical path planning algorithms for a hybrid approach leads to theoretical worst-case guarantees, while maintaining the computational and optimality advantages [94].

However, it should also be noted that long-term planning using NNs, e.g., applying Deep Reinforcement Learning (DRL) to the Linear Temporal Logic (LTL) or Metric Interval Temporal Logic (MITL) specifications [85], may incur computational drawbacks. This can be alleviated through an MC tree search [86], where Model Predictive Control (MPC) is used under a linear program for convex optimization, to achieve further robustness against noise. This is true for scenarios involving a high number of dynamic objects and may be improved by using Long Short-Term Memory (LSTM) [102].

Parameterized probabilistic models of human behavior can be utilized to predict and prepare for future actions. This allows adapting a model to circumstances and observations, based on a distribution over the model parameters, obviating the necessity for prior determination. Other alternatives include Bayesian inference and worst-case forward reachable sets. The additional computational burden may be leveraged by solving a stochastic reachability problem in the joint space of the human and the belief for the model parameters. These can be deterministically approximated using a Hamilton–Jacobi framework. This approach needs to be implemented allowing only a set of actions, as opposed to the aforementioned distribution, while maintaining the belief as an explicit state.

For robots with numerous continuous states, improved scalability can be achieved by decomposing the problem into smaller sub-problems using a lazy satisfiability modulo theory approach. At each iteration, a coarse discretization of the workspace is carried out, taking into account Boolean constraints, while capturing the low-level continuous dynamics, to produce high-level discrete plans. Iterations are performed until a feasible plan is found. Possible infeasibilities include the transitions between the workspace regions [144].

From the point of view of software development, frameworks realized using utilities such as Motion Planning Templates (MPT) take advantage of compile-time polymorphism. Such approaches may help to more efficiently handle the usual lack of computational resources for small robots running on batteries, by producing individual codes tailored to the specifications of the robot and motion planning problem, which can determine the data structure. Although this deprives the robot of runtime flexibility, it provides extended compile-time capabilities, such as storing robot-specific information in the resulting motion planning graphs, creating firmly packed data structures, and modifying the scalar precision [145].

A common practice is to use a unified representation on the basis of a discrete, finite motion space and to split the problem into task and motion planning components. The former decides what to do, and the latter determines the geometric feasibility based on finer details. Width-based search algorithms and AI planning languages that represent the high-dimensional problem in compact form are essential components of such frameworks. Moreover, some problem representations enable the prediction of constraints and the transfer of knowledge from one instance of the problem to another. For example, the score space proposed in [146] provides a representation according to the performance of solutions applied previously, where similarity indices are used sequentially [147].

Although collision detection is widely believed to be the main computational bottleneck in asymptotically optimal SBMP, a nearest-neighbor search may be computationally even more expensive, depending on the experimental scenario. This has motivated the development of more efficient nearest-neighbor search strategies and data structures that take into account the mission specifications [50].

Given an online preparation of a set of random controls that is large enough, the computational cost may be reduced by finding a balance between exploration and exploitation. Machine learning techniques can further contribute by producing choices of maneuvers based on the dynamics, as well as the heuristics and particularities related to obstacles [16].

On the other hand, incorporating loop-closure constraints may help reduce the configuration space to a manifold within the joint ambient space, which is higher-dimensional [61].

In numerous recent studies, using a many-core Graphics Processing Unit (GPU) has been reported to allow fast motion planning [148,149]. Nevertheless, robot-specific circuitry could be built to achieve scalability in motion planning and parallelization of collision detection over the RM edges [70]. Using semi-infinite nonlinear optimization, the constraints and their gradients are calculated simultaneously, and obtaining polynomial approximations over time helps satisfy the constraints continuously. Parallelizability arises from the fact that the constraints and their gradients are analyzed for each time interval independently. The rest of the computational elements, such as kinematics, dynamics, and geometry, are also essentially parallelizable [149]. Continuous-state Partially Observable Markov Decision Process (POMDP) problems can also be solved faster using a GPU or a hybrid of a GPU and Central Processing Unit (CPU), considering a multi-level formulation of the Monte Carlo Value Iteration (MCVI) method. Balancing the workload and interleaving the data and computations constitute the main backbones of parallelization in this context, where the CPU prepares the data, and the GPU handles the simulations [150].

Last but not least, one may resort to a multi-resolution architecture for managing the depth information characterizing the obstacles, where the range measurements are represented using a 2.5-dimensional projective data structure, namely an egospace, within which motion primitives can also be expressed [151].

4. Applications

Motion planning algorithms are useful in many scenarios involving robots or autonomous cars, e.g., trajectory planning for multi-legged walking robots for planetary exploration, or for autonomous cars inside parking garages, as well as for highly actuated service robots in mobile manipulation [49], and even in precision medicine [75]. In the following sections, applications of motion planning algorithms realized using a certain type of robot will be briefly reviewed, to make their importance in practical contexts more tangible, as well as touching upon some of the relevant open problems.

4.1. Social Robots

Social robots have gained increased popularity. In particular, omnidirectional robots are capable of performing social interactions relatively efficiently, due to their ability to reorient their body independently from the direction of motion. Nevertheless, apart from general objectives typically considered while framing the problem, assimilating a robot into the social context necessitates accounting for additional requirements. Prominently, the robot should behave cooperatively and in a similar manner to human beings, e.g., in crowded or narrow passages. Thus, certain constraints may be derived from human subjects, and imposing these while devising the translation and rotation profiles helps make the robot resemble humans in a more realistic fashion [62,152].

In the absence of vocal clues implying the user's intentions or expectations, a robot should preferably be able to understand and handle a situation by observing the users' behavior and predicting their actions [153]. On the other hand, mimicking and trying to replicate the motion profiles adopted by human beings may contribute to developing socially acceptable and cooperative behaviors for mobile robots, small-sized transportation vehicles, and wheelchairs, thereby improving operational efficiency and safety using an acceleration-based *social force model* [154].

A robot's perceptual and decision-making capabilities should enable it to behave well by, e.g., choosing a reasonable trajectory. Statistical models of density and velocity profiles can be made and integrated into the robot's resources. These are produced by analyzing and representing the subjects' attitudes using, e.g., a Boltzmann factor defined based on the comfort attributed to their distance from walls and their possible preference for walking on a certain side within a pedestrian corridor. This, in turn, provides further insights into the authenticity and reliability of theoretical and simulation-based pedestrian distributions and collision avoidance models [154,155].

Modeling human behaviors is among the factors that may contribute to better motion planning for social robots. This may not be easy, due to the underlying stochasticity. Nevertheless, a robot should ideally be able to plan and perform its motions such that they are predictable for nearby pedestrians [156]. More clearly, feature-matching techniques, possibly based on maximum entropy, may fail to model the path followed by humans under scenarios involving different subjects or different instances of the same path being followed by a given subject. Therefore, utilizing DRL may be considered to avoid violations of social norms, as well as for achieving motion at normal human walking speed, while avoiding extended computation times [103].

Assuming that a motion is planned and performed in a human-like manner, the above facilitates the prediction of the robot's motions by the human, thereby enabling more convenient joint navigation. Nevertheless, the principles governing Human-Human Interaction (HHI) are unlikely to be readily applicable to Human-Robot Interaction (HRI), due to the different circumstances arising from the usual interest presented by humans in getting to know more about various aspects of a robot's behavior [156].

HRI may be enhanced in an adaptive and online manner in dynamic and unknown environments, through utilizing communicative means of locomotion, where, based on the Artificial Potential Field (APF), the velocity and orientation information is utilized to take advantage of patterns represented by the motions of household animals [72].

It is worth noting that human-likeness is measured based on social spatial attention models indicating the capture time and direction as a consequence of a cue, e.g., a head turn [157]. When it comes to replicating human motions intended to resemble an emotion, analyzing upper-body motions inspired by human gait, along with the trajectory and the center-of-mass, has been shown to have a significant influence from the vertical oscillations of the human subjects' perceptions of the emotional states and the corresponding confidence levels [158].

On the other hand, the motion diversity produced independently from the subject's intentions and incorporated into the android's behavior could contribute to the soundness of the resulting impressions, in terms of the interpersonal skills required for natural interaction [159]. The congruence based on the balancing motions of the robot, which could be achieved by, e.g., a human-like inverted pendulum mechanism, could greatly affect these impressions and swaying [160]. Last but not least, motion planning for persons with reduced mobility, such as those with visual impairment, may require further considerations; e.g., of the constraints imposed based on their capability for handling crowded areas, wet floors or roadblocks, or avoiding undesirable items, such as stairs.

4.2. *Self-Driving Cars*

In principle, self-driving cars are a type of mobile robot, as they take advantage of similar technology in terms of autonomy. Likewise, the problem of avoiding hitting other cars or pedestrians is, in fact, tantamount to that of avoiding dynamic obstacles dealt with in the motion planning for mobile or humanoid robots [1].

They also need to perceive the world, make maps and update them, and plan their motion with several layers of hierarchy. Nevertheless, unlike mobile robots, which may be utilized under considerably more flexible setups, the motions of self-driving cars are restricted to predefined roads and are further constrained by traffic regulations and good practices of driving, e.g., while navigating intersections [86]. This leads to both advantages and disadvantages in motion planning [1].

Improved convenience, efficiency, accessibility, and safety, as well as reduced overall costs and manpower, constitute the significant advantages of self-driving cars in automotive transportation settings [35], where, from the point of view of urban administration, they also contribute to tackling congestion and reducing emissions [37]. Moreover, due to their capability for utilizing redundant sensors, they reduce the risk of accidents due to possible late reactions, faulty perceptions, or a lack of attentiveness from the human driver [1].

Tasks for a self-driving car can be extremely safety-critical, and performing them may demand making use of feedback control [35]. Real-time speed control of navigating a self-driving car through an occluded crosswalk may be achieved by modeling the problem as a partially observable Markov decision process and optimizing the longitudinal acceleration using dynamic programming, on the basis of the belief of the crosswalk [161].

In order to enhance the speed for reliable performance under real-world scenarios, rule-template sets may be created. These make it possible to determine an appropriate decision for the traffic scene at hand. The CommonRoad benchmark was proposed in [162], where for each real or artificial experimental scenario, constraints, goals, the cost function, and a vehicle model are provided. For example, autonomous parking is modeled as a dynamic optimization problem and solved using an interior-point simultaneous algorithm [163].

Algorithms such as A^*, which are designed to work in unstructured environments, are useful for handling highly constrained environments where on-road trajectory planning may not suffice. Node expansion algorithms are utilized to obtain trajectories that are subsequently refined using a pure-pursuit controller producing edges, as short motion primitives, which ensure the general adherence of the vehicle to the trajectory. Using this strategy, a vehicle can exploit explorative advantages in challenging scenarios, e.g., reversing to pass another vehicle that suddenly stops.

After extracting motion primitives from real-world driving data, they can be connected, followed by applying Expectation-Maximization (EM) and an initial segmentation. The Dynamic Movement Primitives (DMPs) may then be inferred probabilistically, thereby correlating the independent motion plans for a smooth transition and improving the tracking accuracy [164].

Finally, it should be noted that more complex maneuvers, such as changing lanes, parking, and overtaking, also require modeling the reactions of human drivers and are computationally disproportionately heavy. This could be remedied using prior knowledge obtained by training neural networks [95].

4.3. Humanoid Robots

Humanoid robots are characterized by their relatively high DoF. Classically, their high-level motion plans resulted from path planning, and their trajectory optimization used to be transformed into a dynamic locomotion plan for the whole body in a later stage. More recent approaches have tired to take advantage of simultaneous optimization of the low-level and high-level motion specifications where applicable. This will be further elaborated later in this section. While analyzing the generalized inverse kinematics, the joint limits, foot positions, and balance need to be considered. A motion planner should be able to manipulate massive objects, take appropriate actions with the unforeseen appearances of obstacles in cluttered environments, and plan footsteps by means of variable kinematic modeling of the footholds [165]. On the other hand, state space representations are used to apply the existing SBMP techniques with biased sampling to humanoid robots, followed by steering using inverse kinematics [62].

Motion planning for humanoid robots may be performed using a hierarchical strategy, such that, first, a collision-free path is obtained for the End-Effector (EE), followed by searching for collision-free points in the case of each of the via-points for the elbow, resulting in an asymptotically-optimal raw path consisting of straight lines, further refined for kinematic smoothness and feasibility using an online Cartesian calculator and controller [24].

The Contact-Invariant Optimization (CIO) approach aims to optimize whether or not contact forces should be active during each phase of presenting a behavior, simultaneously with the behavior itself. This could be utilized in conjunction with ensembles of perturbed models to achieve robustness against model uncertainties while performing behaviors such as turning, as well as sideways and forward walking [166]. Moreover, human experience may be exploited by considering multiple operational phases and optimizing the specific contact configurations separately [2].

Gait primitives for a humanoid robot may be extracted as limit-cycle behaviors, then being arranged into a switched, discrete-time system. Nevertheless, the frequency of switching needs to be limited, in order to achieve fluent performance. This can then be utilized in conjunction with Hybrid Zero Dynamics (HZD) to enhance the stability through dimensional reduction and sums-of-squares programming [167].

Motions limited to the sagittal plane constitute the main element in pick-and-place tasks [168]. A more challenging scenario would be to move toward a target stance pose, and after reaching it, perform a prescribed manipulation task. The complexity arises from requirements such as finding a stable stance pose that not only is collision-free but also leads to a full-body configuration that is robustly maintained during manipulation. Using inverse dynamic reachability maps and assuming that the feet are positioned close to each other on a flat surface, a solution can be found with a success rate dependent on the coverage density of the sampling space. Obviously, the map size should not exceed the memory capacity.

Nevertheless, the map may not offer samples with a high enough variety from the high-dimensional configuration space to handle real-world applications involving, e.g., uneven terrain. Therefore, a paired forward-inverse dynamic reachability map needs to be utilized to achieve a wider modularity, if the robot's kinematic structure allows. This would help achieve a given number of composed configurations, while requiring storing a smaller number of samples. For a given memory capacity, this would lead to a more diverse set of possible configurations. Consequently, taking advantage of whole-body redundancy, a higher level of flexibility would be achieved in planning the motion to reach an end-pose in the presence of constraining factors such as obstacles and uneven terrain [169].

Humanoid robots are widely used in scenarios involving HRI. One of the challenges typically encountered in haptic interactions arises from the inevitable impact of the robot's own motions on the sensory data supposed to represent the user's tactile input. A possible partial remedy to the latter problem could be to estimate and subtract the effect of the robot's motions based on a sequence of joint values, through linearizing the underlying posture sub-spaces [170].

Various strategies using the safety of the operator may be accommodated within a motion planning framework that can be assessed based on, among other things, the technical specification ISO/TS 15066 [171] and the principle of energy absorption by the operator body. In this context, the regulated body movements are simulated for an experimental setup, bearing in mind the fundamental consideration that the relative velocity needs to be predicted, estimated, and bound to improve the operators' safety, by preventing the robot from harming them, according to the prescribed trajectory and the relevant velocity limits [172].

Multi-limbed robots can take part in missions such as climbing. For example, a six-legged robot could climb up two walls, where numerous constraints on parameters such as posture, contact force, and torque need to be taken into account, in order for the robot to be able to perform the task using the friction between its end-effectors and the wall surfaces. One approach to tackle this problem consists in utilizing a NonLinear Programming (NLP) solver. Alternatively, the problem could be decoupled into parts: one concerning the torso posture, and the other dealing with contact forces. The former could be solved by means of NLP or MICP, while the latter needs to be worked out as a series of convex optimization problems. Introducing angled walls, uneven surfaces, and obstacles could be considered as further challenges to expose the robot to [173].

4.4. Object Manipulation

Multi-arm robotic systems require task- and joint-level coordination, where the former is meant to make the arms coordinated with an object or with each other, and the latter is aimed at preventing self-occlusions. The space of free motion may be learned using sparse non-linear kernel classification approaches. Whether the robots are expected to coordinate with each other to reach a moving object simultaneously or only to perform independent point-to-point motions determines the synchrony requirements, which need to be properly considered, especially for possible transitions between different scenarios [174].

The higher-dimensionality and multi-modality of the configuration and search spaces involved in motion planning problems for multi-object manipulation tasks give rise to additional challenges, due to kinematic and geometric constraints. To tackle different placements of objects, FastForward (FF) planners such as FFRob use multi-query RM structures to assess reachability. They handle rearrangement planning problems using an Extended Action Specification (EAS) representation suitable for conditions involving, e.g., arbitrary predicates. They are capable of solving strip planning problems, e.g., delete-relaxations. More specifically, by means of batch sampling of the manipulation primitives, the motion planning problem is iteratively discretized, resulting in a probabilistically complete planner with the finite runtime required for estimating the distance to goal [71,175]. The primitives may be generated using a graph-search under a linear quadratic minimum time problem [176].

In cluttered environments with objects blocking the path to grasping an object, it may be necessary to move the objects first. This may require robot–object and object–object interactions of a dynamical multibody nature, involving varying object poses, and using algorithms such as KPIECE and p-KPIECE [63,64].

4.5. Cooperation and Multi-Robot Motion Planning

Autonomous systems may benefit greatly from communication with their peers, as well as management centers. This would help them make more efficient decisions. For example, self-driving cars may park close to each other, which saves space, or cooperate with each other to alleviate congestion [1]. Moreover, a self-driving car can be equipped with sensors and processing units enabling it to obtain and share valuable information regarding, e.g., the duration of each color of traffic lights, as well as their locations. Cloud-based networks enabling such communications could help them to calculate an Estimated Time of Arrival (ETA) more efficiently. Similarly, the information could be analyzed to allow an improved understanding of traffic patterns and alternative routes [177].

Although setting up global restrictions on robots' movements based on worst-case scenarios may suffice for ensuring safety, this may result in longer operation times. This could be alleviated by incorporating safety considerations at the motion planning level [172,178].

Multi-Robot (MR) teams are useful where, e.g., a robot's own reachable workspace or payload does not suffice for handling large loads [179], and for missions such as patrolling rooms, or in handling deadlocks, e.g., where a moving obstacle hinders the robot's motion [180]. LTL formulas are widely utilized in this context. To this end, robot dynamics and specifications can be captured for convex and Boolean constraints within a feasibility problem, to be decomposed and solved using, e.g., Satisfiability Modulo Convex (SMC) [181].

Even the aerial transportation of large loads, including cable-suspended, could be achieved using teams of robots, where, e.g., Parametric Dynamic Movement Primitives (PDMPs) are used to coordinate the robots and their manipulators. This could be subsequently transformed into an explicit parameterization of motions using statistical approaches with a Bayesian property, such as Gaussian Process Regression (GPR) [66]. Constraints can be imposed on the relative positions of the robots, where onboard, real-time localization detects the relationships between neighboring robots [42,67].

Hybrid approaches only take advantage of centralized planning at a global level based on a consensus, where distributed controls are applied to individual robots to avoid obstacles and transfer forces to the rest of the robots through the object being manipulated. Typically, a receding horizon planner is employed to achieve velocity control via a convex optimization problem [182]. In cooperative missions, a network topology controller is required to ensure connectivity [73].

Centralized strategies may be adopted under optimal control formulations. Although computationally more expensive, these yield higher quality solutions [183]. They also improve the robustness against possible changes in the environment, avoid deadlocks, and obviate the necessity of decoupling. This problem may be discretized using, e.g., orthogonal collocation direct transcription, resulting in a large-scale NLP [184].

In unlabeled Multi-Robot Motion Planning (MRMP) scenarios, robots need to be placed interchangeably, such that each of the prescribed positions is filled by one of them. In contrast, in labeled cases, each position is associated with a specific robot. For unit-square robots maneuvering within an environment involving polygonal obstacles, this problem is PSPACE-hard [185].

For optimal formation trajectory tracking, finite-term performance can be achieved by optimizing a cost function in the presence of control and state-variable constraints, as well as invertible conditions, over Euclidean rigid body motions, using Pontryagin's maximum principle for Lie groups. The optimality of the performance index is guaranteed if, and only if, the linear dynamic systems are controllable [186,187].

Complexity issues may be partially alleviated by utilizing a priority assignment algorithm to divide a team of robots into smaller groups. This can be more robustly dealt with using incremental approaches, based on, e.g., Satisfiability Modulo Theories (SMT) solvers. Robots with the same priority are in the same group, and, regarding motion planning, handled simultaneously. Robots taking higher priorities are treated as dynamic obstacles, collisions with which are avoided using a relatively small delay [188]. Priorities are also assigned dynamically to avoid collisions between the robots, using a Minimum Linear Ordering Problem (MLOP).

The total length, i.e., the summation of the lengths of the individual paths can be considered as an optimization criterion. This results in a computational complexity of $\tilde{O}(m^4 + m^2n^2)$, with m and n standing for the number of robots and the workspace's total complexity, respectively. The maximum length is OPT$+4m$, where OPT denotes the cost of the optimal solution [189].

Population-based stochastic optimization approaches such as PSO or Biogeography-Based Optimization (BBO), which are inspired by group behaviors observed in wildlife, may be utilized to find a global solution, i.e., to avoid local minima, through position updating strategies applied to diverse populations. The Normalized Step Cost (NSC) can be used to initialize the optimization procedure, which may also benefit from previously optimized parameters. Upon violation of the constraints, the parameters are replaced with those of the solution that is the closest to the query vector [76,77]. Moreover, path-finding through geometrically embedded discrete graphs can be combined with an implicit representation of the RMs for a reliable performance [65].

The metric utilized for finding the nearest neighbors within a certain configuration may have a considerable impact on the perception of connectivity of the RMs, as well as the path quality. Recently, other metrics have been adopted from the field of shape matching. Combining different metrics might lead to a preferable number of vertices within the RMs [190].

When it comes to computations, the cost functions related to collision testing can be enhanced by summarizing the most salient data through probabilistic modeling [44]. Directly searching a dense motion planning RM is run in $O(V \log V + E) \approx O(n^2)$ time, where n is the number of vertices. For a large n, it can be alleviated through successively shortening the path by making the r-disk sub-graphs increasingly dense over a low-dispersion deterministic sequence.

In MRMP, potential targets are captured using, e.g., visual sensors, where probabilities are modeled using, e.g., a Bayesian likelihood ratio tracker, which is recursively updated as a probability density function. Each robot's motion plan is dynamically determined according to the probability of each target for the corresponding location, and its distance from neighboring robots. Considering the inverse log-likelihood ratio as the temperature, the robot moves along the negative gradient of the temperature surface. The interaction with the local agents is performed using a Lennard–Jones potential, practically maximizing the mutual information between the target state and the measurements [191].

Consensus among the agents can be achieved in diverse ways. In the absence of a leader, distributed consensus protocols may be produced using, e.g., Pontryagin's principle, where the relative information is measured at sampling instants. To tackle the inadequacy in bandwidth, the continuity of the information may be improved using a distributed framework, taking advantage of multiple leaders. Distributed containment protocols are followed, such that the remaining robots converge, in terms of position and velocity, to the convex hull resulting from the leaders. Control gains, communication topologies, and sampling periods are devised independently, which provide further freedom for the controller [192].

5. Discussion and Future Work

We have provided a brief review of motion planning frameworks in the context of robotics, along with some of the essential considerations for their efficient practical utilization. In this section, to give a quick overview of the motion planning algorithms discussed throughout the present review, we categorize them from two different points of view. We also list the most prevalent approaches for each, followed by a discussion of which frameworks have been omitted and why, and then the concluding remarks.

5.1. Categorizations

The first categorization is based on the property of being conventional or heuristic and is shown in Figures 5 and 6, respectively. However, based on the studies covered in the present survey, we have clearly shown that with the emergence of Artificial Neural Network (ANN)s and machine learning algorithms, especially Deep Neural Network (DNN)s, there has been a noticeable transition in the trend of research and development activities within the field of motion planning. Owing to their power and efficiency in solving problems such as optimization and route search, etc., AI-based methods, especially DL, have emerged as an independent category of techniques. Since 2015, they have occupied the largest proportion of related works.

The second categorization is on the basis of being global or local, and this is shown in Figure 7. Global path planning uses the global geographical or a static map to find an optimal path [193,194], whereas local path planning requires a constant supply of sensory information to compute a collision-free path within the robot's immediate vicinity. Unlike local planners, global ones may struggle with real-time performance [195–197]. A combinatory interpretation using the two types of categorization shown in Figure 5 and Figure 7 provides an advantageous insight into which choice of motion planning algorithm would best suit a certain application.

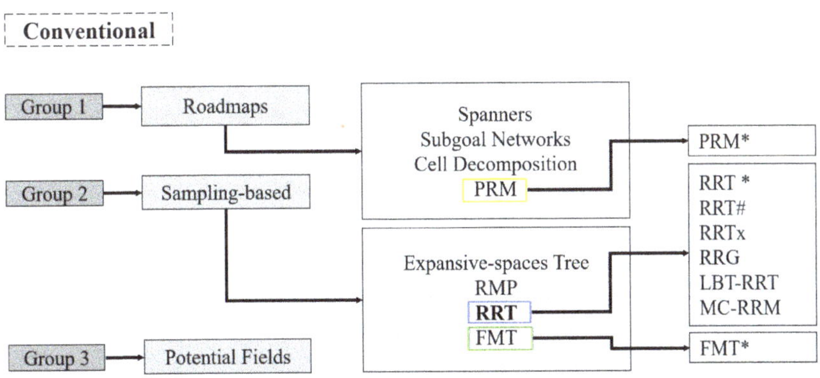

Figure 5. Conventional motion planning frameworks covered in this review.

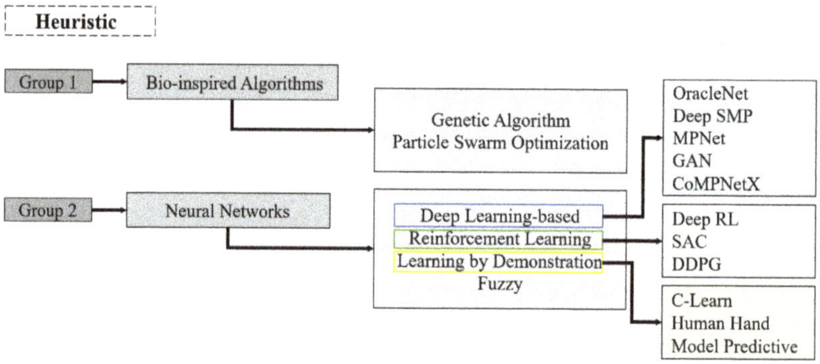

Figure 6. Heuristic motion planning approaches reviewed throughout the present survey.

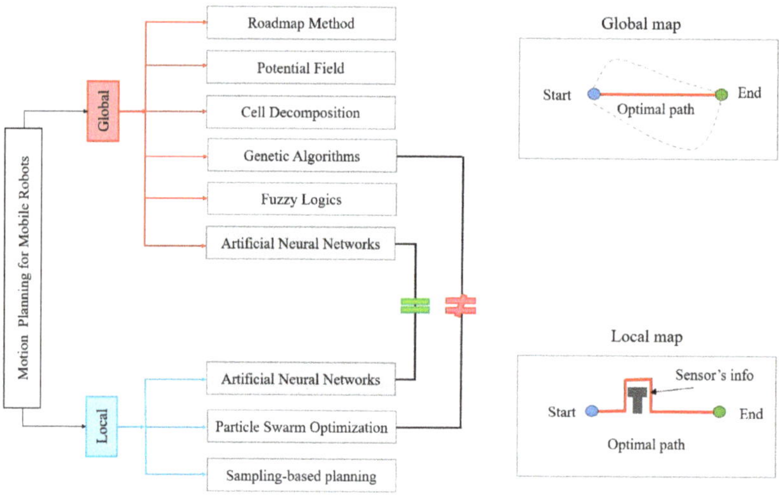

Figure 7. Categorization of the motion planning algorithms discussed in this review based on being global or local.

When it comes to older methods, one may clearly see that over time, heuristic approaches have gained a wider utilization for solving motion design problems. This is shown in the diagrams illustrated in Figure 8, which indicate the percentage usage of conventional and heuristic techniques for three consecutive decades spanning the 1980s through the 2000s. These statistics were taken from [198].

However, in the case of more recent years, the share of motion planning pipelines relying on learning has risen significantly. For DL networks, RL, and DBL, this has changed from 6% to 27%. This has taken place from the period 2010–2015 to 2015–2022 [5]. The ratios of the utilization of these strategies are shown, along with those of heuristic and conventional approaches in Figure 9. A more detailed representation is shown in Figure 10, which indicates the specific ratios for, not only learning-based methods, but also for the bio-inspired, potential field, and sampling-based ones. It should be noted that Figure 10 was adopted from [5], which was also the source of the above statistics.

These figures demonstrate the remarkable pace at which the category of DNN-based motion planning systems has been expanding with respect to the others, and it may well continue doing so in the coming years. While classical methods seem to have maintained their share, classical heuristic structures appear to be rapidly losing their competitiveness against learning-based mechanisms. This can be attributed to learning-based mechanisms' strong robustness and scalability, which enables them to handle the challenging motion-planning requirements arising from the growing complexities of modern robots, as well as those of the environments they are expected to operate within [199].

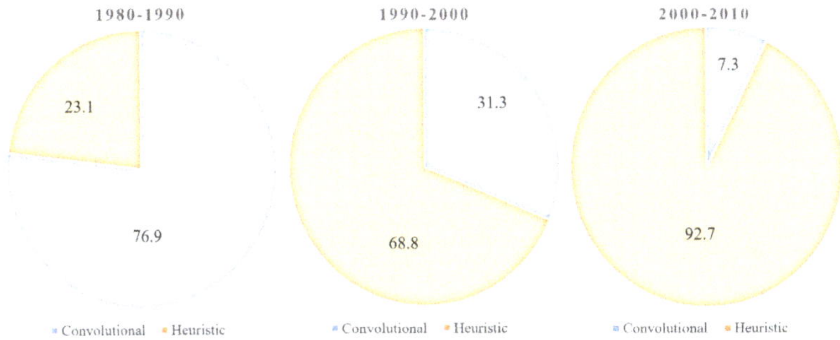

Figure 8. Visual comparison of percentages, indicating the trend of heuristic motion planning methods surpassing conventional ones over time. These values were taken from [198]. The diagrams cover the period from 1980 until 2010.

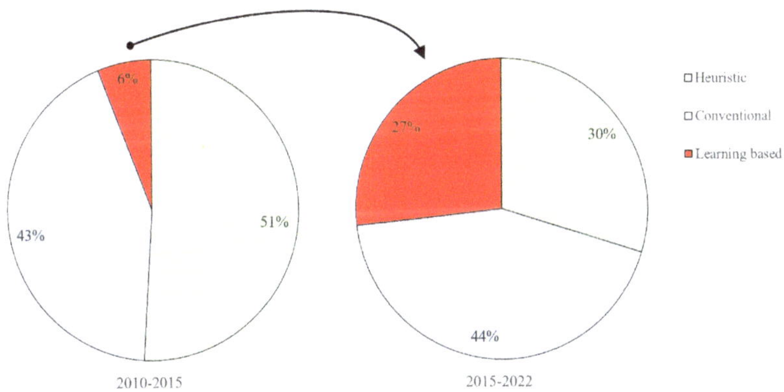

Figure 9. The trend in use of learning-based motion planning frameworks in recent years compared to heuristic and conventional methods, represented as percentages. More detailed statistics are shown in Figure 10.

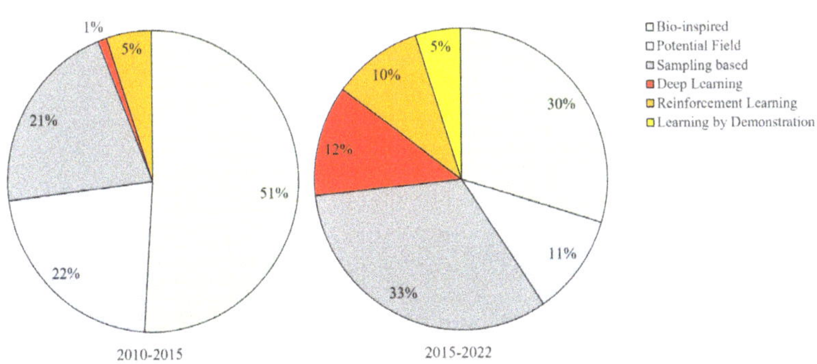

Figure 10. The trend in popularity of bio-inspired, potential field, and sampling-based motion planning frameworks compared to learning-based techniques taking advantage of DL, RL, and DBL, shown as percentages. The figure was adopted from [5].

It should be noted that in this survey, for the sake of coherence and maintaining brevity, not all motion planning pipelines have been covered. More clearly, to enable an in-depth analysis of the current state of the art and its shortcomings, this review has focused on frameworks of ongoing significance in practical contexts, where this judgment was made based on the frequency of practical usage of each technique in the recent robotics literature. More comprehensive categorizations of conventional and heuristic motion planning approaches are shown in Figures 11 and 12, respectively.

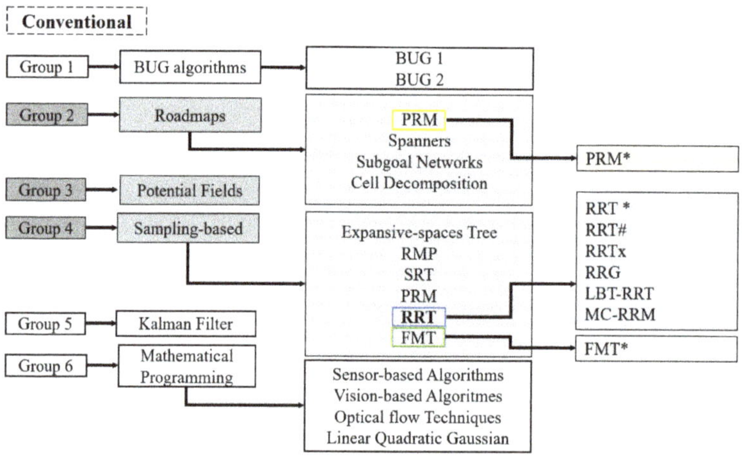

Figure 11. A more comprehensive categorization of conventional motion planning algorithms, with those not covered in the present survey shown within boxes with a plain background. Specifically, BUG algorithms, Kalman filter, mathematical programming, and geometric-based approaches were studied in this survey.

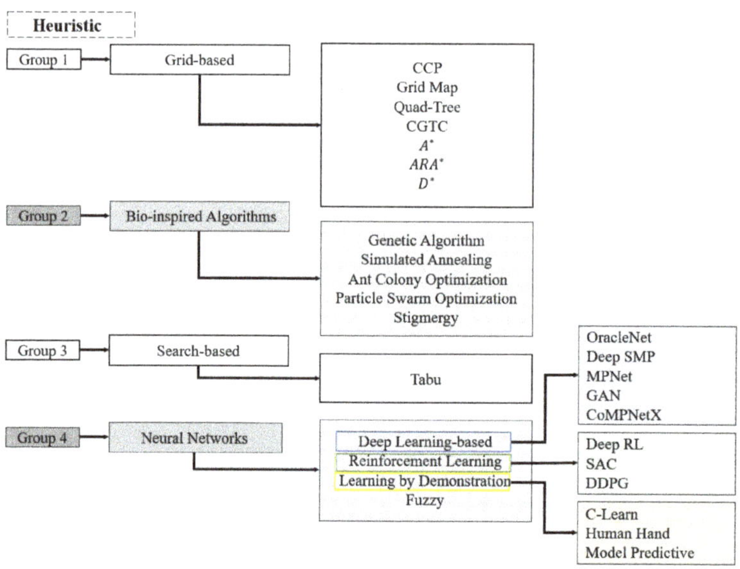

Figure 12. A more inclusive categorization of heuristic motion planning techniques, including grid- and search-based methods, which were not explored in the present review and are placed within boxes with a plain background in the figure.

5.2. Categories Excluded from the Review

For quick reference, each category of methods excluded from the survey will be concisely discussed.

The BUG 1 algorithm is based on the idea of having a robot follow an obstacle's edge, i.e., its contour, until it reaches a full cycle around the obstacle. Meanwhile, the distance to the goal is constantly updated to find the point on the edge which minimizes it. This point will then be used to start the next move toward the goal, which will take place within the next cycle. BUG 2 is faster, as it does not require the robot to undertake a full cycle and instead, makes it depart from the obstacle as soon as the slope of the line from the robot to the goal is the same as that of the line connecting the initial position to the goal [78].

Kalman Filter (KF) is useful in contexts where the tasks of map building and motion planning need to be handled simultaneously through scanning the terrain. Doing so requires obtaining depth information using detectors, where the uncertainties are handled by fusing the data. A fuzzy adaptive KF may help update the map and hierarchically optimize the scanning procedure according to the terrain type by adjusting the filter gain [200].

In mathematical programming, obstacle-avoidance constraints are modeled as inequalities imposed on the configuration parameters, where due to the underlying nonlinearity and the high number of obstacles, typically, a numerical approach is employed to solve the resulting optimization problem [198].

Grid-based motion planning covers a diverse range of algorithms. Complete Coverage Planning (CCP) considers the whole mapping area and the grid-based coverage method is the most popular example, which is suitable for indoor environments. However, its memory usage may be significantly high, due to the fact that the map resolution is not adjusted with respect to the actual environment's complexity, including the shapes and positions of the obstacles. Moreover, it is prone to nonoptimality, due to possible inefficient organization of the cells [201]. Other variants of grid-based algorithms include quadtree [202], Circle Grid Trajectory Cell (CGTC) [203], A* [203,204], Anytime Repairing A* (ARA*) [205,206], and D* [207].

Search-based algorithms work based on a neighborhood search over several waypoints, aiming at real-time, simultaneous map building and path planning. For example, in Tabu search, to avoid becoming stuck in local minima, a Tabu list is made to store nodes that have been evaluated previously [208].

5.3. Future Work

Various applications of motion planning are worthy of further investigation and are still at a primitive stage. Therefore, in what follows, some open challenges that need to be more rigorously addressed in upcoming studies will be concisely discussed, to provide insights into potential future research directions

Although some studies have suggested projecting the problem of motion planning in unstructured environments onto the 2D domain, this may not be practically sufficient. On the other hand, dealing with the problem in the actual 3D domain might fail to accommodate all requirements [20]. Therefore, it is recommended to seek solutions that provide a compromise between the two, using high-level motion primitives in an alternative space, thereby tracing down lower-level characteristics, to devise more detailed configurations over the preliminary outcomes projected onto the other space.

As far as intention-aware and DBL motion planning techniques are concerned, a robot should ideally resemble the intended patterns as closely as possible and perform robustly in the presence of possible errors. This is also the case for self-driving cars [88], which are expected to present a human-like driving behavior. Therefore, larger and more comprehensive representation databases offering more inclusive samples that resemble appropriate reactions under different situations may make of great contribution to a robotic system's fluency.

For example, self-driving cars possess limited sensory equipment but are meant to tackle occlusions in highly cluttered urban areas, while ensuring convenience and driving fluency [25]. Thus, further investigation and development are needed to devise motion planning standards that prevent over-cautiousness but do not undermine safety.

Even in the case of humanoid robots, their performance shows a significant depreciation in practical situations compared to simulated environments, even in the presence of local feedback drawn from an optimizer. In extreme cases, they may be unable to physically meet practical demands, where e.g., trying to apply the simulated trajectory could break the robot apart [166]. This indicates the importance of the balance between the advancements targeted within the two realms, i.e., the theoretical and practical sides.

Avoiding collisions in Human-Robot Collaboration (HRC), e.g., in manufacturing, robotic assembly, and robotic disassembly, requires keeping a distance between humans and robots. This may undermine performance and prevent the robot from reaching its goal quickly enough. Context awareness helps a robot to realize the human operator's intentions and enables it to ensure safety within shorter distances. Human pose recognition and collision sensing calibration are among the factors that could contribute to obtaining such an awareness [209]. On the other hand, the size of dynamic safety zones and bounding volumes may be minimized online according to torque constraints and robot dynamics, where the possibility of collision is assessed and avoided using smooth stop trajectories based on intersections, thereby optimizing the robot's speed [210,211]. These concepts play a critical role in improving efficiency and fluency, while ensuring safety, and are worth further investigation and development in upcoming studies.

Finally, shape and motion both play major roles in contacts between a robot and other objects and agents. Although this could be considered at the design or control stages of motion planning; in the literature, they are not usually considered simultaneously, i.e., through a synergy [212], which would be worth dedicating further effort to in the course of future work.

Funding: This work was partly supported by the Center for Research-based Innovation SmartForest: Bringing Industry 4.0 to the Norwegian forest sector (NFR SFI project no. 309671, smartforest.no); the European Social Fund via IT Academy programme; and the Estonian Research Council (projects no. COVSG24 and PSG605).

Institutional Review Board Statement: Not applicable.

Informed Consent Statement: Not applicable.

Data Availability Statement: Not applicable.

Conflicts of Interest: The authors declare no conflict of interest.

Acronyms

ACE	Approximate Clearance Evaluation
AGV	Autonomous Guided Vehicle
AI	Artificial Intelligence
ANN	Artificial Neural Network
APF	Artificial Potential Field
AQP	Alternating Quadratic Programming
ARA*	Anytime Repairing A*
ATA	Aggressive Turn-Around
BBO	Biogeography-Based Optimization
BIT	Batch-Informed Trees
B-spline	Basis spline
BTT	BoTtleneck Tree

BVP	Boundary Value Problem
CAM	Computer Aided Manufacturing
CBF	Control Barrier Function
CCP	Complete Coverage Planning
CGTC	Circle Grid Trajectory Cell
CIO	Contact-Invariant Optimization
CNC	Computer Numerical Control
CNN	Convolutional Neural Network
CoMPNetX	Constrained Motion planning Networks x
CP	Collision Probability
CPU	Central Processing Unit
CRISP	Continuum Reconfigurable Incisionless Surgical Parallel
CV	Control Variates
CVE	Curvature Variation Energy
DBL	Demonstration-Based Learning
DDPG	Deep Deterministic Policy Gradient
DeepSMP	Deep Sampling-based Motion Planner
DGMP	Demonstration-Guided Motion Planning
DL	Deep Learning
DMPs	Dynamic Movement Primitives
DNN	Deep Neural Network
DoF	Degrees of Freedom
DP	Dynamic Programming
DRL	Deep Reinforcement Learning
EAS	Extended Action Specification
EASE	Exploitation of Abstract Symmetry of Environments
EE	End-Effector
EKF	Extended Kalman Filter
EM	Expectation-Maximization
ETA	Estimated Time of Arrival
FaSTrack	Fast and Safe Tracking
FCN	Fully Convolutional neural Network
FEA	Finite Element Analysis
FF	FastForward
FMT	Fast Marching Tree
GA	Genetic Algorithm
GAN	Generative Adversarial Network
GP	Gaussian Process
GPM	Gauss Pseudospectral Method
GPR	Gaussian Process Regression
GPU	Graphics Processing Unit
HER	Hindsight Experience Replay
HFR	High-Frequency Replanning
HHI	Human-Human Interaction
HRC	Human-Robot Collaboration
HRI	Human-Robot Interaction
HZD	Hybrid Zero Dynamics
ICS	Inevitable Collision State
IDTMP	Iteratively Deepened Task and Motion Planning
i.i.d.	independently and identically distributed
INVM	Infinity-Norm Velocity Minimization
IRL	Inverse Reinforcement Learning
I-RRT	Improved Rapidly-exploring Random Trees
IS	Importance Sampling
KF	Kalman Filter
kNN	k-Nearest-Neighbor
KP	Kinodynamic motion Planning
KPIECE	Kinodynamic motion Planning by Interior-Exterior Cell Exploration

LBT	Lower Bound Tree
LPV	Linear Parameter Varying
LRPP	Learned Reactive Planning Problem
LSPI	Least Squares Policy Iteration
LSTM	Long Short-Term Memory
LTI	Linear Time-Invariant
LTL	Linear Temporal Logic
MC	Monte Carlo
MC-RRM	Multi-Component Rapidly-exploring RoadMap
MCVI	Monte Carlo Value Iteration
MICP	Mixed-Integer Convex Programming
MITL	Metric Interval Temporal Logic
MLOP	Minimum Linear Ordering Problem
MPC	Model Predictive Control
MPNet	Motion Planning Networks
MPT	Motion Planning Templates
MR	Multi-Robot
MRMP	Multi-Robot Motion Planning
MSTTMR	Multi-Steering Tractor-Trailer Mobile Robot
NLP	NonLinear Programming
NN	Neural Network
NSC	Normalized Step Cost
PER	Prioritized Experience Replay
PDMPs	Parametric Dynamic Movement Primitives
POMDP	Partially Observable Markov Decision Process
pose	position and orientation
PPO	Proximal Policy Optimization
PSO	Particle Swarm Optimization
QP	Quadratic Programming
RBF	Radial Basis Function
RKHSs	Reproducing Kernel Hilbert Spaces
RL	Reinforcement Learning
RM	RoadMap
RMP	Repetitive Motion Planning
RRG	Rapidly-exploring Random Graph
PRM	Probabilistic RoadMap
RNN	Recurrent Neural Network
RRT	Rapidly-exploring Random Trees
SAC	Soft Actor Critic
SLAM	Simultaneous Localization And Mapping
SBMP	Sampling-Based Motion Planning
SMC	Satisfiability Modulo Convex
SMT	Satisfiability Modulo Theories
SQP	Sequential Quadratic Programming
STOMP	Stochastic Trajectory Optimizer for Motion Planning
SyCLoP	Synergistic Combination of Layers of Planning
TD	Temporal-Difference
TMP	Task and Motion Planning
UAV	Unmanned Aerial Vehicle
V2I	Vehicle-to-Infrastructure
V2V	Vehicle-to-Vehicle
VRU	Vulnerable Road User

References

1. Kala, R. *On-Road Intelligent Vehicles: Motion Planning for Intelligent Transportation Systems*; Butterworth-Heinemann: Oxford, UK, 2016.
2. Liu, C.; Atkeson, C.G.; Feng, S.; Xinjilefu, X. Full-body motion planning and control for the car egress task of the DARPA robotics challenge. In Proceedings of the 2015 IEEE-RAS 15th International Conference on Humanoid Robots (Humanoids), Seoul, South Korea, 3–5 November 2015; pp. 527–532.

3. Halperin, D.; Salzman, O.; Sharir, M. Algorithmic motion planning. In *Handbook of Discrete and Computational Geometry*; Chapman and Hall/CRC: Boca Raton, FL, USA, 2017; pp. 1311–1342.
4. Mac, T.T.; Copot, C.; Tran, D.T.; De Keyser, R. Heuristic approaches in robot path planning: A survey. *Robot. Auton. Syst.* **2016**, *86*, 13–28. [CrossRef]
5. Tamizi, M.G.; Yaghoubi, M.; Najjaran, H. A review of recent trend in motion planning of industrial robots. *Int. J. Intell. Robot. Appl.* **2023**, *7*, 253–274. [CrossRef]
6. Sun, W.; Torres, L.G.; Van Den Berg, J.; Alterovitz, R. Safe motion planning for imprecise robotic manipulators by minimizing probability of collision. In *Robotics Research*; Springer: Cham, Switzerland, 2016; pp. 685–701.
7. Roozegar, M.; Mahjoob, M.; Jahromi, M. Optimal motion planning and control of a nonholonomic spherical robot using dynamic programming approach: Simulation and experimental results. *Mechatronics* **2016**, *39*, 174–184. [CrossRef]
8. Nguyen, Q.V.; Colas, F.; Vincent, E.; Charpillet, F. Motion planning for robot audition. *Auton. Robot.* **2019**, *43*, 2293–2317. [CrossRef]
9. Mccrackin, D.C.; Johnson, S.W. Syntactic Inferential Motion Planning Method for Robotic Systems. U.S. Patent 9,855,657, 2 January 2018.
10. Yang, L.; Qi, J.; Song, D.; Xiao, J.; Han, J.; Xia, Y. Survey of robot 3D path planning algorithms. *J. Control Sci. Eng.* **2016**, *2016*, 7426913. [CrossRef]
11. Gillani, M.; Akbari, A.; Rosell, J. Physics-based motion planning: Evaluation criteria and benchmarking. In Proceedings of the Robot 2015: Second Iberian Robotics Conference, Lisbon, Portugal, 19–21 November 2015; pp. 43–55.
12. Gasparetto, A.; Boscariol, P.; Lanzutti, A.; Vidoni, R. Path planning and trajectory planning algorithms: A general overview. In *Motion and Operation Planning of Robotic Systems*; Springer: Cham, Switzerland, 2015; pp. 3–27.
13. Contreras-Cruz, M.A.; Ayala-Ramirez, V.; Hernandez-Belmonte, U.H. Mobile robot path planning using artificial bee colony and evolutionary programming. *Appl. Soft Comput.* **2015**, *30*, 319–328. [CrossRef]
14. Holladay, R.; Lozano-Pérez, T.; Rodriguez, A. Force-and-Motion Constrained Planning for Tool Use. In Proceedings of the International Conference on Intelligent Robots and Systems (IROS), Macau, China, 3–8 November 2019.
15. Schmerling, E.; Janson, L.; Pavone, M. Optimal sampling-based motion planning under differential constraints: The driftless case. In Proceedings of the 2015 IEEE International Conference on Robotics and Automation (ICRA), Seattle, WA, USA, 26–30 May 2015; pp. 2368–2375.
16. Sivaramakrishnan, A.; Littlefield, Z.; Bekris, K.E. Towards Learning Efficient Maneuver Sets for Kinodynamic Motion Planning. *arXiv* **2019**, arXiv:1907.07876.
17. Elbanhawi, M. Randomised Parameterisation Motion Planning for Autonomous Cars. Ph.D. Thesis, RMIT University, Melbourne, Australia, 2016.
18. Simon, B.; Riegl, P.; Gaull, A. Motion Planning With Flexible Trajectory Chains. In Proceedings of the 2018 IEEE International Conference on Vehicular Electronics and Safety (ICVES), Madrid, Spain, 12–14 September 2018; pp. 1–8.
19. Geng, J.; Arakelian, V. Design of Partially Balanced Planar 5R Symmetrical Parallel Manipulators via an Optimal Motion Planning. In Proceedings of the IFToMM World Congress on Mechanism and Machine Science, Krakow, Poland, 15–18 July 2019; pp. 2211–2220.
20. Ferri, F. Computing Fast Search Heuristics for Physics-based Mobile Robot Motion Planning. 2018. Available online: https://iris.uniroma1.it/handle/11573/1137741 (accessed on 3 April 2023).
21. Vidal, E.; Moll, M.; Palomeras, N.; Hernández, J.D.; Carreras, M.; Kavraki, L.E. Online multilayered motion planning with dynamic constraints for autonomous underwater vehicles. In Proceedings of the 2019 International Conference on Robotics and Automation (ICRA), Montreal, QC, Canada, 20–24 May 2019; pp. 8936–8942.
22. Choudhury, S.; Srinivasa, S.S. A Bayesian Active Learning Approach to Adaptive Motion Planning. In Proceedings of the International Symposium on Robotics Research, Puerto Varas, Chile, 11–14 December 2017.
23. Bhardwaj, M.; Boots, B.; Mukadam, M. Differentiable Gaussian process motion planning. *arXiv* **2019**, arXiv:1907.09591.
24. Nguyen, P.D.; Hoffmann, M.; Pattacini, U.; Metta, G. A fast heuristic Cartesian space motion planning algorithm for many-DoF robotic manipulators in dynamic environments. In Proceedings of the 2016 IEEE-RAS 16th International Conference on Humanoid Robots (Humanoids), Cancun, Mexico, 15–17 November 2016; pp. 884–891.
25. Naumann, M.; Konigshof, H.; Lauer, M.; Stiller, C. Safe but not Overcautious Motion Planning under Occlusions and Limited Sensor Range. In Proceedings of the 2019 IEEE Intelligent Vehicles Symposium (IV), Paris, France, 9–12 June 2019; pp. 140–145.
26. Ravanbakhsh, H.; Laine, F.; Seshia, S.A. Real-time Funnel Generation for Restricted Motion Planning. *arXiv* **2019**, arXiv:1911.01532.
27. Minguez, J.; Lamiraux, F.; Laumond, J.P. Motion planning and obstacle avoidance. In *Springer Handbook of Robotics*; Springer: Cham, Switzerland, 2016; pp. 1177–1202.
28. Janson, L.; Schmerling, E.; Clark, A.; Pavone, M. Fast marching tree: A fast marching sampling-based method for optimal motion planning in many dimensions. *Int. J. Robot. Res.* **2015**, *34*, 883–921. [CrossRef]
29. Rehan, M.; Reyhanoglu, M.; McClamroch, H. Motion planning for a knife-edge on the surface of a hyperboloid. In Proceedings of the 2017 11th Asian Control Conference (ASCC), Gold Coast, QLD, Australia, 17–20 December 2017; pp. 1326–1330.
30. Marin-Plaza, P.; Hussein, A.; Martin, D.; Escalera, A.D.L. Global and local path planning study in a ros-based research platform for autonomous vehicles. *J. Adv. Transp.* **2018**, *2018*, 6392697. [CrossRef]
31. Ye, G.; Alterovitz, R. Guided motion planning. In *Robotics Research*; Springer: Cham, Switzerland, 2017; pp. 291–307.

32. Palmieri, L.; Arras, K.O. Distance metric learning for RRT-based motion planning with constant-time inference. In Proceedings of the 2015 IEEE International Conference on Robotics and Automation (ICRA), Seattle, WA, USA, 26–30 May 2015; pp. 637–643.
33. Ramanagopal, M.S.; Phu-Van Nguyen, A.; Le Ny, J. A motion planning strategy for the active vision-based mapping of ground-level structures. *IEEE Trans. Autom. Sci. Eng.* **2017**, *15*, 356–368. [CrossRef]
34. Francis, G.; Ott, L.; Ramos, F. Stochastic functional gradient for motion planning in continuous occupancy maps. In Proceedings of the 2017 IEEE International Conference on Robotics and Automation (ICRA), Singapore, 29 May–3 June 2017; pp. 3778–3785.
35. Paden, B.; Čáp, M.; Yong, S.Z.; Yershov, D.; Frazzoli, E. A survey of motion planning and control techniques for self-driving urban vehicles. *IEEE Trans. Intell. Veh.* **2016**, *1*, 33–55. [CrossRef]
36. González, D.; Pérez, J.; Milanés, V.; Nashashibi, F. A review of motion planning techniques for automated vehicles. *IEEE Trans. Intell. Transp. Syst.* **2015**, *17*, 1135–1145. [CrossRef]
37. Katrakazas, C.; Quddus, M.; Chen, W.H.; Deka, L. Real-time motion planning methods for autonomous on-road driving: State-of-the-art and future research directions. *Transp. Res. Part C Emerg. Technol.* **2015**, *60*, 416–442. [CrossRef]
38. Kuffner, J.J.; LaValle, S.M. RRT-connect: An efficient approach to single-query path planning. In Proceedings of the Proceedings 2000 ICRA, Millennium Conference, IEEE International Conference on Robotics and Automation, Symposia Proceedings (Cat. No. 00CH37065), San Francisco, CA, USA, 24–28 April 2000; Volume 2, pp. 995–1001.
39. Boscariol, P.; Gasparetto, A.; Scalera, L. Path Planning for Special Robotic Operations. In *Robot Design: From Theory to Service Applications*; Carbone, G., Laribi, M.A., Eds.; Springer International Publishing: Cham, Switzerland, 2023; Volume 123, pp. 69–95.
40. Karaman, S.; Frazzoli, E. Sampling-based algorithms for optimal motion planning. *Int. J. Robot. Res.* **2011**, *30*, 846–894. [CrossRef]
41. Otte, M.; Frazzoli, E. RRTX: Asymptotically optimal single-query sampling-based motion planning with quick replanning. *Int. J. Robot. Res.* **2016**, *35*, 797–822. [CrossRef]
42. Okadome, Y.; Nakamura, Y.; Ishiguro, H. Sampling-Based Motion Planning with a Prediction Model using Fast Gaussian Process Regression. *Electron. Commun. Jpn.* **2017**, *100*, 24–34. [CrossRef]
43. Ma, L.; Xue, J.; Kawabata, K.; Zhu, J.; Ma, C.; Zheng, N. Efficient sampling-based motion planning for on-road autonomous driving. *IEEE Trans. Intell. Transp. Syst.* **2015**, *16*, 1961–1976. [CrossRef]
44. Knepper, R.A.; Mason, M.T. Realtime informed path sampling for motion planning search. In *Robotics Research*; Springer: Cham, Switzerland, 2017; pp. 401–417.
45. Salzman, O.; Halperin, D. Asymptotically near-optimal RRT for fast, high-quality motion planning. *IEEE Trans. Robot.* **2016**, *32*, 473–483. [CrossRef]
46. Xie, C.; van den Berg, J.; Patil, S.; Abbeel, P. Toward asymptotically optimal motion planning for kinodynamic systems using a two-point boundary value problem solver. In Proceedings of the 2015 IEEE International Conference on Robotics and Automation (ICRA), Seattle, WA, USA, 26–30 May 2015; pp. 4187–4194.
47. Tang, Y.; You, R.; Kan, H.; Tithi, J.J.; Ganapathi, P.; Chowdhury, R.A. Cache-oblivious wavefront: Improving parallelism of recursive dynamic programming algorithms without losing cache-efficiency. In Proceedings of the ACM SIGPLAN Notices, San Francisco, CA, USA, 7–11 February 2015; Volume 50, pp. 205–214.
48. Janson, L.; Ichter, B.; Pavone, M. Deterministic sampling-based motion planning: Optimality, complexity, and performance. *Int. J. Robot. Res.* **2018**, *37*, 46–61. [CrossRef]
49. Klemm, S.; Oberländer, J.; Hermann, A.; Roennau, A.; Schamm, T.; Zollner, J.M.; Dillmann, R. RRT*-Connect: Faster, asymptotically optimal motion planning. In Proceedings of the 2015 IEEE International Conference on Robotics and Biomimetics (ROBIO), Zhuhai, China, 6–9 December 2015; pp. 1670–1677.
50. Kleinbort, M.; Salzman, O.; Halperin, D. Collision detection or nearest-neighbor search? On the computational bottleneck in sampling-based motion planning. *arXiv* **2016**, arXiv:1607.04800.
51. Kleinbort, M.; Solovey, K.; Bonalli, R.; Bekris, K.E.; Halperin, D. RRT2.0 for Fast and Optimal Kinodynamic Sampling-Based Motion Planning. *arXiv* **2019**, arXiv:1909.05569.
52. Solovey, K.; Kleinbort, M. The critical radius in sampling-based motion planning. *Int. J. Robot. Res.* **2020**, *39*, 266–285. [CrossRef]
53. Joshi, S.S.; Tsiotras, P. Relevant Region Exploration On General Cost-maps For Sampling-Based Motion Planning. *arXiv* **2019**, arXiv:1910.05361.
54. Harper, M.Y.; Ordonez, C.; Collins, E.G.; Erlebacher, G. Parallel approach to motion planning in uncertain environments. In Proceedings of the Unmanned Systems Technology XX International Society for Optics and Photonics, Orlando, FL, USA, 15–19 April 2018; Volume 10640, p. 106400H.
55. Sun, W.; Patil, S.; Alterovitz, R. High-frequency replanning under uncertainty using parallel sampling-based motion planning. *IEEE Trans. Robot.* **2015**, *31*, 104–116. [CrossRef]
56. Vonásek, V.; Saska, M.; Winkler, L.; Přeučil, L. High-level motion planning for CPG-driven modular robots. *Robot. Auton. Syst.* **2015**, *68*, 116–128. [CrossRef]
57. Reist, P.; Preiswerk, P.; Tedrake, R. Feedback-motion-planning with simulation-based LQR-trees. *Int. J. Robot. Res.* **2016**, *35*, 1393–1416. [CrossRef]
58. Janson, L.; Schmerling, E.; Pavone, M. Monte Carlo motion planning for robot trajectory optimization under uncertainty. In *Robotics Research*; Springer: Cham, Switzerland, 2018; pp. 343–361.
59. Yang, G.; Vang, B.; Serlin, Z.; Belta, C.; Tron, R. Sampling-based Motion Planning via Control Barrier Functions. *arXiv* **2019**, arXiv:1907.06722.

60. Pendleton, S.D.; Liu, W.; Andersen, H.; Eng, Y.H.; Frazzoli, E.; Rus, D.; Ang, M.H. Numerical approach to reachability-guided sampling-based motion planning under differential constraints. *IEEE Robot. Autom. Lett.* **2017**, *2*, 1232–1239. [CrossRef]
61. Jaillet, L.; Porta, J.M. Path planning with loop closure constraints using an atlas-based RRT. In *Robotics Research*; Springer: Cham, Switzerland, 2017; pp. 345–362.
62. Yang, Y.; Ivan, V.; Merkt, W.; Vijayakumar, S. Scaling sampling-based motion planning to humanoid robots. In Proceedings of the 2016 IEEE International Conference on Robotics and Biomimetics (ROBIO), Qingdao, China, 3–7 December 2016; pp. 1448–1454.
63. Sucan, I.A.; Kavraki, L.E. A sampling-based tree planner for systems with complex dynamics. *IEEE Trans. Robot.* **2011**, *28*, 116–131. [CrossRef]
64. Muhayyuddin; Moll, M.; Kavraki, L.; Rosell, J. Randomized physics-based motion planning for grasping in cluttered and uncertain environments. *IEEE Robot. Autom. Lett.* **2017**, *3*, 712–719. [CrossRef]
65. Solovey, K.; Salzman, O.; Halperin, D. Finding a needle in an exponential haystack: Discrete RRT for exploration of implicit roadmaps in multi-robot motion planning. *Int. J. Robot. Res.* **2016**, *35*, 501–513. [CrossRef]
66. Kim, H.; Lee, H.; Choi, S.; Noh, Y.K.; Kim, H.J. Motion planning with movement primitives for cooperative aerial transportation in obstacle environment. In Proceedings of the 2017 IEEE International Conference on Robotics and Automation (ICRA), Singapore, 29 May–3 June 2017; pp. 2328–2334.
67. Spurny, V.; Petrlik, M.; Vonasek, V.; Saska, M. Cooperative transport of large objects by a pair of unmanned aerial systems using sampling-based motion planning. In Proceedings of the 2019 24th IEEE International Conference on Emerging Technologies and Factory Automation (ETFA), Zaragoza, Spain, 10–13 September 2019; pp. 955–962.
68. Valencia, R.; Andrade-Cetto, J. Path planning in belief space with pose SLAM. In *Mapping, Planning and Exploration with Pose SLAM*; Springer: Cham, Switzerland, 2018; pp. 53–87.
69. Marble, J.D.; Bekris, K.E. Asymptotically near-optimal is good enough for motion planning. In *Robotics Research*; Springer: Cham, Switzerland, 2017; pp. 419–436.
70. Murray, S.; Floyd-Jones, W.; Qi, Y.; Sorin, D.J.; Konidaris, G. Robot Motion Planning on a Chip. In Proceedings of the Robotics: Science and Systems, Ann Arbor, MI, USA, 18–22 June 2016.
71. Garrett, C.R.; Lozano-Pérez, T.; Kaelbling, L.P. Ffrob: An efficient heuristic for task and motion planning. In *Algorithmic Foundations of Robotics XI*; Springer: Cham, Switzerland, 2015; pp. 179–195.
72. Kovács, B.; Szayer, G.; Tajti, F.; Burdelis, M.; Korondi, P. A novel potential field method for path planning of mobile robots by adapting animal motion attributes. *Robot. Auton. Syst.* **2016**, *82*, 24–34. [CrossRef]
73. Di, B.; Zhou, R.; Duan, H. Potential field based receding horizon motion planning for centrality-aware multiple UAV cooperative surveillance. *Aerosp. Sci. Technol.* **2015**, *46*, 386–397. [CrossRef]
74. Roy, A.G.; Rakshit, P. Motion planning of non-holonomic wheeled robots using modified bat algorithm. In *Nature-Inspired Algorithms for Big Data Frameworks*; IGI Global: Hershey, PA, USA, 2019; pp. 94–123.
75. Meng, K.; Jia, Y.; Yang, H.; Niu, F.; Wang, Y.; Sun, D. Motion Planning and Robust Control for the Endovascular Navigation of a Microrobot. *IEEE Trans. Ind. Inform.* **2019**, *16*, 4557–4566. [CrossRef]
76. Kim, J.J.; Lee, J.J. Trajectory optimization with particle swarm optimization for manipulator motion planning. *IEEE Trans. Ind. Inform.* **2015**, *11*, 620–631. [CrossRef]
77. Mo, H.; Xu, L. Research of biogeography particle swarm optimization for robot path planning. *Neurocomputing* **2015**, *148*, 91–99. [CrossRef]
78. Campbell, S.; O'Mahony, N.; Carvalho, A.; Krpalkova, L.; Riordan, D.; Walsh, J. Path planning techniques for mobile robots a review. In Proceedings of the 2020 6th International Conference on Mechatronics and Robotics Engineering (ICMRE), Barcelona, Spain, 12–15 February 2020.
79. Ghorbani, A.; Shiry, S.; Nodehi, A. Using genetic algorithm for a mobile robot path planning. In Proceedings of the 2009 International Conference on Future Computer and Communication, Kuala Lumpar, Malaysia, 3–5 April 2009.
80. Gerke, M. Genetic path planning for mobile robots. In Proceedings of the 1999 American Control Conference (Cat. No. 99CH36251), San Diego, CA, USA, 2–4 June 1999; Volume 4.
81. Nonoyama, K.; Liu, Z.; Fujiwara, T.; Alam, M.; Nishi, T. Energy-efficient robot configuration and motion planning using genetic algorithm and particle swarm optimization. *Energies* **2022**, *15*, 2074. [CrossRef]
82. Pamosoaji, A.K.; Piao, M.; Hong, K.-S. PSO-based minimum-time motion planning for multiple vehicles under acceleration and velocity limitations. *Int. J. Control Autom. Syst.* **2019**, *17*, 2610–2623. [CrossRef]
83. Ordonez, C.; Gupta, N.; Reese, B.; Seegmiller, N.; Kelly, A.; Collins, E.G., Jr. Learning of skid-steered kinematic and dynamic models for motion planning. *Robot. Auton. Syst.* **2017**, *95*, 207–221. [CrossRef]
84. Li, Y.; Cui, R.; Li, Z.; Xu, D. Neural network approximation based near-optimal motion planning with kinodynamic constraints using RRT. *IEEE Trans. Ind. Electron.* **2018**, *65*, 8718–8729. [CrossRef]
85. Lindemann, L.; Dimarogonas, D.V. Robust motion planning employing signal temporal logic. In Proceedings of the 2017 American Control Conference (ACC), Seattle, WA, USA, 24–26 May 2017; pp. 2950–2955.
86. Paxton, C.; Raman, V.; Hager, G.D.; Kobilarov, M. Combining neural networks and tree search for task and motion planning in challenging environments. In Proceedings of the 2017 IEEE/RSJ International Conference on Intelligent Robots and Systems (IROS), Vancouver, BC, Canada, 24–28 September 2017; pp. 6059–6066.

87. Abbeel, P.; Ng, A.Y. Apprenticeship learning via inverse reinforcement learning. In Proceedings of the Twenty-First International Conference on Machine Learning, Banff, AB, Canada, 4–8 July 2004; p. 1.
88. Wulfmeier, M.; Wang, D.Z.; Posner, I. Watch this: Scalable cost-function learning for path planning in urban environments. In Proceedings of the 2016 IEEE/RSJ International Conference on Intelligent Robots and Systems (IROS), Daejeon, South Korea, 9–14 October 2016; pp. 2089–2095.
89. Pfeiffer, M.; Schaeuble, M.; Nieto, J.; Siegwart, R.; Cadena, C. From perception to decision: A data-driven approach to end-to-end motion planning for autonomous ground robots. In Proceedings of the 2017 IEEE International Conference on Robotics and Automation (ICRA), Singapore, 29 May–3 June 2017; pp. 1527–1533.
90. Mainprice, J.; Hayne, R.; Berenson, D. Predicting human reaching motion in collaborative tasks using inverse optimal control and iterative replanning. In Proceedings of the 2015 IEEE International Conference on Robotics and Automation (ICRA), Seattle, WA, USA, 26-30 May 2015; pp. 885–892.
91. Park, J.S.; Park, C.; Manocha, D. Intention-Aware Motion Planning Using Learning Based Human Motion Prediction. In Proceedings of the Robotics: Science and Systems, Cambridge, MA, USA, 12–16 July 2017.
92. Hernández, J.D.; Moll, M.; Kavraki, L.E. Lazy Evaluation of Goal Specifications Guided by Motion Planning. In Proceedings of the 2019 International Conference on Robotics and Automation (ICRA), Montreal, QC, Canada, 20–24 May 2019.
93. Wilde, N.; Kulić, D.; Smith, S.L. Learning user preferences in robot motion planning through interaction. In Proceedings of the 2018 IEEE International Conference on Robotics and Automation (ICRA), Brisbane, QLD, Australia, 21–25 May 2018; pp. 619–626.
94. Qureshi, A.H.; Simeonov, A.; Bency, M.J.; Yip, M.C. Motion planning networks. In Proceedings of the 2019 International Conference on Robotics and Automation (ICRA), Montreal, QC, Canada, 20–24 May 2019; pp. 2118–2124.
95. Lenz, D. Motion Planning for Highly-Automated Vehicles under Uncertainties and Interactions with Human Drivers. Ph.D. Thesis, Technische Universität München, München, Germany, 2018.
96. Tian, Y.; Chang, Q. Motion Planning for Human-Robot Collaboration based on Reinforcement Learning. In Proceedings of the 2022 IEEE 18th International Conference on Automation Science and Engineering (CASE), Mexico City, Mexico, 20–24 August 2022.
97. Qureshi, A.H.; Yip, M.C. Deeply informed neural sampling for robot motion planning. In Proceedings of the 2018 IEEE/RSJ International Conference on Intelligent Robots and Systems (IROS), Madrid, Spain, 1–5 October 2018.
98. Qureshi, A.H.; Miao, Y.; Simeonov, A.; Yip, M.C. Motion planning networks: Bridging the gap between learning-based and classical motion planners. *IEEE Trans. Robot.* **2020**, *37*, 48–66. [CrossRef]
99. Bency, M.J.; Qureshi, A.H.; Yip, M.C. Neural path planning: Fixed time, near-optimal path generation via oracle imitation. In Proceedings of the 2019 IEEE/RSJ International Conference on Intelligent Robots and Systems (IROS), Macau, China, 3–8 November 2019.
100. Lembono, T.S.; Pignat, E.; Jankowski, J.; Calinon, S. Learning constrained distributions of robot configurations with generative adversarial network. *IEEE Robot. Autom. Lett.* **2021**, *6*, 4233–4240. [CrossRef]
101. Qureshi, A.H.; Dong, J.; Baig, A.; Yip, M.C. Constrained motion planning networks x. *IEEE Trans. Robot.* **2021**, *38*, 868–886. [CrossRef]
102. Everett, M.; Chen, Y.F.; How, J.P. Motion planning among dynamic, decision-making agents with deep reinforcement learning. In Proceedings of the 2018 IEEE/RSJ International Conference on Intelligent Robots and Systems (IROS), Madrid, Spain, 1–5 October 2018; pp. 3052–3059.
103. Chen, Y.F.; Everett, M.; Liu, M.; How, J.P. Socially aware motion planning with deep reinforcement learning. In Proceedings of the 2017 IEEE/RSJ International Conference on Intelligent Robots and Systems (IROS), Vancouver, BC, Canada, 24–28 September 2017; pp. 1343–1350.
104. Sangiovanni, B.; Incremona, G.P.; Piastra, M.; Ferrara, A. Self-configuring robot path planning with obstacle avoidance via deep reinforcement learning. *IEEE Control Syst. Lett.* **2020**, *5*, 397–402. [CrossRef]
105. Zhou, D.; Jia, R.; Yao, H.; Xie, M. Robotic arm motion planning based on residual reinforcement learning. In Proceedings of the 2021 13th International Conference on Computer and Automation Engineering (ICCAE), Melbourne, Australia, 20–22 March 2021.
106. Prianto, E.; Park, J.-H.; Bae, J.-H.; Kim, J.-S. Deep reinforcement learning-based path planning for multi-arm manipulators with periodically moving obstacles. *Appl. Sci.* **2021**, *11*, 2587. [CrossRef]
107. Guo, M.; Wang, Y.; Liang, B.; Chen, Z.; Lin, J.; Huang, K. Robot Path Planning via Deep Reinforcement Learning with Improved Reward Function. In *Proceedings of 2021 Chinese Intelligent Systems Conference: Volume III*; Springer: Singapore, 2022.
108. Gupta, K.; Najjaran, H. Exploiting Abstract Symmetries in Reinforcement Learning for Complex Environments. In Proceedings of the 2022 International Conference on Robotics and Automation (ICRA), Philadelphia, PA, USA, 23–27 May 2022.
109. Incremona, G.P.; Sacchi, N.; Sangiovanni, B.; Ferrara, A. Experimental Assessment of Deep Reinforcement Learning for Robot Obstacle Avoidance: A LPV Control Perspective. *IFAC-PapersOnLine* **2021**, *54*, 89–94. [CrossRef]
110. Wang, S.; Cao, Y.; Zheng, X.; Zhang, T. An end-to-end trajectory planning strategy for free-floating space robots. In Proceedings of the 2021 40th Chinese Control Conference (CCC), Shanghai, China, 26–28 July 2021.
111. Yang, J.; Peng, G. DDPG with Meta-Learning-Based Experience Replay Separation for Robot Trajectory Planning. In Proceedings of the 2021 7th International Conference on Control, Automation and Robotics (ICCAR), Singapore, 23–26 April 2021.
112. Jurgenson, T.; Tamar, A. Harnessing reinforcement learning for neural motion planning. *arXiv* **2019**, arXiv:1906.00214.
113. Chen, P.; Pei, J.; Lu, W.; Li, M. A deep reinforcement learning based method for real-time path planning and dynamic obstacle avoidance. *Neurocomputing* **2022**, *497*, 64–75. [CrossRef]

114. Hockman, B.; Pavone, M. Stochastic motion planning for hopping rovers on small solar system bodies. In Proceedings of the International Symposium on Robotics Research (ISRR), Puerto Varas, Chile, 11–14 December 2017.
115. Phiquepal, C.; Toussaint, M. Combined task and motion planning under partial observability: An optimization-based approach. In Proceedings of the 2019 International Conference on Robotics and Automation (ICRA), Montreal, QC, Canada, 20–24 May 2019; pp. 9000–9006.
116. Dantam, N.T.; Chaudhuri, S.; Kavraki, L.E. The Task-Motion Kit: An Open Source, General-Purpose Task and Motion-Planning Framework. *IEEE Robot. Autom. Mag.* **2018**, *25*, 61–70. [CrossRef]
117. Farber, M. Configuration spaces and robot motion planning algorithms. *arXiv* **2017**, arXiv:1701.02083.
118. Elhafsi, A.; Ivanovic, B.; Janson, L.; Pavone, M. Map-Predictive Motion Planning in Unknown Environments. *arXiv* **2019**, arXiv:1910.08184.
119. Dong, J.; Mukadam, M.; Dellaert, F.; Boots, B. Motion Planning as Probabilistic Inference using Gaussian Processes and Factor Graphs. In Proceedings of the Robotics: Science and Systems, Ann Arbor, MI, USA, 18–22 June 2016; Volume 12, p. 4.
120. Marinho, Z.; Dragan, A.; Byravan, A.; Boots, B.; Srinivasa, S.; Gordon, G. Functional gradient motion planning in reproducing kernel hilbert spaces. *arXiv* **2016**, arXiv:1601.03648.
121. Mercy, T.; Van Loock, W.; Pipeleers, G. Real-time motion planning in the presence of moving obstacles. In Proceedings of the 2016 European Control Conference (ECC), Aalborg, Denmark, 29 June–1 July 2016; pp. 1586–1591.
122. Mukadam, M.; Yan, X.; Boots, B. Gaussian process motion planning. In Proceedings of the 2016 IEEE International Conference on Robotics and Automation (ICRA), Stockholm, Sweden, 16–21 May 2016; pp. 9–15.
123. Bitar, G.; Martinsen, A.B.; Lekkas, A.M.; Breivik, M. Two-Stage Optimized Trajectory Planning for ASVs Under Polygonal Obstacle Constraints: Theory and Experiments. *IEEE Access* **2020**, *8*, 199953–199969. [CrossRef]
124. Fan, W.; Lee, C.H.; Chen, J.H. A realtime curvature-smooth interpolation scheme and motion planning for CNC machining of short line segments. *Int. J. Mach. Tools Manuf.* **2015**, *96*, 27–46. [CrossRef]
125. LeCun, Y.; Bengio, Y.; Hinton, G. Deep learning. *Nature* **2015**, *521*, 436–444. [CrossRef]
126. Guo, D.; Xu, F.; Yan, L.; Nie, Z.; Shao, H. A new noise-tolerant obstacle avoidance scheme for motion planning of redundant robot manipulators. *Front. Neurorobot.* **2018**, *12*, 51. [CrossRef] [PubMed]
127. Majumdar, A.; Tedrake, R. Funnel libraries for real-time robust feedback motion planning. *Int. J. Robot. Res.* **2017**, *36*, 947–982. [CrossRef]
128. Cohen, D.C.; Vandembroucq, L. Motion planning in connected sums of real projective spaces. *arXiv* **2018**, arXiv:1807.09947.
129. Hubicki, C.M.; Aguilar, J.J.; Goldman, D.I.; Ames, A.D. Tractable terrain-aware motion planning on granular media: An impulsive jumping study. In Proceedings of the 2016 IEEE/RSJ International Conference on Intelligent Robots and Systems (IROS), Daejeon, South Korea, 9–14 October 2016; pp. 3887–3892.
130. Mastalli, C.; Havoutis, I.; Focchi, M.; Caldwell, D.; Semini, C. Motion planning for challenging locomotion: A study of decoupled and coupled approaches. 2017. Available online: https://cmastalli.github.io/publications/locomotion20tro.pdf (accessed on 3 April 2023).
131. Ghosh, S.; Otsu, K.; Ono, M. Probabilistic Kinematic State Estimation for Motion Planning of Planetary Rovers. In Proceedings of the 2018 IEEE/RSJ International Conference on Intelligent Robots and Systems (IROS), Madrid, Spain, 1–5 October 2018; pp. 5148–5154.
132. Aceituno-Cabezas, B.; Mastalli, C.; Dai, H.; Focchi, M.; Radulescu, A.; Caldwell, D.G.; Cappelletto, J.; Grieco, J.C.; Fernández-López, G.; Semini, C. Simultaneous contact, gait, and motion planning for robust multilegged locomotion via mixed-integer convex optimization. *IEEE Robot. Autom. Lett.* **2017**, *3*, 2531–2538.
133. Sieverling, A.; Eppner, C.; Wolff, F.; Brock, O. Interleaving motion in contact and in free space for planning under uncertainty. In Proceedings of the 2017 IEEE/RSJ International Conference on Intelligent Robots and Systems (IROS), Vancouver, BC, Canada, 24–28 September 2017; pp. 4011–4073.
134. Akbari, A.; Diab, M.; Rosell, J. Contingent task and motion planning under uncertainty for human–robot interactions. *Appl. Sci.* **2020**, *10*, 1665. [CrossRef]
135. Xu, X.; Yang, Y.; Pan, S. Motion Planning for Mobile Robots. In *Advanced Path Planning for Mobile Entities*; IntechOpen: Rijeka, Croatia, 2018; p. 145.
136. Wang, H.; Huang, Y.; Khajepour, A.; Zhang, Y.; Rasekhipour, Y.; Cao, D. Crash mitigation in motion planning for autonomous vehicles. *IEEE Trans. Intell. Transp. Syst.* **2019**, *20*, 3313–3323. [CrossRef]
137. Simon, B.; Franke, F.; Riegl, P.; Gaull, A. Motion Planning for Collision Mitigation via FEM–Based Crash Severity Maps. In Proceedings of the 2019 IEEE Intelligent Vehicles Symposium (IV), Paris, France, 9–12 June 2019; pp. 2187–2194.
138. Zhao, Y.; Wang, Y.; Zhou, M.; Wu, J. Energy-Optimal Collision-Free Motion Planning for Multiaxis Motion Systems: An Alternating Quadratic Programming Approach. *IEEE Trans. Autom. Sci. Eng.* **2018**, *16*, 327–338. [CrossRef]
139. Herbert, S.L.; Chen, M.; Han, S.; Bansal, S.; Fisac, J.F.; Tomlin, C.J. FaSTrack: A modular framework for fast and guaranteed safe motion planning. In Proceedings of the 2017 IEEE 56th Annual Conference on Decision and Control (CDC), Melbourne, VIC, Australia, 12–15 December 2017; pp. 1517–1522.
140. Dantam, N.T.; Kingston, Z.K.; Chaudhuri, S.; Kavraki, L.E. Incremental Task and Motion Planning: A Constraint-Based Approach. In Proceedings of the Robotics: Science and Systems, Ann Arbor, MI, USA, 18–22 June 2016; Volume 12, p. 00052.

141. Becker, A.T.; Fekete, S.P.; Keldenich, P.; Konitzny, M.; Lin, L.; Scheffer, C. Coordinated Motion Planning: The Video (Multimedia Exposition). In Proceedings of the 34th International Symposium on Computational Geometry (SoCG 2018), Schloss Dagstuhl-Leibniz-Zentrum fuer Informatik, Budapest, Hungary, 11–14 June 2018.
142. Sayre-McCord, T.; Karaman, S. Perception-driven sparse graphs for optimal motion planning. In Proceedings of the 2018 IEEE/RSJ International Conference on Intelligent Robots and Systems (IROS), Madrid, Spain, 1–5 October 2018; pp. 8110–8117.
143. Vasquez, D. Novel planning-based algorithms for human motion prediction. In Proceedings of the 2016 IEEE International Conference on Robotics and Automation (ICRA), Stockholm, Sweden, 16–21 May 2016; pp. 3317–3322.
144. Shoukry, Y.; Nuzzo, P.; Saha, I.; Sangiovanni-Vincentelli, A.L.; Seshia, S.A.; Pappas, G.J.; Tabuada, P. Scalable motion planning using lazy smt-based solving. In Proceedings of the 55th IEEE Conference on Decision and Control (IEEE CDC), Las Vegas, NV, USA, 12–14 December 2016; pp. 6683–6688.
145. Ichnowski, J.; Alterovitz, R. Motion Planning Templates: A Motion Planning Framework for Robots with Low-power CPUs. In Proceedings of the 2019 International Conference on Robotics and Automation (ICRA), Montreal, QC, Canada, 20–24 May 2019.
146. Kim, B.; Kaelbling, L.P.; Lozano-Pérez, T. Learning to guide task and motion planning using score-space representation. In Proceedings of the 2017 IEEE International Conference on Robotics and Automation (ICRA), Singapore, 29 May–3 June 2017; pp. 2810–2817.
147. Ferrer-Mestres, J.; Frances, G.; Geffner, H. Combined task and motion planning as classical AI planning. *arXiv* **2017**, arXiv:1706.06927.
148. Lin, C.J.; Chen, C.S.; Yu, S. A GPU-based evolution algorithm for motion planning of a redundant robot. *Int. Robot. Autom. J.* **2017**, *2*, 1–14. [CrossRef]
149. Chretien, B.; Escande, A.; Kheddar, A. GPU robot motion planning using semi-infinite nonlinear programming. *IEEE Trans. Parallel Distrib. Syst.* **2016**, *27*, 2926–2939. [CrossRef]
150. Lee, T.; Kim, Y.J. Massively parallel motion planning algorithms under uncertainty using POMDP. *Int. J. Robot. Res.* **2016**, *35*, 928–942. [CrossRef]
151. Fragoso, A.T.; Matthies, L.H.; Murray, R.M. A fast motion planning representation for configuration flat robots with applications to micro air vehicles. In Proceedings of the 2017 American Control Conference (ACC), Seattle, WA, USA, 24–26 May 2017; pp. 4195–4202.
152. Kitagawa, R.; Liu, Y.; Kanda, T. Human-inspired Motion Planning for Omni-Directional Social Robots. In Proceedings of the 2021 ACM/IEEE International Conference on Human-Robot Interaction, New York, NY, USA, 8–11 March 2021; pp. 34–42.
153. Jung, J.; Kanda, T.; Kim, M.S. Guidelines for contextual motion design of a humanoid robot. *Int. J. Soc. Robot.* **2013**, *5*, 153–169. [CrossRef]
154. Saiki, L.Y.M.; Kanda, T. Social Group Motion in Robots. In *Social Robotics: 9th International Conference, ICSR 2017, Tsukuba, 22–24 Japan, November 2017, Proceedings*; Springer: Cham, Switzerland, 2017; Volume 10652, p. 474.
155. Zanlungo, F.; Chigodo, Y.; Ikeda, T.; Kanda, T. Experimental study and modelling of pedestrian space occupation and motion pattern in a real world environment. In *Pedestrian and Evacuation Dynamics 2012*; Springer: Cham, Switzerland, 2014; pp. 289–304.
156. Pfeiffer, M.; Schwesinger, U.; Sommer, H.; Galceran, E.; Siegwart, R. Predicting actions to act predictably: Cooperative partial motion planning with maximum entropy models. In Proceedings of the 2016 IEEE/RSJ International Conference on Intelligent Robots and Systems (IROS), Daejeon, South Korea, 9–14 October 2016; pp. 2096–2101.
157. Li, X.A.; Florendo, M.; Miller, E.L.; Ishiguro, H.; Saygin, P.A. Robot form and motion influences social attention. In Proceedings of the 2015 10th ACM/IEEE International Conference on Human-Robot Interaction (HRI), New York, NY, USA, 2–5 March 2015; pp. 43–50.
158. Yagi, S.; Ise, N.; Yu, S.; Nakata, Y.; Nakamura, Y.; Ishiguro, H. Perception of emotional gait-like motion of mobile humanoid robot using vertical oscillation. In Proceedings of the Companion of the 2020 ACM/IEEE International Conference on Human-Robot Interaction, New York, NY, USA, 23–26 March 2020; pp. 529–531.
159. Minato, T.; Ishiguro, H. Construction and evaluation of a model of natural human motion based on motion diversity. In Proceedings of the 3rd ACM/IEEE International Conference on Human Robot Interaction, New York, NY, USA, 12–15 March 2008; pp. 65–72.
160. Masayuki, K.; Takahiro, M.; Noriaki, M.; Hiroshi, I.; Norihiro, H. How does a Balancing Motion of a Humanoid Robots Affect a Human Motion and Impression? In Proceedings of the 2006 6th IEEE-RAS International Conference on Humanoid Robots, Genova, Italy, 4–6 December 2006; pp. 252–257.
161. Thornton, S.M.; Lewis, F.E.; Zhang, V.; Kochenderfer, M.J.; Gerdes, J.C. Value sensitive design for autonomous vehicle motion planning. In Proceedings of the 2018 IEEE Intelligent Vehicles Symposium (IV), Changshu, China, 26–30 June 2018; pp. 1157–1162.
162. Althoff, M.; Koschi, M.; Manzinger, S. CommonRoad: Composable benchmarks for motion planning on roads. In Proceedings of the 2017 IEEE Intelligent Vehicles Symposium (IV), Los Angeles, CA, USA, 11–14 June 2017; pp. 719–726.
163. Li, B.; Shao, Z. A unified motion planning method for parking an autonomous vehicle in the presence of irregularly placed obstacles. *Knowl.-Based Syst.* **2015**, *86*, 11–20. [CrossRef]
164. Wang, B.; Gong, J.; Chen, H. Motion Primitives Representation, Extraction and Connection for Automated Vehicle Motion Planning Applications. *IEEE Trans. Intell. Transp. Syst.* **2019**, *21*, 3931–3945. [CrossRef]
165. Yoshida, E.; Kanehiro, F.; Laumond, J.P. Whole-body Motion Planning. In *Humanoid Robotics: A Reference*; Springer: Dordrecht, The Netherlands, 2017.

166. Mordatch, I.; Lowrey, K.; Todorov, E. Ensemble-cio: Full-body dynamic motion planning that transfers to physical humanoids. In Proceedings of the 2015 IEEE/RSJ International Conference on Intelligent Robots and Systems (IROS), Hamburg, Germany, 28 September–2 October 2015; pp. 5307–5314.
167. Motahar, M.S.; Veer, S.; Poulakakis, I. Composing limit cycles for motion planning of 3D bipedal walkers. In Proceedings of the 2016 IEEE 55th Conference on Decision and Control (CDC), Las Vegas, NV, USA, 12–14 December 2016; pp. 6368–6374.
168. Zielinska, T.; Zimin, L.; Szumowski, M.; Ge, W. Motion Planning for a Humanoid Robot with Task Dependent Constraints. In Proceedings of the IFToMM World Congress on Mechanism and Machine Science, Krakow, Poland, 15–18 July 2019; pp. 1681–1690.
169. Yang, Y.; Merkt, W.; Ferrolho, H.; Ivan, V.; Vijayakumar, S. Efficient humanoid motion planning on uneven terrain using paired forward-inverse dynamic reachability maps. *IEEE Robot. Autom. Lett.* **2017**, *2*, 2279–2286. [CrossRef]
170. Tajika, T.; Miyashita, T.; Ishiguro, H.; Hagita, N. Reducing influence of robot's motion on tactile sensor based on partially linear model. In Proceedings of the 2008 IEEE/RSJ International Conference on Intelligent Robots and Systems, Nice, France, 22–26 September 2008; pp. 512–517.
171. Rosenstrauch, M.J.; Krüger, J. Safe human-robot-collaboration-introduction and experiment using ISO/TS 15066. In Proceedings of the 2017 3rd International Conference on Control, Automation and Robotics (ICCAR), Nagoya, Japan, 24–26 April 2017; pp. 740–744.
172. Vysocký, A.; Wada, H.; Kinugawa, J.; Kosuge, K. Motion Planning Analysis According to ISO/TS 15066 in Human–Robot Collaboration Environment. In Proceedings of the 2019 IEEE/ASME International Conference on Advanced Intelligent Mechatronics (AIM), Hong Kong, China, 8–12 July 2019; pp. 151–156.
173. Lin, X.; Zhang, J.; Shen, J.; Fernandez, G.; Hong, D.W. Optimization Based Motion Planning for Multi-Limbed Vertical Climbing Robots. *arXiv* **2019**, arXiv:1909.06339.
174. Mirrazavi Salehian, S.S.; Figueroa, N.; Billard, A. A unified framework for coordinated multi-arm motion planning. *Int. J. Robot. Res.* **2018**, *37*, 1205–1232. [CrossRef]
175. Garrett, C.R.; Lozano-Perez, T.; Kaelbling, L.P. FFRob: Leveraging symbolic planning for efficient task and motion planning. *Int. J. Robot. Res.* **2018**, *37*, 104–136. [CrossRef]
176. Liu, S.; Atanasov, N.; Mohta, K.; Kumar, V. Search-based motion planning for quadrotors using linear quadratic minimum time control. In Proceedings of the 2017 IEEE/RSJ International Conference on Intelligent Robots and Systems (IROS), Vancouver, BC, Canada, 24–28 September 2017; pp. 2872–2879.
177. Singhal, A.; Aharony, N.; Liang, A.T.; Muralidhar, G. Adaptive Route and Motion Planning Based on Learned External and Internal Vehicle Environment. U.S. Patent 10,234,302, 19 March 2019.
178. Dragan, A.D.; Bauman, S.; Forlizzi, J.; Srinivasa, S.S. Effects of robot motion on human-robot collaboration. In Proceedings of the Tenth Annual ACM/IEEE International Conference on Human-Robot Interaction, New York, NY, USA, 2–5 March 2015; pp. 51–58.
179. Sina Mirrazavi Salehian, S.; Figueroa, N.; Billard, A. Coordinated multi-arm motion planning: Reaching for moving objects in the face of uncertainty. In Proceedings of the Robotics: Science and Systems, Ann Arbor, MI, USA, 18–22 June 2016.
180. Alonso-Mora, J.; DeCastro, J.A.; Raman, V.; Rus, D.; Kress-Gazit, H. Reactive mission and motion planning with deadlock resolution avoiding dynamic obstacles. *Auton. Robot.* **2018**, *42*, 801–824. [CrossRef]
181. Shoukry, Y.; Nuzzo, P.; Balkan, A.; Saha, I.; Sangiovanni-Vincentelli, A.L.; Seshia, S.A.; Pappas, G.J.; Tabuada, P. Linear temporal logic motion planning for teams of underactuated robots using satisfiability modulo convex programming. In Proceedings of the 2017 IEEE 56th Annual Conference on Decision and Control (CDC), Melbourne, VIC, Australia, 12–15 December 2017; pp. 1132–1137.
182. Alonso-Mora, J.; Knepper, R.; Siegwart, R.; Rus, D. Local motion planning for collaborative multi-robot manipulation of deformable objects. In Proceedings of the 2015 IEEE International Conference on Robotics and Automation (ICRA), Seattle, WA, USA, 26–30 May 2015; pp. 5495–5502.
183. Ma, X.; Jiao, Z.; Wang, Z.; Panagou, D. Decentralized prioritized motion planning for multiple autonomous UAVs in 3D polygonal obstacle environments. In Proceedings of the 2016 International Conference on Unmanned Aircraft Systems (ICUAS), Arlington, VA, USA, 7–10 June 2016; pp. 292–300.
184. Li, B.; Liu, H.; Xiao, D.; Yu, G.; Zhang, Y. Centralized and optimal motion planning for large-scale AGV systems: A generic approach. *Adv. Eng. Softw.* **2017**, *106*, 33–46. [CrossRef]
185. Solovey, K.; Halperin, D. On the hardness of unlabeled multi-robot motion planning. *Int. J. Robot. Res.* **2016**, *35*, 1750–1759. [CrossRef]
186. Liu, Y.; Zhao, Y.; Chen, G. Finite-time formation tracking control for multiple vehicles: A motion planning approach. *Int. J. Robust Nonlinear Control* **2016**, *26*, 3130–3149. [CrossRef]
187. Liu, Y.; Geng, Z. Finite-time formation control for linear multi-agent systems: A motion planning approach. *Syst. Control Lett.* **2015**, *85*, 54–60. [CrossRef]
188. Saha, I.; Ramaithitima, R.; Kumar, V.; Pappas, G.J.; Seshia, S.A. Implan: Scalable incremental motion planning for multi-robot systems. In Proceedings of the 2016 ACM/IEEE 7th International Conference on Cyber-Physical Systems (ICCPS), Vienna, Austria, 11–14 April 2016; pp. 1–10.
189. Solovey, K.; Yu, J.; Zamir, O.; Halperin, D. Motion planning for unlabeled discs with optimality guarantees. *arXiv* **2015**, arXiv:1504.05218.

190. Atias, A.; Solovey, K.; Salzman, O.; Halperin, D. Effective metrics for multi-robot motion-planning. *Int. J. Robot. Res.* **2018**, *37*, 1741–1759. [CrossRef]
191. Sydney, N.; Paley, D.A.; Sofge, D. Physics-inspired motion planning for information-theoretic target detection using multiple aerial robots. *Auton. Robot.* **2017**, *41*, 231–241. [CrossRef]
192. Liu, Y.; Zhao, Y.; Chen, G. Sampled-data-based consensus and containment control of multiple harmonic oscillators: A motion-planning approach. *Chaos Interdiscip. J. Nonlinear Sci.* **2016**, *26*, 116303. [CrossRef]
193. Mohanty, P.K.; Singh, A.K.; Kumar, A.; Mahto, M.K.; Kundu, S. Path Planning Techniques for Mobile Robots: A Review. In *Proceedings of the International Conference on Soft Computing and Pattern Recognition, 2021*; Springer International Publishing: Cham, Switzerland, 15–17 December 2021.
194. Lu, Y.; Xue, Z.; Xia, G.-S.; Zhang, L. A survey on vision-based UAV navigation. *Geo-Spat. Inf. Sci.* **2018**, *21*, 21–32. [CrossRef]
195. Wang, Q.; Wieghardt, C.S.; Jiang, Y.; Gong, J.; Wagner, B. Generalized path corridor-based local path planning for nonholonomic mobile robot. In Proceedings of the 2015 IEEE 18th International Conference on Intelligent Transportation Systems, Gran Canaria, Spain, 15–18 September 2015.
196. Huy, Q.; Mita, S.; Nejad, H.T.N.; Han, L. Dynamic and safe path planning based on support vector machine among multi moving obstacles for autonomous vehicles. *IEICE Trans. Inf. Syst.* **2013**, *96*, 314–328.
197. Liu, L.; Wang, X.; Yang, X.; Liu, H.; Li, J.; Wang, P. Path planning techniques for mobile robots: Review and prospect. *Expert Syst. Appl.* **2023**, *227*, 120254. [CrossRef]
198. Tang, S.H.; Khaksar, W.; Ismail, N.; Ariffin, M. A review on robot motion planning approaches. *Pertanika J. Sci. Technol.* **2012**, *20*, 15–29.
199. Xiao, X.; Liu, B.; Warnell, G.; Stone, P. Motion planning and control for mobile robot navigation using machine learning: A survey. *Auton. Robot.* **2022**, *46*, 569–597. [CrossRef]
200. Najjaran, H.; Goldenberg, A. Real-time motion planning of an autonomous mobile manipulator using a fuzzy adaptive Kalman filter. *Robot. Auton. Syst.* **2007**, *55*, 96–106. [CrossRef]
201. Aydemir, H.; Tekerek, M.; Gök, M. Complete coverage planning with clustering method for autonomous mobile robots. *Concurr. Comput. Pract. Exp.* **2023**, e7830. [CrossRef]
202. Hwang, J.Y.; Kim, J.S.; Lim, S.S.; Park, K.H. A fast path planning by path graph optimization. *IEEE Trans. Syst. Man Cybern.-Part A Syst. Hum.* **2003**, *33*, 121–128. [CrossRef]
203. Zhu, M.; Xiao, C.; Gu, S.; Du, Z.; Wen, Y. A circle grid-based approach for obstacle avoidance motion planning of unmanned surface vehicles. *Proc. Inst. Mech. Eng. Part J. Eng. Marit. Environ.* **2023**, *237*, 132–152. [CrossRef]
204. Song, R.; Liu, Y.; Bucknall, R. Smoothed A* algorithm for practical unmanned surface vehicle path planning. *Appl. Ocean Res.* **2018**, *83*, 9–20. [CrossRef]
205. Likhachev, M.; Gordon, G.J.; Thrun, S. ARA*: Anytime A* with provable bounds on sub-optimality. *Adv. Neural Inf. Process. Syst.* **2003**, *16*, 767–774.
206. Zhang, Y.; Wang, H.; Yin, M.; Wang, J.; Hua, C. Bi-AM-RRT*: A Fast and Efficient Sampling-Based Motion Planning Algorithm in Dynamic Environments. *arXiv* **2023**, arXiv:2301.11816.
207. Stentz, A. The focussed d* algorithm for real-time replanning. In Proceedings of the International Joint Conference on Artificial Intelligence (IJCAI), Montreal, QC, Canada, 20–25 August 1995; Volume 95.
208. Balan, K.; Luo, C. Optimal trajectory planning for multiple waypoint path planning using tabu search. In Proceedings of the 2018 9th IEEE Annual Ubiquitous Computing, Electronics & Mobile Communication Conference (UEMCON), New York, NY, USA, 8–10 November 2018.
209. Liu, H.; Wang, L. Collision-free human-robot collaboration based on context awareness. *Robot. Comput.-Integr. Manuf.* **2021**, *67*, 101997. [CrossRef]
210. Scalera, L.; Vidoni, R.; Giusti, A. Optimal scaling of dynamic safety zones for collaborative robotics. In Proceedings of the 2021 IEEE International Conference on Robotics and Automation (ICRA), Xi'an, China, 30 May–5 June 2021.
211. Scalera, L.; Giusti, A.; Vidoni, R.; Gasparetto, A. Enhancing fluency and productivity in human-robot collaboration through online scaling of dynamic safety zones. *Int. J. Adv. Manuf. Technol.* **2022**, *121*, 6783–6798. [CrossRef]
212. Taylor, O.; Rodriguez, A. Optimal shape and motion planning for dynamic planar manipulation. *Auton. Robot.* **2019**, *43*, 327–344. [CrossRef]

machines

MDPI

Technical Note

Dual Cone Continuously Variable Transmission Model Controlled by LabVIEW

Šimon Berta [1,*], Vladimír Goga [1], Kristián Ondrejička [1], Erik Kučera [2] and Vladimír Kutiš [1]

[1] Department of Applied Mechanics and Mechatronics, Faculty of Electrical Engineering and Information Technology, Slovak University of Technology in Bratislava, Ilkovičova 3, 841 04 Bratislava, Slovakia; vladimir.goga@stuba.sk (V.G.); kristian.ondrejicka@stuba.sk (K.O.); vladimir.kutis@stuba.sk (V.K.)

[2] Department of Information, Communication and Control Systems, Faculty of Electrical Engineering and Information Technology, Slovak University of Technology in Bratislava, Ilkovičova 3, 841 04 Bratislava, Slovakia; erik.kucera@stuba.sk

* Correspondence: simon.berta@stuba.sk

Abstract

This paper outlines the design, development, and practical implementation of a rubber belt-driven dual cone continuously variable transmission (CVT) model. This model enables a demonstration of stepless changes in the transmission ratio between input and output shafts. Although the model can be operated manually via a control panel, enhanced functionality, such as automated measurement, proportional-integral-derivative (PID) speed control, and data measurement and storage, is achieved through a control application created within the LabVIEW virtual instrument environment. This work also includes a partial comparison between the practical implementation and its simulation model created in MATLAB-Simulink.

Keywords: continuously variable transmission; transmission ratio; cone CVT; model; LabVIEW; MATLAB-Simulink; PID control; VISA

1. Introduction

Continuously variable transmission systems provide significant advantages in both automotive and machine tool applications, enabling infinitely adjustable and seamless transfer of power while enhancing efficiency compared to traditional geared mechanical transmissions [1]. Furthermore, CVT technology can contribute to improved fuel efficiency and a reduction in emissions in vehicles equipped with internal combustion engines [2,3]. Numerous scientific publications, monographs, and dissertations [1–6] have examined the broader challenges associated with CVT systems, analyzing their benefits and limitations in comparison to conventional mechanical and automatic transmissions in both vehicles and machinery.

Among all types of CVT systems, the conventional push belt types are the most widely utilized and researched, with various elements of their dynamics assessed and modeled [2–4,7–10]. The academic paper by Carbone et al. [11] delineates a comparison between theoretical proposes and empirical validation of phenomena related to the functioning of a belt CVT system during a variation in the output gear ratio. Furthermore, a paper by Srivastava and Haque [12] summarizes the current status of research in dynamic modeling and control of friction-limited continuously variable transmissions. The fundamental principles, mathematical models, and computing approaches for such systems are thoroughly discussed in both these papers.

The aforementioned push belt type CVTs use a transmission belt or chain as the torque transmission element and moving pulleys that change shape during operation [13]. The change in rotation speed of the input (primary) and output (secondary) shafts is caused by a change in diameter of the transmission belt acting via friction on both pulleys (Figure 1) [3]. The two boundary states of transmission ratios (low and overdrive) are shown in Figure 1.

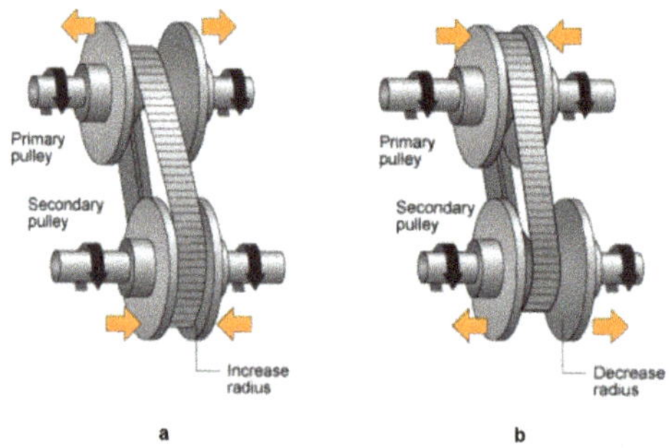

Figure 1. Push belt CVT—(**a**) low, (**b**) overdrive [13].

This fact indicates the primary disadvantage and limiting aspect of these systems: the maximum transmitted torque constrained by friction, along with the consequent friction-induced mechanical losses and wear of the pulley-belt system. Nevertheless, this was also one of the contributing factors that compelled researchers and engineers to engage in technical enhancements and develop the subsequent methodology of performance evaluation of the CVT concept.

Numerous studies show a variety of approaches for analyzing and demonstrating the performance of CVT systems, including the use of test rigs, laboratory testbeds, and prototype creation accompanied by experimental validation. Duong et al. [14] describe the design and development of a CVT test bench for evaluating the performance and characteristics of CVT. The test bench enables the assessment of key parameters influencing CVT performance, including pulley pressure, belt force, and torque under varying load conditions. Wong et al. [15] introduced a dual-belt CVT prototype and test rig for heavy-duty vehicles, validating their analytical model and studying basic characteristics. Spanoudakis and Tsourveloudis [16] explored the efficiency of an Electronic Shift Variable Transmission in a hydrogen fuel cell-powered urban vehicle using a laboratory testbed. Centeno et al. [17] designed and tested a CVT with an inertia-regulating system, consisting of three subsystems that convert engine rotation into oscillating movement, regulate it, and rectify it to unidirectional angular velocity. Supriyo et al. [18] proposed the development and experimental study of an electro-mechanical CVT system for motorcycle applications, which uses a DC motor and cam mechanism to control the primary pulley position and thereby change the transmission ratio. Work by Mohamed and Albatlan [19] presents a dynamic analysis and experimental work on a CVT system, focusing on the influence of loading conditions on the slip behavior and torque-transmitting ability of the CVT.

In addition to the push belt design, a significant number of experimental CVT systems are currently under research and development. These include magnetic, wheel, spherical, toroidal and cone systems [20–26]. The cone CVTs, which have a constant shape of the moving parts but a variable timing belt or chain location, adjust the transmission ratio by moving a wheel or belt down the axis of one or more bevel-shaped cylinders (Figure 2) [27]. The most basic design, known as the single-cone CVT, works by enabling a wheel to traverse the inclined surface of the cone, changing the effective diameter from a narrower

 https://doi.org/10.3390/machines14020141

to a broader segment. Notably, certain cone CVT systems use dual-cylinder topologies to improve performance and transmission ratio variability. The first mention of this kind of system was in 1903, when Evans and Knauf applied for a patent for a continuously variable transmission that uses two parallel conical cylinders facing opposite directions and connected by belts that can be moved along the cones to vary the gear ratio [28,29]. Added to that, the Evans variable speed gearbox, manufactured in the 1920s, is a more straightforward design [30]. It features two bevel cylinders arranged with a small gap of constant width between them, with the position of the so-called leather sleeve that runs between the cylinders determining the gear ratio.

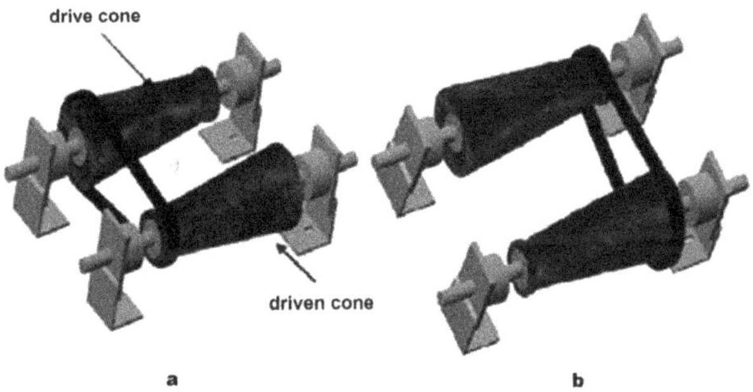

Figure 2. Cone type CVT—(**a**) overdrive, (**b**) low [27].

Cone-based CVTs have been explored for various purposes and applications. The design, optimization, and functionality of such systems have been examined in various academic papers. One design integrates a dynamic vibration absorber to create an adaptively tuned vibration absorber based on the principle of a twin-cone CVT system, capable of suppressing resonance in primary systems over a wide frequency range [31]. Lin et al. presented an innovative design of a twin-cone CVT gearbox based on a discretely adjustable radius pulley system [32]. Another innovative approach by Mobedi and Dede [33] proposed a novel two-sphere cone CVT design instead of friction wheels to enable independent output torque and position control, as well as bidirectional power transmission. Researchers have also applied system-driven product development and topology optimization approaches to develop cone CVT systems, highlighting their advantages over conventional transmissions [34]. The findings of their research have also been disseminated by Bazios et al. [35], who explored the utilization of computer-controlled topological optimization in the design of a twin-cone CVT transmission system. These systems offer benefits such as higher fuel efficiency, wider torque range, and smoother acceleration in vehicles [16,35]. The cone CVTs have been investigated for their slip and speed ratio characteristics in [36], with potential applications in both the automotive and machine tool industries. Additionally, a purely mechanical continuous automatic variable transmission (CAVT) system for bicycles was designed using a centrifugal governor and slotted cone, addressing durability issues associated with electronic components in other CVT systems [37]. These studies highlight the adaptability and potential of cone-based CVT systems across a wide range of applications. Beyond their industrial utility, these mechanisms also serve as powerful pedagogical tools for illustrating complex kinematic principles.

Mechanical engineering students gain a deeper understanding and increased motivation to learn about transmission systems when they can physically interact with such systems [38]. Consequently, incorporating a simple, demonstrative device into the educational process can further enhance the students' learning experience. This work details the design and implementation of a dual cone CVT physical model with dual-mode control

 https://doi.org/10.3390/machines14020141

capabilities. The system integrates manual adjustment with a LabVIEW-based PC interface, providing a versatile tool for educational demonstrations and research prototyping (Figure 3).

The development of CVT test rigs for educational and research purposes has been explored through various configurations. Previous work has focused on performance analysis using customized rigs [39] and the use of LabVIEW for monitoring control signals in belt-driven systems [40]. Specifically, automated clutch control [41] and the fundamental mechanics of cone-to-cone (CTC) transmissions [42] have been documented. While these didactic models [43] provide a foundation for understanding CVT behavior, this paper contributes a unique dual-mode architecture that allows for a seamless transition between manual and LabVIEW-based PID speed regulation, specifically optimized for HIL (Hardware-in-the-Loop) pedagogical applications (see Table 1).

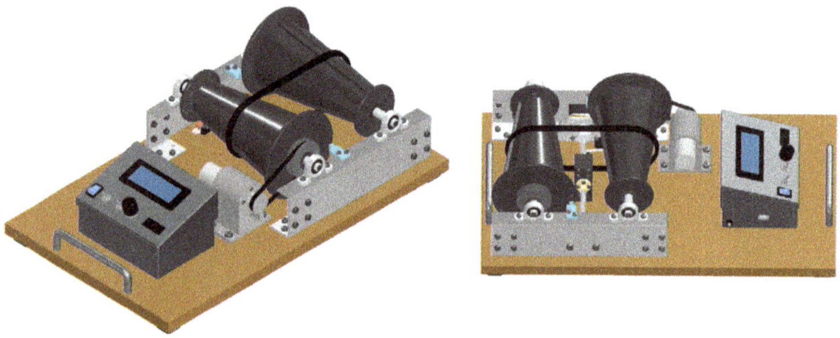

a b

Figure 3. Overview of the proposed dual cone CVT design model—(**a**) isometric view, (**b**) side view.

Table 1. Comparison of various CVT test rigs and didactic implementations.

Reference	Configuration	Control Interface	Primary Focus
Srivastava et al. [39]	V-Belt CVT	Manual/Mechanical	Performance analysis
Bieniek et al. [40]	Industrial CVT	LabVIEW	Signal characteristics
Younes et al. [41]	Centrifugal Clutch	LabVIEW	Automation of engagement
Bertini et al. [42]	Cone-to-Cone	Mechanical/Experimental	Fundamental mechanics
Kushwaha et al. [43]	Pedagogical Model	Manual	Educational fabrication
This Work	Dual Cone CVT	Manual + LabVIEW	Pedagogical HIL & PID

While several CVT demonstrators are documented in the literature, existing publications often emphasize theoretical outcomes rather than the granular implementation of hardware–software interfaces. This paper provides a comprehensive framework for the integration of dual-mode control (manual and LabVIEW), a technical detail frequently omitted in broader research articles. The proposed design utilizes a unique cost-effective configuration of LabVIEW-based data acquisition and PID regulation, specifically optimized for pedagogical reproducibility. Consequently, this work serves as a practical reference for researchers and educators seeking to develop functional, low-cost HIL platforms for CVT characterization.

2. Design of the Proposed Dual Cone CVT Model

The proposed design is derived from previously referenced studies on cone type CVTs, specifically influenced by references [27,31,34,35]. The cone geometry is inspired by the

 https://doi.org/10.3390/machines14020141

work of Patil et al. [34], and the mechanism for adjusting the belt position is conceptually based on ideas presented in [31,35]. The overall system architecture and configuration are primarily influenced by works [27–30].

The system is mounted on a main base plate equipped with side handles to facilitate transport (Figures 3 and 4). Main structural crossbars are affixed to this base plate and support the radial bearings and shafts of the rotating cones. A DC motor with a planetary gearbox, also mounted on the base plate, transmits rotational motion to the driving cone via a toothed belt and gear assembly. The driving cone, in turn, transfers torque to the driven cone through rubber belts. The rotational speeds of both cones are measured using Hall effect sensors, with twelve embedded neodymium magnets in each cone to ensure precise detection. The entire structure is assembled using standardized metric bolts and nuts, ensuring mechanical stability and allowing for straightforward disassembly in the event of component failure (Figure 3).

A lead screw, positioned between the cones, is driven by a NEMA 17 stepper motor. A nut attached to a linear slider travels along the lead screw. The slider moves across the full span between the cones, from one structural crossbar to the other (Figure 4). Limit switches are installed at both ends to control and restrict the slider's travel range. Rubber belts are threaded through an opening in the slider. By moving the slider left or right, the belt positions on the surfaces of both rotating cones can be adjusted smoothly and synchronously.

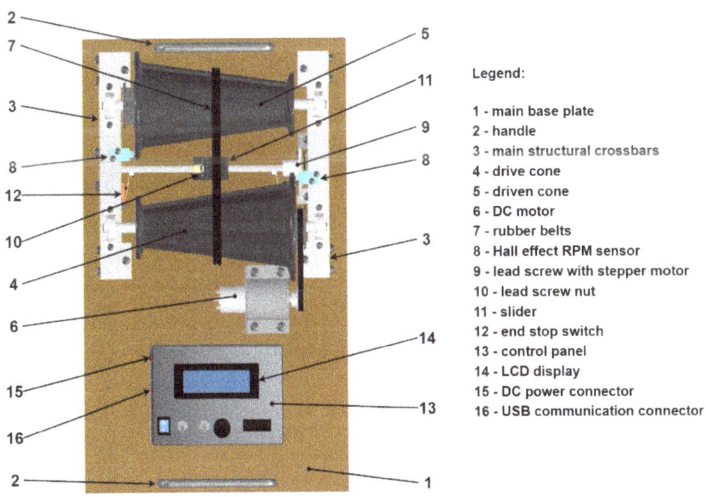

Figure 4. Detailed view of the proposed dual cone CVT design—top view.

3. Device Operation

The device enables the measurement and practical demonstration of continuous (stepless) variation in the transmission ratio between the driving and driven cones by adjusting the position of rubber belts (see Figure 2). The entire system can be operated manually via toggle switches and control elements located on the control panel (see Figures 4–6). A DC power connector is located on the side of the control panel to supply the system with power from a 15 V, 4 A power source. Additionally, a USB-C connector is provided for communication with and programming of the integrated Arduino Nano microcontroller (Figures 4 and 5). The utilized microcontroller is programmed using the Arduino IDE programming environment (version 2.3.0).

https://doi.org/10.3390/machines14020141

Figure 5. Constructed dual cone CVT (**left**) with control PC (**right**).

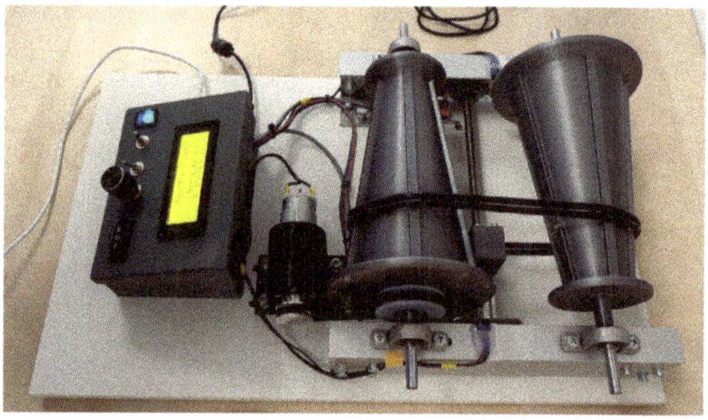

Figure 6. Constructed dual cone CVT—detailed side view.

3.1. Operating Principle

The concurrent variation in the rotational radii of the rubber belts on both cones (denoted as d and D, Figure 7) is governed by their lateral displacement. This produces an immediate and continuous adjustment of the transmission ratio, while the constant inter-cone distance c is maintained by radial bearings and crossbars (Figure 6). The transmission ratio (TR) can be expressed in terms of the rotational speeds of the drive cone (ω_1) and the driven cone (ω_2) as:

$$TR = \frac{\omega_2}{\omega_1} = \frac{\frac{2v}{d}}{\frac{2v}{D}} = \frac{D}{d}, \tag{1}$$

where ω_1 and ω_2 are measured in revolutions per minute (RPM). Alternatively, the transmission ratio (TR) may be determined from the effective diameters of the cones beneath the rubber belts, assuming a constant tangential velocity v of the elements of the belts. However, because accurately measuring the effective diameters is challenging, TR is instead calculated directly from the rotational speeds, which are measured using Hall-effect speed sensors.

Considering the construction constraints (cone taper angle, positions of the limit switches and stepper motor, and slider shape design), the proposed device enables transmission ratio adjustment within the range of 0.60 (low) to 1.55 (overdrive). The slider's movement is limited to 125 mm from its initial position on the stepper motor side of the lead screw (see Figure 4). At present, the slider's displacement speed is predefined and set to 4.9 mm/s, determined by the frequency preset of the pulse-width modulation (PWM) generation pin of the microcontroller. It should be noted, however, that this value may be

https://doi.org/10.3390/machines14020141

modified in future iterations of the device through the integration of an additional rotary knob for user adjustment.

The dual cone-and-belt system is designed for a maximum input speed of 2000 RPM at the DC motor shaft (at a nominal input voltage of 15 V DC under no-load conditions at the output shaft of the driven cone), which corresponds to a maximum of 450 RPM at the drive cone following gear reduction via the flat transmission belt (see Figure 6). Consequently, the maximum output speed of the driven cone reaches 700 RPM in the overdrive gear state. Under no-load conditions, the system achieves a peak angular acceleration of approximately 75 rad/s^2. This dynamic performance is primarily governed by the rotational inertia of the mechanical components and the frictional limits of the cone-and-belt interface. These performance benchmarks were established through preliminary characterization tests conducted during the initial system commissioning.

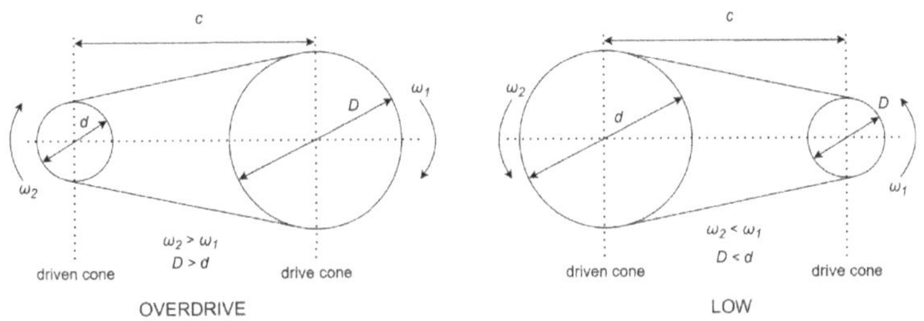

Figure 7. Boundary states of the transmission ratio.

3.2. Manual Control

The first switch from the left (Figure 4) serves as the main power switch, allowing the user to activate or deactivate the device. The second switch enables the selection of the control mode, either manual or LV. The third switch determines the DC motor's rotation direction (clockwise—CW or counterclockwise—CCW), which, in turn, sets the rotation direction of the cones. A rotary potentiometer with a control knob is provided to regulate the power supplied to the DC motor within a range of 0 to 255. As a result, the rotational speed of the drive cone can be adjusted from 0 to 450 RPM. The final control element is a three-position switch that controls slider movement—left, right, or stop (neutral center position). At the end positions, slider travel is programmatically constrained through inputs from the limit switches.

An LCD display (4 lines × 20 characters) mounted on the control panel (Figure 4) provides real-time operational data, including the rotational speeds of both cones (in RPM), the calculated transmission ratio (*TR*), the cone rotation direction (CW or CCW), the DC motor RPM setting, and the active operating mode (manual or LV). When the operating mode is switched to LV, enhanced functionality and control are enabled through a proprietary application running on the control PC (Figure 5).

Microcontroller Code

The detailed block functionality of the microcontroller code is illustrated in the block diagram shown in Figure 8. During initialization (function *setup()*), the program establishes serial communication via a universal serial bus (USB) port with the control PC, configures all input/output pins, sets the presets for the PWM output pin, initializes the LCD display, and presents the startup screen. At the conclusion of the *setup()* function, interrupts on the selected input pins are enabled to measure the rotational speeds of the cones. Within the interrupt service routines —specialized subprograms executed in response to hardware events—the input pulses from the Hall effect sensors are counted.

 https://doi.org/10.3390/machines14020141

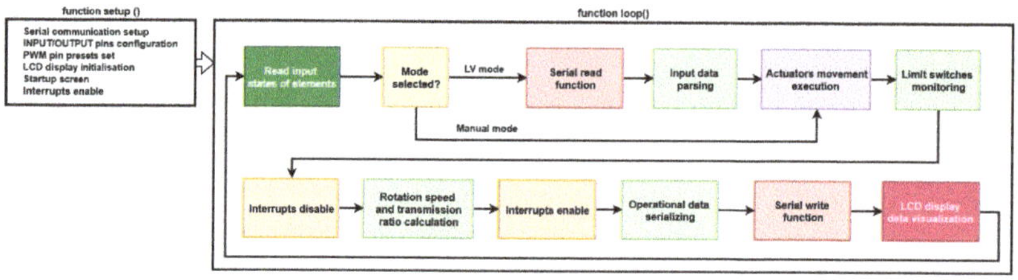

Figure 8. Block diagram of the microcontroller code.

The main program executes within an infinite loop (function *loop()*). First, the input states of the limit switches and all manual control elements on the control panel are read. Depending on the mode switch setting, the program either continues in manual mode or processes serial input from the LV application.

In manual mode, the DC motor and stepper motor are controlled according to the values provided by the control panel inputs. In LV mode, the incoming data string from the LV application—transmitted via USB serial communication—is programmatically parsed using comma delimiters into discrete numerical control variables—namely, the DC motor rotation direction, RPM setpoint, and left/right stepper motor commands—which are subsequently executed by the actuators. An example of the received control string can be found in Figure 9 in the *DATA Sent* box. This example indicates CW rotation (first 0), no left/right movement of the belts (a combination of 1 and 1), and no movement of the DC motor (setpoint set to 0 RPM).

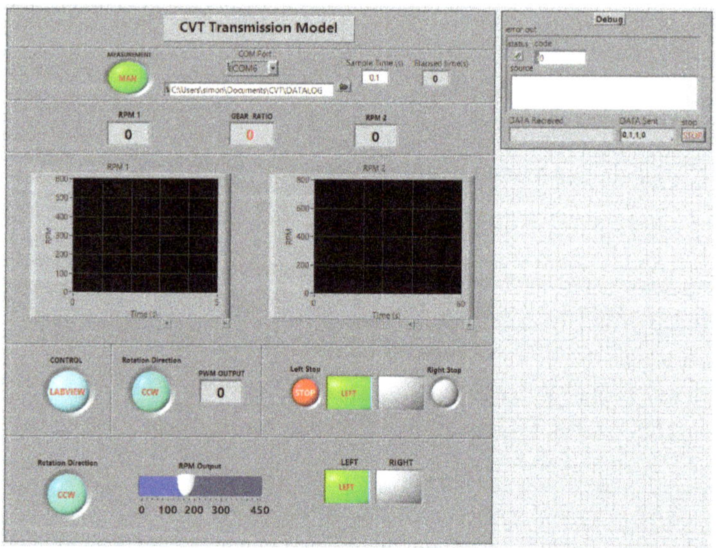

Figure 9. LabVIEW application control panel.

Next, the states of the limit switches are continuously monitored, and the stepper motor is halted if the slider reaches either of the limit positions.

After the control commands for the actuators are executed and the limit switches are checked, the interrupts for speed measurements are temporarily disabled, and the rotational speeds of both cones are calculated based on the number of detected pulses and the elapsed time. Using these values, the transmission ratio is determined, after which the interrupts are re-enabled. All relevant data—including measured rotational speeds, panel control states, and limit switch states—are serialized into a comma-separated string

 https://doi.org/10.3390/machines14020141

and transmitted using the USB serial communication protocol to the LV application via the serial write function. Finally, the real-time operational data is displayed on the device's LCD panel. The cycle of the *loop()* function then repeats.

3.3. *LabVIEW Control*

In addition to manual operation, the device can be switched to LV mode using the mode selection switch on the control panel. When connected to a control computer through a USB programming and communication interface, the LV application (version 2021 SP1) can be launched (Figure 9). This software application primarily enables direct operation of individual actuators (DC motor, stepper motor) and provides visualization of sensor data (limit switches, magnetic speed sensors), as illustrated in Figure 9.

The detailed functionality of the application is illustrated in the block diagram in Figure 10. At startup, the application initializes the USB serial communication protocol with the device's microcontroller (using the virtual instrument software architecture—VISA) and prepares a comma-separated values (CSV) log file by generating a timestamped filename and defining column headers. Data exchange with the device (microcontroller) is conducted via USB serial communication (at the port named COM6). Within the main *while()* loop, the program continuously reads byte streams from the USB communication buffer (using the VISA string read function), which contain serialized and comma-separated string data transmitted from the microcontroller in the serial write function (as detailed in the Microcontroller code subsubsection). The received data string, formatted as comma-separated values, can be monitored in the debug window (Figure 9, *DATA Received* box). An example of a formatted data string is represented by the sequence: "175, 420, 0, 0, 1, 0, 1, 1, 250". In this sequence, the first two variables correspond to the rotational speeds of the respective cones. The subsequent value indicates the selection of control mode (0 represents manual mode); the next value indicates the rotation direction of the cones (0 represents CCW rotation), while the following binary pair (1, 0) denotes the directional command for belt translation (to the right). The next two values represent the status of the limit switches (both off), and the final parameter conveys the target RPM setpoint as defined by the rotary potentiometer on the device's control panel. This string is parsed into individual numerical parameters and displayed using LV visualization components (RPM1, RPM2, CONTROL, Rotation Direction, RPM Output, Left Stop, Right Stop, Left, Right; see Figure 9). Conversely, setpoints from LV control components (Rotation Direction, RPM Output, Left, Right; see Figure 9, bottom) are serialized into a comma-separated data string and transmitted to the device through the VISA string write function and USB serial communication protocol. All measured and control values are simultaneously stored at the specified sampling interval (*Sample Time*) in the CSV file (Figure 9) created at the beginning of the measurement. When the application is stopped or closed, both the communication channel and the CSV file are properly terminated (closed).

In the LabVIEW control application, two measurement modes can be selected using the switch (Figure 10, center; Figure 9, top left). In manual measurement mode, the DC motor and stepper motor are operated directly through the LV control components. In automatic measurement mode, the DC motor speed and stepper motor movement are governed by predefined arbitrary timed functions (RPM1, Right, Left; see Table 2). These timed functions are specifically designed to enable measurements across the full range of the device's transmission ratio. Furthermore, the generated input functions serve as inputs for the MATLAB-Simulink (version R2023b) model of the device, allowing validation of the simulation model against the experimental data gathered on the physical model.

https://doi.org/10.3390/machines14020141

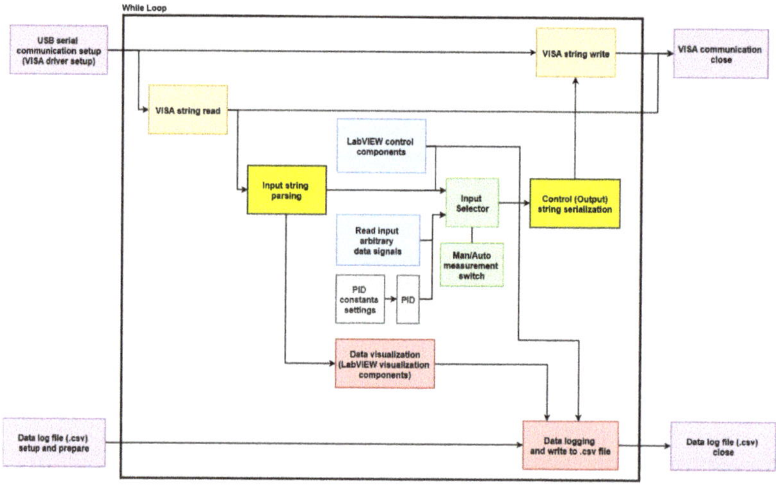

Figure 10. LabVIEW functionality block diagram.

In automatic mode, the rotational speed of the drive cone is regulated by a PID controller to maintain the RPM setpoint. The controller parameters were tuned using the heuristic Ziegler–Nichols method, yielding the following constants: P = 0.1, I = 0.1667, and D = 0.03. The implementation of the PID speed controller is essential, as the movement of the rubber belts along the cone surfaces introduces non-negligible load variations, reducing the cone speed by approximately 10 to 15 RPM. To ensure continuous and smooth adjustment of the transmission ratio during automatic measurements, the PID control structure was therefore applied.

Table 2. LabVIEW automatic measurement control signals (RPM1, Right, Left).

Time (s)	RPM1	Right Signal	Left Signal
0	0	0	0
2	0	0	0
10	350	0	0
15	350	1	0
39	350	1	0
49	350	0	−1
73	350	0	−1
78	350	0	0
86	0	0	0
88	0	0	0

4. MATLAB-Simulink Model

A simplified first-order MATLAB-Simulink model (Figure 11) was developed based on the primary motion characteristics (1) and the experimentally measured motion behavior of the physical system (see Figure 12). The rotational output speed of the driven cone (*RPM2*) is computed from the predefined rotational speed of the drive cone (*RPM1*) and the instantaneous transmission ratio. The transmission ratio is defined by the slider position (x), which is determined from its direction of motion (left to right or right to left, see Figure 4) and elapsed time, according to experimentally derived lookup curves (Figure 12).

https://doi.org/10.3390/machines14020141

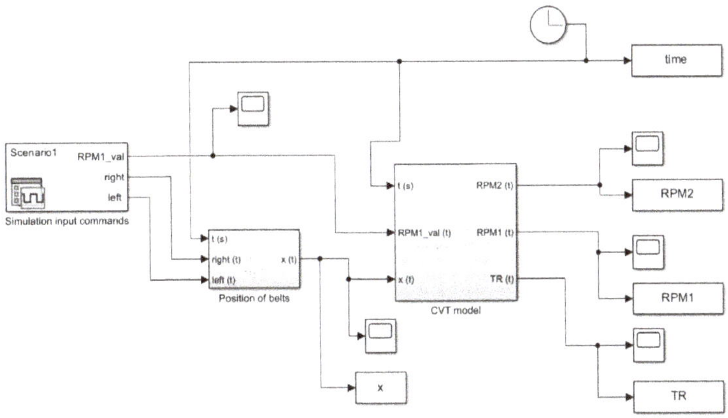

Figure 11. Simulation model of the CVT created in MATLAB-Simulink.

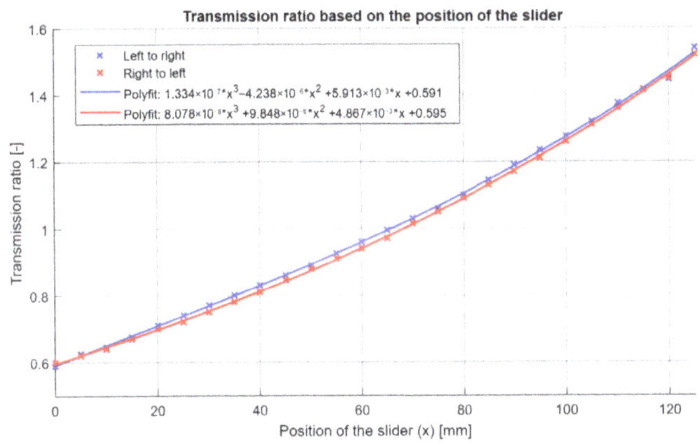

Figure 12. Slider position (x) functions.

The two functions were obtained by measuring the slider position with a precision ruler and determining the corresponding transmission ratio as the quotient of the rotational speeds of the two cones (with the drive cone fixed at 350 RPM). The functions and the resulting graph (Figure 12) indicate that the transmission ratio differs depending on whether the slider moves from left to right or from right to left (top view—Figure 4). This asymmetry arises from the slightly larger opening in the slider through which the rubber belts pass, leading to a minor delay and hysteresis in the transmission ratio response.

The *Simulation input commands* block (Figure 11) defines the timed control signals for slider motion (*left/right*) and the desired drive cone speed (*RPM1_val*). To validate the simulation model, identical arbitrary signal inputs from the LV application (automatic measurement mode) were applied (Table 2). The *Position of belts* block computes the slider position based on the timed direction movement inputs. The *CVT Model* block then calculates the driven cone speed (*RPM2*) and the corresponding transmission ratio (*TR*) from the slider position (x) and the prescribed drive cone speed (*RPM1_val*). These calculations are derived from experimentally obtained functions in Figure 12.

All simulation parameters (*time*, slider position x, drive cone speed *RPM1*, driven cone speed *RPM2*, and transmission ratio *TR*) are stored in the MATLAB workspace and plotted alongside the experimental data obtained from automatic measurements in the LV application (in CSV format). A comparison of simulation and experimental results is presented in the Results Comparison section.

https://doi.org/10.3390/machines14020141

Discussion

To ensure a computationally efficient model suitable for real-time control, the following first-order modeling assumptions were made:

- Inextensible belt drive: The round connecting belt is considered a longitudinally rigid member. While belt creep and longitudinal elasticity are present in high-torque industrial applications, they are considered negligible within the current experimental envelope (max 500 RPM, no-load conditions).
- Constant friction coefficient: A uniform friction coefficient (μ) is assumed across the contact interface of the cones. Potential variations due to local heating or surface wear were omitted, as the short duration of the test cycles prevents significant thermal gradients.
- Negligible parasitic losses: Energy losses attributed to aerodynamic drag and bearing rolling resistance were excluded. At the current rotational velocities, these factors account for less than 2% of the system's total energy expenditure during transmission change, thus not compromising the transmission-ratio accuracy.
- Ideal geometry: The contact between the round belt and the cones is simplified to a mean-radius line contact. This omission of the 'seating' effect (belt thickness compensation) is justified by the high correlation between the simulation results and empirical measurements.

The validity of these simplifications is confirmed by the experimental results, where the maximum deviation between the model and the physical prototype remains within an acceptable margin of error for pedagogical and prototype-development frameworks.

5. Results Comparison

Figures 13–15 provide a graphical comparison of the time-dependent courses of the rotational speeds of the drive and driven cones, along with the corresponding transmission ratio. Throughout the measurement, the rotational speed of the drive cone was maintained at a constant value of 350 RPM using the implemented PID control. At both the beginning and the end of the automatic measurement, the rotational speed was linearly increased or decreased over a short interval of eight seconds to minimize sharp peaks and abrupt transitions in the recorded data (Figure 13). The small dips and rises observed in the rotational speed (Figure 13) during the measurement can be attributed to the action of the PID controller, which regulated the rotational speed of the DC motor during the stop–start motion of the slider at its end positions (see Figure 15).

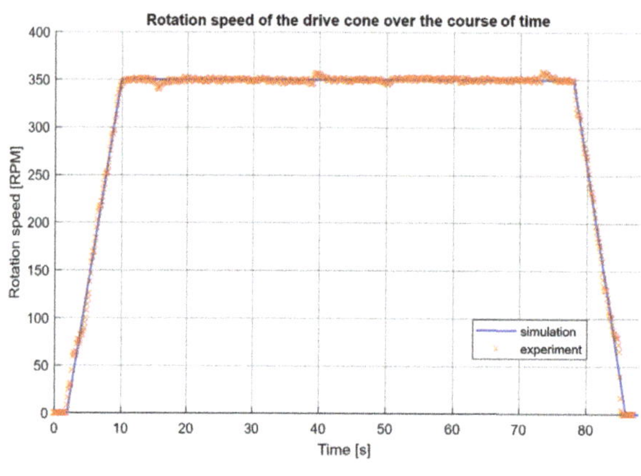

Figure 13. Rotation speed of the drive cone (*RPM1*).

https://doi.org/10.3390/machines14020141

Figure 14 illustrates the time-dependent course of the rotational speed of the driven cone. The figure shows that, at specific intervals (10–15 s, 39–49 s, and 73–78 s), the slider remained stationary, resulting in a constant rotational speed of the driven cone. These intervals correspond to the slider being positioned at its end limits, as indicated by the Right/Left movement signals in Table 1. The slight discrepancies observed between the simulated and measured rotational speeds can be attributed to dynamic factors not incorporated into the simplified model; these are discussed in detail in the Discussion subsection of the previous section. These effects introduced minor, unpredictable slippage, leading to a small reduction in RPM, which is particularly evident in the middle of Figure 14.

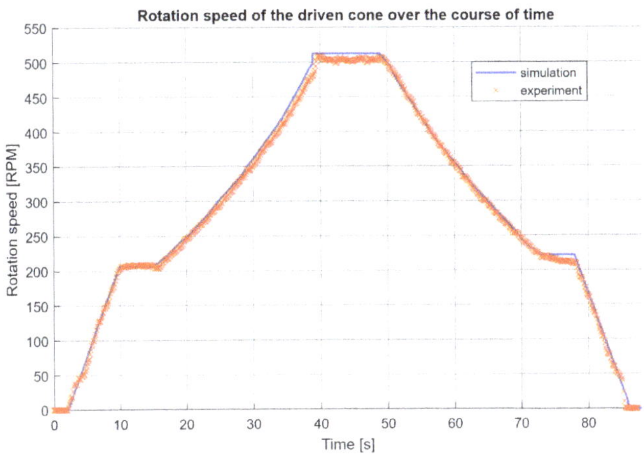

Figure 14. Rotation speed of the driven cone (*RPM2*).

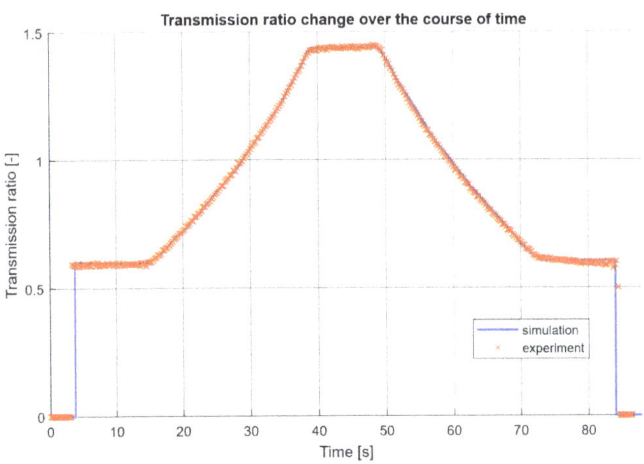

Figure 15. Transmission ratio change over time (*TR*).

During the measurements, the slider executed nearly a full stroke from its initial position to 120 mm and back (Figure 4). This movement produced the expected variation of the transmission ratio from the low value of 0.6 to the overdrive value of 1.45. The measured change in the transmission ratio shows strong agreement with the simulated (calculated) values, as illustrated in Figure 15. This high level of predictive accuracy is essential for establishing a reliable plant model, which serves as the foundation for implementing advanced control strategies, such as Model Predictive Control (MPC) or robust

https://doi.org/10.3390/machines14020141

state-estimation frameworks. Furthermore, high-fidelity tracking is indispensable for specific industrial applications, including: (i) proactive slip detection to ensure component longevity; (ii) efficiency optimization along the motor's ideal operating line; and (iii) the regulation of transient dynamics to enhance drivetrain comfort.

6. Conclusions

In conclusion, the developed rubber belt-driven dual cone continuously variable transmission model successfully demonstrates a stepless variation of transmission ratio and provides a reliable, robust, and portable platform for hybrid manual and LabVIEW-based control. The system's microcontroller-based design offers a flexible architecture that facilitates both the straightforward integration of additional software functionalities and the seamless incorporation of supplementary hardware. The integration of the LabVIEW environment facilitates advanced functionalities—including a custom user interface, PID speed regulation, and automated data acquisition and logging—thereby extending the system's utility beyond basic manual demonstration and enabling advanced HIL capabilities.

While this work establishes a robust framework for pedagogical and prototyping applications, the authors acknowledge certain limitations regarding mechanical optimization and comparative benchmarking. Currently, further design optimization, including rigorous tolerance analysis and belt-mechanism improvements, is ongoing to enhance the system's operational efficiency. Furthermore, a direct quantitative benchmarking against other CVT prototypes was precluded by the absence of standardized performance markers and granular data in the existing literature. Nevertheless, by providing a detailed roadmap for low-cost HIL integration, this work establishes the necessary baseline for future comparative studies. Subsequent research will focus on utilizing this platform to generate the standardized datasets required for benchmarking advanced control strategies and mechanical refinements.

The results presented in this study demonstrate that the developed first-order MATLAB-Simulink model maintains high fidelity with experimental measurements obtained from the physical prototype. Despite the deliberate omission of specific higher order dynamics—such as belt slippage, bearing-related rotational losses, and simplified contact geometries—the model's predictive accuracy is confirmed, establishing its suitability for further analytical refinement. The behavioral characteristics observed during experimental validation underscore the necessity of integrating hysteresis and slip-loss modeling in future iterations to further enhance the system's precision. Ultimately, these findings substantiate the platform as a robust experimental framework for investigating CVT behavior, facilitating educational applications, and serving as a benchmark for subsequent research, including the characterization of transmission efficiency and torque-load dynamics.

Future revisions of the physical model are expected to incorporate several enhancements: (i) measurement of slider position using a laser time-of-flight (TOF) sensor, (ii) adjustable slider movement speed control through both manual operation and the LabVIEW application, (iii) implementation of incremental encoders to enable more accurate and reliable cone speed measurements, (iv) implementation of a see-through enclosure for the entire system to ensure operator safety, and (v) design optimizations of the cone and belt system to enhance transmission efficiency.

Author Contributions: Conceptualization, Š.B. and V.G.; methodology, Š.B.; software, Š.B.; validation, V.G., Š.B. and K.O.; formal analysis, V.G. and V.K.; investigation, Š.B.; resources, Š.B.; data curation, K.O.; writing—original draft preparation, Š.B.; writing—review and editing, K.O. and E.K.; visualization, Š.B.; supervision, V.G. and V.K.; project administration, Š.B. and E.K.; funding acquisition, Š.B., E.K., V.G. and V.K. All authors have read and agreed to the published version of the manuscript.

 https://doi.org/10.3390/machines14020141

Funding: This work was supported by Grant Agency KEGA, grant No. 006STU-4/2023, and by the Scientific Grant Agency of the Ministry of Education, Research, Development and Youth of the Slovak Republic No. 1/0637/23 and 1/0045/25. APC was funded by the Institute of Automotive Mechatronics, Faculty of Electrical Engineering and Information Technology, Slovak University of Technology in Bratislava.

Data Availability Statement: The data presented in this study are openly available in Zenodo at https://zenodo.org/records/17734363 (accessed on 21 January 2026).

Conflicts of Interest: The authors declare no conflicts of interest.

Abbreviations

The following abbreviations are used in this article:

CSV	comma-separated values
CVT	continuous variable transmission
DC	direct current
HIL	hardware-in-loop
IDE	integrated development environment
LCD	liquid crystal display
LV	LabVIEW
PC	personal computer
PID	proportional, integral, derivative
RPM	rotations per minute
USB	universal serial bus
VISA	virtual instrument software architecture

References

1. Seelan, V. Analysis, Design and Application of Continuously Variable Transmission (CVT). *Int. J. Eng. Res. Appl.* **2015**, *5*, 99–105.
2. Del Castillo, J.M.; Pintado, P.; Benitez, F.G. A procedure for determining the efficiency of a continuously variable transmission. *SAE Trans.* **2002**, *111*, 2104–2108.
3. Chen, Y. *Automotive Transmissions: Design, Theory and Applications*; Springer Nature: Berlin/Heidelberg, Germany, 2020.
4. Mane, L.J.; Kekan, S.N.; Salunke, M.S.; Mane, S.S.; Adhapure, D.U. Continuous Variable Transmission. *Int. J. Recent Res. Civ. Mech. Eng.* **2016**, *2*, 51–55.
5. Lang, K.R. Continuously variable transmissions: An Overview of CVT Research Past, Present, and Future. *W* **2000**, *21*, 1–15.
6. van der Meulen, S.H. High-Performance Control of Continuously Variable Transmissions. Ph.D. Thesis, Technische Universiteit Eindhoven, Eindhoven, The Netherlands, 2010. [CrossRef]
7. Shabrov, N.; Ispolov, Y.; Orlov, S. Simulations of continuously variable transmission dynamics. *ZAMM J. Appl. Math. Mech.* **2014**, *94*, 917–922. [CrossRef]
8. Bonsen, B. Efficiency Optimization of the Push-Belt CVT by Variator Slip Control. Ph.D. Thesis, Technische Universiteit Eindhoven, Eindhoven, The Netherlands, 2006. [CrossRef]
9. Klaassen, T.W.G.L. The Empact CVT: Dynamics and Control of an Electromechanically Actuated CVT. Ph.D. Thesis, Technische Universiteit Eindhoven, Eindhoven, The Netherlands, 2007. [CrossRef]
10. Mascenik, J.; Murcinkova, Z. Experimental Determination of the Belt Transmission Slip. *Manag. Syst. Prod. Eng.* **2019**, *27*, 205–210. [CrossRef]
11. Carbone, G.; Mangialardi, L.; Bonsen, B.; Tursi, C.; Veenhuizen, P.A. CVT dynamics: Theory and experiments. *Mech. Mach. Theory* **2007**, *42*, 409–428. [CrossRef]
12. Srivastava, N.; Haque, I. A review on belt and chain continuously variable transmissions (CVT): Dynamics and control. *Mech. Mach. Theory* **2009**, *44*, 19–41. [CrossRef]
13. MathWorks. Continuously Variable Transmission Block. Available online: https://www.mathworks.com/help/autoblks/ref/continuouslyvariabletransmission.html (accessed on 21 January 2026).

https://doi.org/10.3390/machines14020141

14. Duong, T.T.; Huynh, P.S.; Tran, N.H.; Nguyen, T.H. Research on Designing the Continuously Variable Transmission Test Bench. *J. Tech. Educ. Sci.* **2023**, *18*, 15–24. [CrossRef]
15. Wong, P.K.; Xie, Z.; Chen, Y.; Tam, L.M. Development of a novel Dual-Belt Van Doorne's Continuously Variable Transmission for Automobiles—A preliminary study. *Int. J. Struct. Stab. Dyn.* **2017**, *18*, 1850016. [CrossRef]
16. Spanoudakis, P.; Tsourveloudis, N.C. Prototype variable transmission system for electric vehicles: Energy consumption issues. *Int. J. Automot. Technol.* **2015**, *16*, 525–537. [CrossRef]
17. Centeno, G.; Morales, F.; Perez, F.B.; Benitez, F.G. Continuous Variable Transmission With an Inertia-Regulating System. *J. Mech. Des.* **2010**, *132*, 051004-1. [CrossRef]
18. Supriyo, B.; Ariyono, S.; Sihono, S.; Sumiyarso, B.; Tjahyono, B. Experimental study of transmission ratio changing mechanism for motorcycle applications. *IOP Conf. Ser. Mater. Sci. Eng.* **2021**, *1108*, 012044. [CrossRef]
19. Mohamed, E.S.; Albatlan, S.A. Experimental investigation and theoretical model approach on transmission efficiency of the vehicle continuously variable transmission. *Am. J. Veh. Des.* **2014**, *2*, 43–52.
20. Atallah, K.; Wang, J.; Calverley, S.D.; Duggan, S. Design and operation of a magnetic continuously variable transmission. *IEEE Trans. Ind. Appl.* **2012**, *48*, 1288–1295. [CrossRef]
21. Belter, J.T.; Dollar, A.M. A passively adaptive rotary-to-linear continuously variable transmission. *IEEE Trans. Robot.* **2014**, *30*, 1148–1160. [CrossRef]
22. Chen, X.; Hang, P.; Wang, W.; Li, Y. Design and analysis of a novel wheel type continuously variable transmission. *Mech. Mach. Theory* **2017**, *107*, 13–26. [CrossRef]
23. Kim, J.; Park, F.C.; Park, Y.; Shizuo, M. Design and analysis of a spherical continuously variable transmission. *J. Mech. Des.* **2002**, *124*, 21–29. [CrossRef]
24. Ghariblu, H.; Behroozirad, A.; Madandar, A. Traction and efficiency performance of ball type CVTs. *Int. J. Automot. Eng.* **2014**, *4*, 738–748.
25. Raghavan, M. Kinematics of the Full-Toroidal Traction Drive Variator. *J. Mech. Des.* **2002**, *124*, 448–455. [CrossRef]
26. Spanoudakis, P.; Tsourveloudis, N.C. On the efficiency of a prototype Continuous Variable Transmission system. In Proceedings of the 21st Mediterranean Conference on Control and Automation, Crete, Greece, 25–28 June 2013; pp. 290–295.
27. Srinath, N. Cone Type CVT with High Speed Variations. *Power* **2015**, *1*, T2.
28. Knauff, P.W.; Evans, W. Variable-Speed-Transmission Device. U.S. Patent 759,872, 17 October 1904.
29. Knauff, P.W.; Evans, W. Power-Transmission Device. U.S. Patent 759,873, 17 May 1904.
30. Evans, G.P. Speed Regulating Device. U.S. Patent 390,216, 2 October 1888.
31. Punyakaew, S. Adaptively tuned vibration absorber using cone based continuously variable transmission. *Cogent Eng.* **2022**, *9*, 2102044. [CrossRef]
32. Lin, X.; Peng, Y.; Hong, R.; Wang, Y. Research on a novel discrete adjustable radiuses type continuously variable transmission. *Meccanica* **2022**, *57*, 1155–1171. [CrossRef]
33. Mobedi, E.; Dede, M.İ.C. A continuously variable transmission system designed for human–robot interfaces. In *Mechanism and Machine Science: Select Proceedings of Asian MMS 2018*; Springer: Singapore, 2021; pp. 29–41.
34. Patil, N.S.; Malekipour, E.; El-Mounayri, H. *Development of a Cone CVT by SDPD and Topology Optimization*; Technical Report 2019-01-0775, SAE Technical Paper; SAE: Warrendale, PA, USA, 2019.
35. Bazios, P.; Bitzas, G.; Stavroulakis, G. Topology optimization of a Continuous Variable Transmission (CVT) system. In Proceedings of the 11th HSTAM International Congress on Mechanics, Athens, Greece, 27–30 May 2016.
36. Garud, V.U.; Awadhani, L.V.; Vivekanandan, N.; Panchawadkar, A.A.; Lakade, S.S.; Tiwari, H.U. Investigation of slip and speed ratio in axially displaceable cone variator CVT drive. In Proceedings of the 2016 International Conference on Electrical, Electronics, and Optimization Techniques (ICEEOT), Chennai, India, 3–5 March 2016; pp. 3710–3714.
37. Sonkhaskar, Y.M.; Rao, G.S.; Khade, K.V.; Bhute, R.L.; Mahant, S.P. Design and analysis of continuously automatic variable transmission (CAVT) system for bicycle. *Aip Conf. Proc.* **2024**, *2974*, 020023. [CrossRef]
38. Dennis, S.; Bowe, M.; Ball, J.; Jensen, D. A Student Developed Teaching Demo Of An Automatic Transmission. In Proceedings of the 2001 Annual Conference, Rome, Italy, 16–21 July 2001; pp. 6–104.
39. Srivastava, S.D.; Haque, S. Design and Development of a Continuously Variable Transmission (CVT) Test Rig for Performance Analysis. *Int. J. Eng. Res. Technol.* **2016**, *5*, 112–117.
40. Bieniek, A.; Graba, M.; Mamala, J.; Burdzik, R. Research on the Influence of Control Signals on the Characteristics of a CVT Transmission in a LabVIEW Environment. *Sensors* **2021**, *21*, 5842. [CrossRef]
41. Younes, T.M.; Al-Nasser, A. Automation and Control of a Centrifugal Clutch for CVT Applications using LabVIEW. *Int. J. Control. Autom. Syst.* **2014**, *3*, 25–34.

 https://doi.org/10.3390/machines14020141

42. Bertini, L.; Carmignani, C.; Frendo, F.; Arnone, L. Experimental Analysis of a Cone-to-Cone Continuously Variable Transmission. *Proc. Inst. Mech. Eng. Part D J. Automob. Eng.* **2004**, *218*, 1321–1329.
43. Kushwaha, A.K.; Dwivedi, S. Design and Fabrication of a Pedagogical CVT Model for Educational Purposes. *J. Mech. Eng. Technol.* **2018**, *6*, 45–52.

 https://doi.org/10.3390/machines14020141

MDPI AG
Grosspeteranlage 5
4052 Basel
Switzerland
Tel.: +41 61 683 77 34

Machines Editorial Office
E-mail: machines@mdpi.com
www.mdpi.com/journal/machines

www.ingramcontent.com/pod-product-compliance
Lightning Source LLC
Chambersburg PA
CBHW061457210926
51850CB00008B/1743
* 9 7 8 3 7 2 5 8 7 5 1 2 2 *